ECOLOGY OF PLANKTON

Editor
Professor (Dr.) Arvind Kumar
Environmental Science Research Unit
Post Graduate Department of Zoology
S.K.M. University, Dumka – 814 101 (Jharkhand)

2024

Daya Publishing House®

A Division of

Astral International Pvt. Ltd.

New Delhi – 110 002

First Published, 2005

Reprinted, 2024

ISBN: 978-81-7035-374-4

Published by : **Daya Publishing House**®
A Division of
Astral International Pvt. Ltd.
– ISO 9001:2015 Certified Company –
4736/23, Ansari Road, Darya Ganj
New Delhi-110 002
Ph. 011-43549197, 23278134
E-mail: info@astralint.com
Website: www.astralint.com

Preface

Plankton are the weakly swimming but mostly drifting small organisms that inhabit the water column of the ocean, and bodies of freshwater. The name comes from the Greek term, plankton-meaning "wanderer"or "drifter". While some forms of plankton can move several hundreds of meters vertically in a single day (a behavior called vertical migration), their horizontal position is mostly determined by movement (currents) of the body of water they inhabit. Larger organisms, such as squid, fish, and marine mammals that can control their horizontal movement and swim against the average flow of the water environment, are called nekton. The study of plankton is termed planktology. The term holoplankton refers to organisms that spend their entire life cycle as part of the plankton, such as krill, copepods, salps, and jellyfish. Meroplankton, in contrast, are only planktonic for part of their lives (usually the larval stage). Examples of meroplankton include the larvae of sea urchins, starfish, crustaceans, marine worms, and most fish.

Plankton concentration and distribution are sensitive to chemical and physical changes in the water. Plankton are often described in terms of size, such as Megaplankton, 20-200 cm; Macroplankton, 2-20 cm; Mesoplankton, 0.2 mm-2 cm; Microplankton, 20-200 µm; Nanoplankton, 2-20 µm; Picoplankton, 0.2-2 µm, mostly bacteria; Femtoplankton, smaller than 0.2 µm, consisting of marine viruses. However, some of these terms may be used with very different boundaries, especially on the larger end of the scale. The existence and importance of nano- and even smaller plankton was only discovered during the 1980s, but they are thought to make up the largest proportion of all plankton in number and diversity. Plankton are also divided into three broad functional groups such as, Phytoplankton (from Greek *phyton* or plant), algae that live near the water surface where there is sufficient light to support photosynthesis Zooplankton (from Greek *zoon* or animal), small protozoa, crustaceans, and various other animals that feed on other plankton. Some of the eggs and larvae of larger animals, such as fish, crustaceans, and annelids, are included here. Bacterioplankton, bacteria and archaea, which play an important role in absorbing nutrients dissolved in the water.

Above all, some plankton serve the purpose of monitoring the environmental pollution as they are tolerant to adverse environmental conditions, that's why these are also termed as bioindicators and are capable of measuring the actual response of organisms or populations to the environmental qualities. The knowledge of their seasonal qualitative and quantitative fluctuations have been considered essential for proper manipulation of the factors influencing biological productivity of the wetlands. Hence, the ecology of plankton is helpful in opening a new vista in the field aquatic ecology. Keeping this in view, the book has been compiled in order to know the hydrological spectrum of the wetlands.

My special thanks and appreciation go to the scientists whose contributions have enriched this volume. I wish to express my sincere gratitude to Dr. P. C. Hembram, Hon'ble Vice Chancellor, S. K. M. University, Dumka who has been a source of constant inspiration. I am especially thankful to Professor (Dr.) B. N. Jha, Hon'ble Pro Vice-Chancellor, S. K. M. University, Dumka for his encouragement. I owe my special thanks to Professor M. C. Dash, Hon'ble Vice Chancellor of Sambalpur University, Professor N. C. Datta of Calcutta University, Professor S. K. Konar of Kalyani University, Professor P. S. Murthy of Bangalore University, Professor P. Natarajan of Trivandrum University, Professor S. P. Roy of Bhagalpur University, Professor A. K. Quereshi of Bhopal University, Professor Tanmay Bhattacharya of Midnapore University, Professor P. C. Mishra of Sambalpur University, Professor B. D. Joshi of Hardwar University, Professor G. C. Pandey of Faizabad University, Professor K. C. Sharma of Ajmer University, Professor M. Raziuddin of Hazaribag University, Professor U. S. Bagde of Mumbai University, Dr. M. P. Sinha of Ranchi University, Professor Gurdeep Singh of I. S. M., Dhanbad, Dr. P. K. Goel of Karad, Dr. M. C. Varma and Shri T. Poddar of Bhagalpur University. I also acknowledge the incentives provided by one of my research scholars, Dr. Chandan Bohra, Head of Zoology, B. S. K. College, Barharwa. for helping me actively in bringing out this book.

I also express my deep sense of gratitude to my parents whose blessings have always prompted me to pursue academic activities deeply. I am also thankful to my sweet wife, Professor Kumari Bimla and my two lovely sons, Kumar Pallav Shivshankaran and Kumar Prasun Ramakrishnan whose natural smiles extended to me relief all through this tiresome endeavour.

Last but not the least, I am also thankful to Mr. Anil Mittal, Proprietor, Daya Publishing House, Delhi for taking keen interest in bringing out of this book. Finally, I will always remain a debtor to all my well-wishers for their blessings, without which this book would not have come into existence.

Dumka **Professor Arvind Kumar**

Contents

Chapter 1

Dynamics of Phytoplankton Productivity of Certain Lentic Ecosystems of Jharkhand, India

Arvind Kumar and Chandan Bohra

Environmental Science Research Unit, Post Graduate Department of Zoology, S.K.M. University, Dumka–814 101, India

Introduction

Phytoplankton is the primary producer of organic matter in aquatic ecosystem and constitutes the first trophic level in the pelagic food web and through the process of photosynthesis provides food and oxygen to other organisms.

These organisms are the important biotic component of an aquatic ecosystem and provide the main food of fishes directly or indirectly and can be used as indicator of the trophic phase of the water body (Verma and Datta Munshi, 1987 and Bohra and Kumar, 1999). The quality and quantity of phytoplankton population in a biotope is altered by any change in the prevailing environmental conditions. The magnitude and dynamics of phytoplankton becomes an essential parameter to assess the state of pollution as the biodiversity is stressed and the environment becomes poor. Certain living organisms serve the purpose of monitoring the environmental pollution, as they are tolerant to adverse environmental condition. These are known as bioindicators and are capable of measuring the actual response of organisms or populations to the environmental qualities.

The density and diversity of phytoplankton are controlled by several physico-chemical factors of water. Light plays a significant role as an energy source for photosynthesis of phytoplankton. A number of researchers suggested that light is often a limiting factor for phytoplankton growth (Tailing,

1971 and Kaufman, 1980). Temperature has been considered as a major environmental factor to control the phytoplankton growth (Verma and Datta Munshi, 1987; Kaushik *et al.,* 1991 and Bohra and Kumar, 1999). Alkalinity also plays an important role to monitor the potential of the phytoplankton (Jackson, 1961). In addition to carbon, hydrogen and oxygen, phytoplankton require some other essential inorganic nutrients to grow well. It is generally observed that if the concentrations of nitrogen and phosphorus are low, they limit the phytoplankton productivity. However, phosphorus limitation of productivity is more common in the freshwaters. Wetzel (1975), Trivedy *et al.* (1985), Outridge *et al.* (1989). Agarwal *et al.* (1990) etc. have developed a relationship between nutrients and algal growth. Several workers have used the different phytoplankton groups to evaluate the trophic status and pollution potential of the freshwater bodies all over the world (Bais *et al.,* 1993 and 1995).

In Jharkhand, Kumar (1996); Gupta and Kumar (1998); Kumar (1998); Kumar and Bohra (1998); Bohra and Kumar (1999); Bohra and Kumar (2002a and b); Kumar (2003); Gupta *et al.* (2003) and Bohra and Kumar (2004) have done considerable work on phytoplankton diversity.

Freshwater bodies are important resources of production. The exploitation of these resources by man in the present time is due to the increasing cultural eutrophication which further increases their productivity. Phytoplankton through their photosynthesis apparatus absorb radiant energy and package it as bond energy in the form of high energy compound ATP. In majority of the studies, algal photosynthesis is used as a measure of primary productivity. However, productivity figures are meaningful only if the most important factors controlling them are identified. There are a number of important physico-chemical factors of water bodies like temperature, light, pH, alkalinity, organic matter, inorganic nutrients etc. which control the algal productivity (Tailing, 1971; Vijaya Raghwan, 1971; Richardson *et al.,* 1993; Laal, 1989; Rai and Sharma. 1991; Vijay Kumar, 1994; Agrawal *et al.,* 1995; Kumar, 1996 and 1999; Pandit, 1999; Harikrishnan and Aziz, 2000; Kumar, 2002 and Bohra and Kumar, 2002b). Phytoplankton productivity can be used as indices of trophic status and fisheries resources potential of water bodies.

In the present investigation, seasonal and spatial variations of different phytoplankton groups as well as their primary productivity have been studied and correlated their with hydrological factors of water to evaluate the trophic status and pollution potential of two lentic ecosystems of Barharwa (Santal Pargana).

Materials and Methods

The present investigation was conducted during January to December, 2003 in the Munshi and Raja Dighi ponds of Barharwa. Four representative sampling stations, two at littoral zone and two at limnetic zone were selected for the water, phytoplankton and primary productivity analysis in both the ponds. Samples for water analysis and for qualitative and quantitative study of phytoplankton were collected from sub-surface using Ruttner's water sampler. Physico-chemical factors of water were done adopting the methods given in APHA (1985) and Trivedy and Goel (1986). For the quantitative enumeration of phytoplankton, a microtransact method as described by Lackey (1938) and simplified by Edmondson (1974) was used. Identification of phytoplankton was done with the help of work of Smith (1950), Fritsch (1935), Edmondson (1959), Desikachary (1959), Randhwa (1959), Ward and Whipple (1959) and Phillipose (1967). Phytoplankton productivity was estimated adopting the Garder and Gran's (1927) light and dark bottles method. Statistical tools like "F" test, "t" test and coefficients of correlation have been used to obtain the valuable information from the data (APHA, 1985; Trivedy and Goel, 1986; and Misra and Misra, 1989).

Results and Discussion

In the Munshi pond, the phytoplankton groups: Chlorophyceae, Bacillariophyceae, Cyanophyceae and Euglenophyceae were identified while in the Raja Dighi pond apart from these groups, the group Dinophyceae was also recorded. The number of genera of each phytoplankton group identified in both the ponds are as follows:

Phytoplankton	*Group Munshi Pond*	*Raja Dighi Pond*
Chlorophyceae	24 genera	16 genera
Bacillariophyceae	21 genera	20 genera
Cyanophyceae	18 genera	17 genera
Euglenophyceae	02 genera	02 genera
Dinophyceae	Not Identified	02 genera

In the Munshi pond, all phytoplankton groups comprised the maximum number of organisms during summer whereas the minimum was identified during winter season. In the Raja Dighi pond, all phytoplankton groups contained maximum number of organisms during rainy and summer season respectively (Table 1.3).

In the Munshi pond, minimum and maximum number of phytoplankton population during winter and summer seasons respectively might be due to the low and high water temperature (Table 1.1) and short and long photoperiod respectively. Several researchers have proposed the temperature as a vital factor responsible for the growth of algae (Ramkrishnaiah and Sarkar, 1982: Verma and Datta Munshi, 1987; Bohra and Kumar, 1999). Wisharad and Mehrotra (1988) reported that proliferation of phytoplankton from winter to summer could be attributed to the progressively increasing water temperature and photoperiod. High carbonate and low bicarbonate alkalinity during summer in the Munshi pond also reveal the high productivity (Table 1.1). During high photosynthesis, free carbondioxide in the water is completely utilized by the algae and then it is derived from the bicarbonates resulting in the formation of carbonates in large quantity. High pH also indicates the high productivity during summer (Table 1.1). The above inferences are in accordance with the findings of Wetzel (1957); Vyas (1968): Paramasivam and Sree Nivasan (1981) and Golterman and Meyer (1985). Cabecadas and Brogueira (1987) emphasized that the growth and photosynthesis of algae is mainly influenced by the pH and alkalinity of the water. The minimum and maximum Secchi transparency during summer and winter seasons respectively also prove the low and high growth of algae respectively (Table 1.1).

In the Raja Dighi pond, the minimum number of phytoplankton population during summer might have been seen due to the high turbidity of water observed throughout the year which reduced the light penetration resulting in the low algal productivity. It was because the Raja Dighi pond is shallow in nature, during high wind velocity water is mixed with sediment and it becomes turbid. According to Oglesby and Schaffer (1975) and Carlson (1977), water transparency affects the growth of algae directly. Kaufman (1980) reported that the phytoplankton density is declined due to the turbidity of water. In the Raja Dighi pond, during summer, pH and carbonates were minimum and bicarbonates were maximum which also indicate the low photosynthesis (Table 1.1). Several workers have reported that during the low photosynthesis, the pH of water is reduced due to the formation of carbonic acid (Hannan and Young, 1974; Cabecadas and Brogueira, 1987). Although, water temperature and some environmental factors were high during summer, the productivity did not increase due to the non-algal turbidity prevailing in the water.

Table 1.1: Seasonal Mean Values of Physico-chemical Factors of Water

Physico-chemical Factors	*Munshi Pond*			*Raja Dighi Pond*		
	Winter	*Summer*	*Rainy*	*Winter*	*Summer*	*Rainy*
Water Temp. (°C)	19.15	25.73	26.45	18.23	24.52	27.16
Secchi Transp. (cm)	105.75	30.15	39.16	40.45	17.15	40.85
pH	8.70	9.50	8.35	8.75	8.42	8.85
Free CO_2 (mg/l)	1.30	1.75	18.52	0.72	–	0.21
Carbonate Alk (mg/l)	18.35	49.15	3.75	15.20	10.15	8.15
Biocarbonate Alk (mg/l)	99.35	76.85	175.25	121.15	165.35	110.35
Ammonia Alk (mg/l)	0.285	0.153	0.022	0.093	0.285	0.069
Nitrate (mg/l)	0.087	0.182	0.168	0.059	0.072	0.077
Phosphate (mg/l)	0.055	0.132	0.029	0.095	0.037	0.025
Silicate (mg/l)	0.482	1.273	1.639	1.258	1.612	1.515

Table 1.2: Seasonal Variation of Certain Physico-chemical Factors of Water in Littoral and Limnetic Zone

Factors	*Seaons*	*Munshi Pond*				*Raja Dighi Pond*			
		Littoral Station		*Limnetic Station*		*Littoral Station*		*Limnetic Station*	
		A	*B*	*C*	*D*	*A*	*B*	*C*	*D*
Water Temp. (°C)	Winter	19.52	19.31	18.85	19.10	18.41	17.95	18.05	18.45
	Summer	25.40	25.65	26.10	25.70	24.15	24.42	24.75	24.72
	Rainy	26.10	26.75	26.85	26.15	26.90	27.25	23.45	27.80
Secchi Transp. (cm)	Winter	93.15	104.10	120.50	105.70	37.75	40.15	41.25	40.10
	Summer	22.50	30.15	34.40	31.17	16.10	16.85	17.10	17.90
	Rainy	43.75	40.10	39.52	35.10	41.15	39.20	41.35	40.10
pH	Winter	8.93	8.90	8.85	8.41	8.60	8.69	8.65	8.63
	Summer	9.35	9.65	9.69	9.55	8.58	8.56	8.57	8.53
	Rainy	8.35	8.29	8.35	8.25	8.72	8.75	8.76	8.70
Dissolve Oxygen (mg/l)	Winter	7.89	9.20	10.05	9.06	10.35	9.95	9.85	9.35
	Summer	4.95	7.93	8.65	7.05	6.15	5.90	5.45	5.57
	Rainy	4.82	5.37	5.75	5.61	5.67	5.85	5.77	5.64
Nitrate (mg/l)	Winter	0.094	0.125	0.121	0.108	0.072	0.065	0.055	0.057
	Summer	0.152	0.217	0.156	0.182	0.069	0.072	0.079	0.059
	Rainy	0.145	0.232	2.141	0.156	0.067	0.065	0.069	0.082
Phosphate (mg/l)	Winter	0.065	0.048	0.062	0.049	0.041	0.025	0.017	0.015
	Summer	0.151	0.098	0.153	0.125	0.057	0.025	0.032	0.025
	Rainy	0.041	0.025	0.028	0.029	0.035	0.032	0.016	0.019
Silicate (mg/l)	Winter	0.485	0.493	0.501	0.483	1.257	1.259	1.262	1.251
	Summer	1.275	1.268	1.273	1.276	1.615	1.609	1.625	1.600
	Rainy	1.635	1.642	1.625	1.650	1.512	1.518	1.525	1.502

In the Raja Dighi pond, phytoplankton population increased during rainy season. Water temperature was maximum during rainy season, which was the most favourable factor to increase the phytoplankton productivity. Maximum pH and low concentration of free carbon dioxide also indicate the high algal growth during rainy season. Minimum values of bicarbonate, ammonia, nitrate and phosphate during rainy season are indicative of comparatively high consumption of nutrients during high photosynthesis (Table 1.1).

In the Munshi pond, the most dominant group was cyanophyceae (Table 1.3). Dense cyanophycean population might have reduced the species density and diversity of the other groups of phytoplankton by shading them (Talling, 1971). Bais *et al.* (1993a) have also reported the shading effect of Cyanophyceae. According to Ganapati (1957) and Cairns *et al.* (1972), during the formation of cyanophyceae algal bloom, the species of other groups are represented poorly. In the Raja Dighi pond, the most dominant group of phytoplankton was Bacillariophyceae (Table 1.3). It was because Bacillariophyceae prefers loss polluted water. According to Trivedy *et al.* (1985), clean water bodies consist the diatoms dominated flora. Ramanibai and Ravichandran (1987) reported that the occurrence of diatoms in good number indicates the healthy and unpolluted conditions of the water bodies. Diatoms use the silicate in large quantity to build up their cell wall. In the Raja Dighi pond, silicate was also recorded in appreciable amount, which favoured the high production of diatoms as compared to the other group of phytoplankton (Table 1.1). The Dinophyceae group could not be identified in Munshi pond. It was because Dinophyceae prefers only oligotrophic water while the Munshi, pond is eutrophic in nature. Less number of Dinophyceaen population indicates that the Raja Dighi pond is healthy but comparatively very less polluted than the Munshi pond.

Table 1.3: Seasonal Variations (Mean Values) of Phytoplankton Groups by Number at Littoral and Limnetic Zones in Munshi and Raja Dighi Ponds

Phytoplankton Groups	*Seasons*	*Munshi Pond*		*Raja Dighi Pond*	
		Littoral	*Limnetic*	*Littoral*	*Limnetic*
Chlorphyceae	Winter	138	225	33	39
	Summer	891	677	53	32
	Rainy	552	463	105	82
Bacillariophyceae	Winter	405	520	115	95
	Summer	2345	2195	135	61
	Rainy	1290	1585	360	195
Cyanophyceae	Winter	595	1140	72	62
	Summer	5075	5725	79	45
	Rainy	3085	4193	242	129
Euglenophyceae	Winter	33	41	7	6
	Summer	287	189	25	4
	Rainy	175	159	47	27
Dinophyceae	Winter	–	–	5	3
	Summer	–	–	9	2
	Rainy	–	–	20	17

Analysis of variance for phytoplankton groups showed insignificant values between stations in both the ponds. The significant values were observed between seasons in both the ponds for all phytoplankton groups except for Cyanophyceae, which was insignificant in Raja Dighi pond (Table 1.4). The above results reveal that seasonal variations of phytoplankton groups were pronounced in both the ponds and regulated with the seasonal changes in physico-chemical factors of the water. Since both the ponds are shallow in nature, therefore, due to high wind velocity, phytoplankton did not show significant variations among the stations.

Table 1.4: Analysis of Variance for Phytoplankton Groups

Phytoplankton Groups	*Source of Variations*	*Degree of Freedom*	*Significance Level*	
			Munshi Pond	*Raja Dighi Pond*
Chlorophyceae	Between Stations	3 & 6	NS	NS
	Between Seasons	2 & 6	*	**
Bacillariophyceae	Between Stations	3 & 6	NS	NS
	Between Seasons	2 & 6	**	*
Cyanophyceae	Between Stations	3 & 6	NS	NS
	Between Seasons	2 & 6	**	NS
Euglenophyceae	Between Stations	3 & 6	NS	NS
	Between Seasons	2 & 6	*	*
Dinophyceae	Between Stations	3 & 6	–	NS
	Between Seasons	2 & 6	–	**

NS: Not significant; *: Significant at 1% P; **: Significant at 0.1% P.

When t-Test was applied on phytoplankton groups between the littoral and between the limnetic zones of both the ponds, all the values were found significant in which Munshi pond comprised the higher number of phytoplankton than in the Raja Dighi pond at both the zones (Table 1.5).

Table 1.5: Significant t-Test for Phytoplankton Groups

Phytoplankton Groups	*Combinations*	*Degree of Freedom*	*Significance Level*
Chlorophyceae	Lt (MP) Vs Lt (RDP)	10	*
	Lm (MP) Vs Lm (RDP)	10	***
Bacillariophyceae	Lt (MP) Vs Lt (RDP)	10	*
	Lm (MP) Vs Lm (RDP)	10	**
Cyanophyceae	Lt (MP) Vs Lt (RDP)	10	**
	Lm (MP) Vs Lm (RDP)	10	**
Euglenophyceae	Lt (MP) Vs Lt (RDP)	10	*
	Lm (MP) Vs Lm (RDP)	10	**

*: Significant at 5% P; **: Significant at 1% P; ***: Significant at 0.1% P;
Lt: Littoral; Lm: Limnetic; MP: Munshi Pond, RDP: Raja Dighi Pond.

In the Munshi pond, all phytoplankton groups were negatively correlated with Secchi transparency while in the Raja Dighi pond, these all were positively correlated. All phytoplankton groups were found to be positively correlated with water temperature in the Munshi pond while the insignificant correlation were obtained in the Raja Dighi pond. Except euglenophyceae, phytoplankton groups showed significant positive correlation, with pH in Raja Dighi pond while in the Munshi pond, correlations were insignificant. No phytoplankton had any significant correlation with free carbon dioxide bicarbonates and ammonia in both the ponds. Bacillariophyceae in the Munshi pond and Chlorophyceae, Euglenophyceae and Dinophyceae in Raja Dighi pond were correlation significantly with carbonates. Except Chlorophyceae, all other groups showed positive significant correlation with nitrate in Munshi pond while in Raja Dighi pond, correlations were insignificant. Chlorophyceae and Bacillariophyceae were positively correlated with phosphate in Munshi pond. All phytoplankton groups were found significantly correlated with silicate in the Munshi pond while the correlations were insignificant in Raja Dighi pond (Table 1.6).

Table 1.6: Coefficient of Correlation Between Phytoplankton Groups and Physico-chemical Factors of Water

Phytoplankton Groups	*Secchi Transp.*	*Water Temp.*	*pH*	*Free CO_2*	*Carbo-nate Alk.*	*Bicarbo-nate Alk.*	*Ammonia*	*Nitrate*	*Phos-phate*	*Sili-cate*
				Munshi Pond						
Chlorophyceae	– 0.84***	0.72*	0.36	0.16	0.40	0.02	– 0.43	0.52	0.66*	0.65*
Bacillariophyceae	– 0.93***	0.83**	0.46	0.04	0.58*	– 0.18	*0.45	0.73*	0.59*	0.68*
Cyanophyceae	– 0.94***	0.87***	0.41	0.06	0.56*	– 0.13	– 0.53	0.69*	0.57*	0.75**
Euglenophyceae	– 0.88***	0.75**	0.27	0.23	0.38	0.06	– 0.45	0.68*	0.48	0.69*
				Raja Dighi Pond						
Chlorophyceae	– 0.75***	0.41	0.72**	– 0.09	– 0.62*	– 0.51	– 0.53	0.30	– 0.11	0.27
Bacillariophyceae	– 0.59*	0.44	0.69*	– 0.02	– 0.52	– 0.54	– 0.57	0.25	0.13	0.15
Cyanophyceae	– 0.60*	0.46	0.72**	– 0.05	– 0.054	– 0.55	– 0.57	0.20	0.05	0.16
Euglenophyceae	– 0.71*	0.22	*0.56*	– 0.12	– 0.65*	– 0.34	– 0.36	0.30	0.18	0.40
Dinophyceae	0.72**	0.38	0.71*	– 0.11	– 0.61*	– 0.51	– 0.49	0.33	0.07	0.36

In the Munshi pond, phytoplankton productivity ranged from 0.48 to 6.21 $gC/m^3/d$ while in the Raja Dighi pond, it varied from 0.48 to 3.35 $gC/m^3/d$ (Table 1.7). In both the ponds, the minimum and maximum values of phytoplankton productivity were observed during winter and summer seasons respectively (Table 1.8). Analysis of variance for phytoplankton productivity provided insignificant differences among the stations in both the ponds (Table 1.9). It might be due to the mixing of water at high wind velocity, because both the ponds are shallow. Seasonal variation was pronounced and found significant in the Munshi pond ($F = 38.13$. $P < 0.001$) as well as in the Taja Dighi pond $F = 18.31$. $P < 0.05$).

In the Munshi pond, highest productivity during summer might be due to abundant growth of phytoplankton (Singh and Saha, 1987; Takamura *et al.*, 1989 and Bohra and Kumar, 2002b). Productivity was also found maximum during summer in Raja Dighi pond. However, phytoplankton population was comparatively low in summer than in rainy season. Therefore, it can be concluded that a large part of the productivity might have been contributed by the nanoplankton during summer. Many researchers like Khan and Zutshi (1979); Galvao (1985) and Munawar and Munawar (1986) have also reported that the nanoplankton generally assimilate more carbon per unit of biomass than net plankton due to high surface ratio which favours rapid exchange of materials between protoplasm and surrounding environment. Low water level, high solar radiation, longer photoperiod, high water temperature and appreciable amount of nitrate and phosphate could have been responsible for the high productivity during summer in both the ponds (Table 1.2). In both the ponds, the productivity was minimum during winter. It could be attributed to the low temperature, low intensity of light, short photoperiod, low concentration of nutrients and less number of phytoplankton (Khan *et al.*, 1988; Rai and Sharma, 1991; Pati and Sahu, 1993; Vijay Kumar, 1994; Kumar, 1996 and 1999; Mishra and Nayak, 2000; Kumar, 2001 and Bohra and Kumar, 2002b).

Table 1.7: Average Ranges of the Phytoplankton Productivity (gc/m³/d) in Munshi and Raja Dighi Pond

Zones	*Stations*	*Productivity*	*Munshi Pond*	*Raja Dighi Pond*
Littoral	A	GPP	0.78–5.15	0.77–1.69
		NPP	– 1.52–4.39	0.32–1.24
		CRR	0.48–3.49	0.17–0.78
	B	GPP	0.48–4.99	0.63–2.15
		NPP	– 0.47–2.91	0.32–1.54
		CRR	0.32–4.57	0.18–0.79
Limnetic	C	GPP	0.65–5.48	0.64–3.07
		NPP	– 0.47–4.83	0.32–2.75
		CRR	0.32–3.19	0.17–0.64
	D	GPP	0.94–6.21	0.48–3.35
		NPP	0.32–5.56	0.32–3.08
		CRR	0.32–3.55	0.17–0.79

GPP: Gross Primary Productivity; NPP: Net Primary Productivity; CRR: Community Respiration Rate.

Table 1.8: Seasonal Variation in GPP (gc/m³/d) and Phytoplankton Population (organism/l) in Munshi and Raja Dighi Pond

Parameter	*Seaons*	*Munshi Pond*				*Raja Dighi Pond*			
		Littoral Stations		*Limnetic Stations*		*Littoral Stations*		*Limnetic Stations*	
		A	*B*	*C*	*D*	*A*	*B*	*C*	*D*
GPP	Winter	1.46	1.58	1.47	2.03	0.93	0.74	0.85	0.78
	Summer	3.08	3.10	0.69	3.41	1.14	1.27	1.54	1.55
	Rainy	2.55	2.93	3.35	3.38	0.85	0.99	1.19	1.11
Phytoplankton	Winter	1310	995	1692	2115	230	219	198	140
	Summer	8470	8640	8530	8995	290	297	138	14
	Rainy	4190	5992	6085	6680	733	785	435	437

GPP: Gross Primary Productivity.

Table 1.9: Analysis of Variance of Phytoplankton Productivity

Ponds	Source of Variation	Degree of Freedom	Calculated "F"	Tabulated "F"	Significance Level
Munshi Pond	Between Stations	3 & 6	2.35	4.8	NS
	Between Seasons	2 & 6	38.13	27.0	***
Raja Dighi Pond	Between Stations	3 & 6	1.97	4.8	NS
	Between Seasons	2 & 6	18.31	10.9	*

NS: Not significant; *: Significant at 5% P; ***: Significant at 0.1% P

Sometime, the values of NPP were found negative at stations A, B, and C in the Munshi pond (Table 1.7). It might be due to the abundant growth of blue-green algae blooms (*Microcystis* sp.) which shaded themselves and also to the other algae of different groups resulting in the productivity below nil recording due to the failure in using the incident light in required level by phytoplankton. Due to the presence of dense plankton population at optimum temperature, CRR was also high which may also be the reason for negative values of productivity. A part of oxygen might have also used to degrade the organic matter. These results are in accordance with the findings of Tailing (1960); Hawes (1985) and Groegger and Kimmel (1989).

Phytoplankton productivity was higher in the Munshi pond than in the Raja Dighi pond. It could be due to the dense algal growth in the Munshi pond. It was also found that phytoplankton productivity was higher at limnetic zone that the littoral zone in both the ponds. It was probably due to the comparatively high light penetration at the limnetic zone. Statistical t-Test showed significant differences in the values of productivity between the littoral and between the limnetic zones of both the ponds (Table 1.10).

Table 1.10: t-Test for Phytoplankton Productivity

Combinations	Degree of Freedom	Calculated "F"	Tabulated "F"	Significance Level
Lt (MP) Vs Lt (RDP)	10	4.63	4.59	***
Lm (MP) Vs Lm (RDP)	10	4.31	3.17	**

MP: Munshi Pond; RDP: Raja Dighi Pond; Lt: Littoral zone; Lm: Limnetic zone; **: Significant at 1% P;
***: Significant at 0.1% P

Coefficient of correlation between phytoplankton productivity and hydrobiological factors of water in both the ponds have been presented in Table 1.11. Productivity was negatively correlated with Secchi transparency ($r = 0.95$. $P < 0.001$) and positively correlated with temperature ($r = 0.93$. $P < 0.001$) in Munshi pond. These correlations reveal that with the increase in temperature, productivity also increased simultaneously resulting in the dense algal growth which reduced, the light penetration. Sreenivasan (1964); Vijayaraghavan (1971); Purushottaman and Bhatnagar (1976) and Bohra and Kumar (2002b) also obtained the direct correlation between temperature and productivity. Morris *et al.* (1974) and Li *et al.* (1980) concluded that with the increase in algal population Secchi depth was reduced. In the Raja Dighi pond, productivity was also found correlated negatively with Secchi transparency ($r = -0.82$. $P < 0.01$); while the phytoplankton population was very less. Non-algal turbidity was observed throughout the investigation period, which might be main reason for presenting the light penetration in the water (Bandyopadhyaya and Datta, 1986 and Agarwal *et al.*. 1995). A significant direct correlation between productivity and phytoplankton ($r = 0.90$; $P < 0.001$) was found in the Munshi pond, which support the high phytoplankton productivity due to dense algal population. In the Munshi pond, productivity was positively correlated with nitrate ($r = 0.68$; $P < 0.05$) while in the Raja Dighi pond, the correlation was insignificant. This relationship indicates that nitrate was the nutrient mainly responsible to control the phytoplankton productivity. Productivity was negatively correlated with dissolved oxygen in the Munshi pond. The water of the Munshi pond was highly loaded with organic matter due to direct discharge of untreated sewage and wastewater effluents. Thus, a large part of oxygen produced during photosynthesis might have been used to degrade the organic matter. A part of oxygen may also be used in maintenance of plankton themselves. Significant negative correlation of productivity with pH ($r = -0.66$; $P < 0.05$) in the Raja Dighi pond throws light on its low productive nature.

Table 1.11: Coefficient of Correlation Between Phytoplankton Productivity and Hydrobiological Factors of Water in the Munshi Pond and Raja Dighi Pond

Ponds	*Water Temp.*	*Secchi Transp.*	*pH*	*Dissolved Oxygen*	*Nitrate*	*Phosphate*	*Phyto-plankton*
Munshi Pond	0.93***	– 0.95***	0.12	– 0.72*	0.68*	0.23	0.90***
Raja Dighi Pond	0.51	– 0.82**	– 0.66*	– 0.73*	0.39	0.10	– 0.27

*: Significant at 5% P; **: Significant at 1% P; ***: Significant at 0.1% P

From the above discussion, it could be concluded that the trophic status of the Munshi pond is comparatively high than that of Raja Dighi pond. It can also be concluded that Munshi pond in highly productive than that of Raja dighi pond. In both the ponds, the littoral zone is much productive than the limnetic zone. Though, the physico-chemical factors were also favourable for better production but it showed less productivity in the Raja Dighi pond. Thus, non-algal turbidity is the main limiting factor in the Raja Dighi pond which reduced the light penetration.

References

Agarwal, N.C., Bais, V.S. and Shukla, S.N. (1990). Effects of nitrate and phosphate enrichment on primary productivity in the Sagar lake, Sagar. *Poll. Res.,* 9: 29–32.

Agarwal, N.C., Bais, V.S. and Arasta, T. (1995). Comparatively account of phytoplankton productivity in the Sagar lake and the Military Engineering lake: A statistical approach. *J. Freshwat. Biol.,* 7: 27–31.

APHA, AWWA and WPCF (1985). *Standard Methods for the Examination of Water and Wastewater,* 16th ed. Washington D.C. p. 1268.

Bais, V.S., Agarwal, N.C. and Shukla, S.N. (1993). Effects of *Microcystis* bloom on the density and diversity of the Cyanophycean population in the Sagar lake. *Proc. Acad. Environ. Biol.,* 2: 17–23.

Bais, V.S., Agarwal, N.C. and Arasta, T. (1995). Comparative study on seasonal changes in phytoplankton community in the Sagar lake and Military Engineering lake. *J. Freshwat. Biol.,* 7: 19–25.

Bandyopadhyay, B.K. and Datta, N.C. (1986). Studies on the primary productivity of a brackish water ecosystem of Hooghly-Matlah estuarine Complex West Bengal. *Poll. Res.,* 5: 7–11.

Bohra, C. and Kumar, A. (1999). Comparative studies of phytoplankton in two ecologically different lentic freshwater ecosystem. In: *Modern Trends in Environmental Pollution and Ecoplanning,* (Ed.) A. Kumar. ABP Publishers, Jaipur, pp. 220–242.

Bohra, C. and Kumar, A. (2002a). Phytoplankton dynamics in Udhuwa lake, Jharkhand (India). In: *Ecology and Ethology of Aquatic Biota,* Vol. II, (Ed.) A. Kumar. Daya Publ. House, Delhi, pp. 163–177.

Bohra, C. and Kumar, A. (2002b). Primary productivity of a sewage fed aquatic ecosystem. In: *Ecology and Ethology of Aquatic Biota,* Vol. I., (Ed.) A. Kumar. Daya Publ. House, Delhi, pp. 373–392.

Bohra, C. and Kumar, A. (2004). Plankton diversity in the wetlands of Jharkhand. In: *Biodiversity and Environment,* (Ed.) A. Kumar. APH Publ. Corp., New Delhi, pp. 91–123.

Cabecadas, G. and Brogueira, M.J. (1987). Primary production and pigments in three low alkalinity connected reservoirs receiving mine wastes. *Hydrobiol.,* 144: 173–182.

Cairns, J., Lanza, G.R. and Parker, B.C. (1972). Pollution related structural and functional changes in aquatic communities with emphasis on freshwater algae and protozoa. *Proc. Acad. Nat. Sci. Phil.,* 124: 79–127.

Carlson, R.E. (1977). A Trophic State Index for lakes. *Limnol. Oceanogr.,* 22: 361–369.

Desikachary, T.V. (1959). *Cyanophyta.* Indian Council of Agricultural Research (I.C.A.R.), New Delhi, pp. 183.

Edmondson, W.T. (1959). *Freshwater Biology,* 2nd ed. John Wiley and Sons, Inc. N.Y., pp. 412.

Edmondson, W.T. (1974). A simplified method for counting phytoplankton. In: *A Manual on Method for Measuring Primary Production in Aquatic Environment,* (Ed.) Richard, A.V., pp.14.

Fristch, F.E. (1935). *The Structure and Reproduction of Algae.* Vol. 1. Cambridge, pp. 791.

Garder, T. and Gran, H.H. (1927). Investigation of the production of plankton in the Osicfierd. *J. Cons. Perm. Int. Explor. Mer.,* 42: 1–18.

Galvao, S.M.F.G. (1985). Primary production in the reservoirs in Southern Brazil. *Hydrobiologia,* 122: 81–88.

Ganapati, S.V. (1957). Limnological studies of the upland waters in Madras State. *Arch. F. Hydrobiol.,* 53: 30–61.

Golterman, H.L. and Meyer, S. (1985). The geochemistry of Rhina and Rone rivers: The relation between pH, Calcium and Hardness. *Hydrobiol.,* 126: 6–13.

Grogger, A.W. and Kimmel, B.L. (1989). Relationship between photosynthetic and respiratory carbon metabolism in freshwater phytoplankton. *Hydrobiologia,* 173: 107–117.

Gupta, H. P. and Kumar, A. (1998). Impact of domestic sewage on the limnology of a waterbody of Santal Pargana (Bihar). Proceeding of the National Seminar on Environmental Pollution and Ecoplanning held at S.K. University, Dumka (20–21st March, 1998).

Gupta, H.P., Roy, S.S. and Kumar, A. (2003). Ecobiological monitoring of sewage pollution using phytoplankton as a biotool. In: *Environment Pollution and Management,* (Eds.) Kumar, A., Bohra, C. and Singh, L.K. APH Publ. Corp., New Delhi, pp. 429–452.

Hannan, H.H. and Young, W.J. (1974). The influence of a deep storage reservoir on the physico-chemical limnology of a central Texas river. *Hydrobiol.,* 44: 177–207.

Hawes, I. (1985). Light climate and phytoplankton photosynthesis in Maritime Antarctik lakes. *Hydrobiologia*, 123: 69–79.

Harikrishnan and Abdu1 Aziz, P.K. (2000). Primary production studies in a freshwater temple tank in Kerala. *Indian J. Environ. and Ecoplan.*, 3: 127–130.

Jackson, D.F. (1961). Comparative studies on phytoplankton population in relation to total alkalinity. *Verh. Int. Ver. Limnol.*, 14: 125–133.

Kaufman, H.L. (1980). Stream Aufwuchs accumulation process: Effects of Ecosystem Depopulation. *Hydrobiol.*, 70: 75–81.

Kaushik, S., Saksena, M.N. and Saksena, D.N. (1991. Phytoplankton population dynamics in relation to environmental parameters in Matsya Sarovar at Gwalior (M.P.). *Acta. Botanica.* 19: 113–116.

Khan, I.A., Khan, A.A. and Haque, N. (1988). Primary production in Sheikha jheel at Aligarh. *Environ. Ecol.*, 6: 858–862.

Khan, M.A. and Zutshi, D.P. (1979). Relative contribution by Nanno and Net plankton towards primary production of two Kashmir Himalayan lakes. *J. Ind. Bot. Soc.*, 58: 263–267.

Kumar, A. (1996). Impact of sewage pollution on chemistry and primary productivity of two freshwater bodies in Santal Pargana (Bihar). *Indian J. Ecol.*, 23: 86–92.

Kumar, A. (1998). Diversification of phytoplankton and its correlation with certain physico-chemical properties of a tribal pond in Santal Pargana (Bihar). *J. Environ. Poll.*, 5: 83–95.

Kumar, A. (1999). Impact of coal mining effluents on the primary productivity of water bodies and fish farming in and around the Chitra Coal fields in Santal Pargana, Bihar. India. In: *Environmental Management in Coal Mining and Thermal Power Plants,* (Eds.) Mishra, P.C. and Naik, A.. Technoscience Publ., Jaipur, pp. 188–196.

Kumar, A. (2002). The ecology of aquatic biota in thermal springs. In: *Ecology and Ethology of Aquatic Biota*, Vol. II, (Ed.) A. Kumar. Daya Publ. House, New Delhi, pp. 1–53.

Kumar, A. (2003). Investigation on primary productivity and energy structure of a thermal spring. In: *Environment. Pollution and Management*, (Eds.) A. Kumar, Bohra, C. and Singh, L.K. APH. Publ. Corp., New Delhi, pp. 1–14.

Kumar, A. and Bohra, C. (1998). Impact of human activities on the limnology of the religious pond at Basukinath Dham (Dumka), Bihar. *Environ. and Ecol.*, 16: 809–812.

Laal, A.K. (1989). Seasonal variation in primary productivity and photosynthetic efficiency in two tropical swamps. *J. Freshwater Biol.*, 1: 55–59.

Lackey, J.B. (1938). The manipulation and counting of river plankton and changes in some organisms due to formation preservation. *U.S. Public Health Reports*, 53: 2080–2093.

Li, W.K.W., Glover, H.E. and Morris, I. (1980). Physiology of carbon photo assimilation by *Oscillatoria thiebautii* in the Caribbean sea. *Limnol. Oceanogr.*, 25: 447–456.

Misra, B.N. and Misra, N.K. (1989). *Introductory Practical Biostatistics.* Naya Prakash, Calcutta, p. 190.

Misra, Y.K. and Nayak, B.K. (2000). Biomass productivity of energy plantations on Himalayan Foot Hill. *Environ. and Ecol.*, 18: 13–16.

Morris, I., Glover, H.E. and Yentsch, C.S. (1974). Products of photosynthesis by marine phytoplankton. The effect of environmental factors on the relative rates of protein synthesis. *Mar. Biol.*, 27: 1–9.

Munawar, M. and Munawar, I.F. (1986). The seasonality of phytoplankton in North American Great lakes: A comparative study. *Hydrobiologia,* 138: 85–115.

Oglesby, R.T. and Schaffer, W.R. (1975). The response of lake to phosphorus. In: *Nitrogen and Phosphorus Food Production and Environment,* (Ed.) Porter, K.S. Ann. Arbor. Sci., pp. 23–57.

Outridge, P.M., Arthigton, A.H. and Miller, G.J. (1989). Limnology of naturally acidic oligotrophic Dune lake in sub-tropical Australia including chlorophyll-phosphorus relationships. *Hydrobiol.,* 179: 39–51.

Pandit, B. (1999). Dynamics of primary productivity of certain ponds at Barharwa (Santal Pargana): A tribal division of Bihar (India). In: *Modern Trends in Environmental Pollution and Ecoplanning,* (Ed.) A. Kumar. ABD Publ., Jaipur, pp. 97–99.

Paramsivam, M. and Sreenivasan, A. (1981). Changes in the algal flora due to pollution in Cauvery river. *Indian J. Environ. Hlth.,* 23: 222–238.

Pati, S. and Sahu, B.K. (1993). Studies on primary production in Rengali reservoir (Orissa). *J. Ecobiol.,* 5: 85–88.

Phillipose, M.T. (1967). *Chlorococcales.* ICAR, New Delhi, pp. 1–365.

Purushottaman, A. and Bhatanagar, G.P. (1976). Primary production studies in porto Novo waters, South India. *Arch. Hydrobiol.,* 77: 37–50.

Rai, D.N. and Sharma, U.P. (1991). Energy transfer efficiency and primary productivity of tropical wetlands in north Bihar (India). *J. Freshwater Biol.,* 3: 89–97.

Ramanibai, P.S. and Ravichandran, S. (1987). Limnology of a urban pond at Madras, India. *Poll. Res.,* 6: 77–81.

Ramkrishnaiah, M. and Sarkar, S.K. (1982). Plankton productivity in relation to certain hydrobiological factors in Konar reservoir (Bihar). *J. Inland Fish. Soc. India,* 14: 58–68.

Randhwa, M.S. (1959). *Zygnamataceae.* ICAR, New Delhi.

Richardson, K., Beardall, J. and Raven, J.A. (1983). Adaptation of cellular algae to irradiance, an analysis of strategies. *New Phytol.,* 93: 157–191.

Singh, B. and Saha, P.K. (1987). Primary productivity in a composite fish culture pond at Kulia Fish Farm, Kalyani, W.B. Proc. Nat. Acad. Sci., India, 57: 11–25.

Smith, G.M. (1950). *The Freshwater Algae of United States.* McGraw Hill Book Co., N.Y.

Sreenivasan, A. (1964). Limnology, primary production and fish production in a tropical pond. *Limnol. and Oceanogr.,* 9: 391–396.

Takamura, K., Sugaya, Y., Takamura, N., Hanazato, T., Yasuno, M. and Iwakuma, T. (1989). Primary production of phytoplankton and standing crops of zooplankton and zoobenthos in hypertrophic lake, Teganuma. *Hydrobiologia,* 173: 173–184.

Tailing, J.F. (1960). Self shading effects in natural population of a planktonic diatom. *Watter and Leben,* 12: 133–149.

Tailing, J.F. (1971). The underwater light climate as a controlling factor in the production ecology of freshwater plankton. *Limnologia,* 19: 214–243.

Trivedy, R.K. and Goel, P.K. (1986). *Chemical and Biological Methods for Water Pollution Studies*. Environ. Pub., Karad, p. 250.

Trivedy, R.K., Garud, J.M. and Goel, P.K. (1985). Studies on chemistry and phytoplankton of few freshwater bodies in Kolhapur with special reference to human activity. *Poll. Res.*, 4: 25–44.

Verma, P.K. and Datta Munshi, J.S. (1987). Plankton community structure of Badua reservoir, Bhagalpur (Bihar). *Trop. Ecol.*, 28: 200–207.

Vijayakumar, K. (1994). Seasonal variation in the primary productivity of a tropical pond. *J. Ecobiol.*, 6: 207–211.

Vijayaraghavan, S. (1971). Seasonal variation in primary productivity of three tropical ponds. *Hydrobiologia*, 38: 395–408.

Vyas, L.N. (1968). Studies in phytoplankton ecology of Picchola lake, Udaipur. Proc. Symp. Recent Adv. Trop. Ecol., pp. 334–347.

Ward, H.B. and Whipple, G.C. (1959). *Freshwater Biology*, (Ed.) Edmondson, W.T. John Wiley and Sons, Inc., N.Y., p. 1248.

Wetzel, R.G. (1957). Variation in productivity in Goose and hypertrophic sylvan lakes. *Indiana Invest. Indiana Lakes Streams*, 7: 147–184.

Wetzel, R.G. (1975). *Limnology*. W.B. Saunders Co., Philadelphia, p. 743.

Wisharad, S.K. and Mehrotra, S.N. (1988). Periodicity and abundance of plankton in Gulasia reservoir in relation to certain physico-chemical conditions. *J. Ind. and Fish. Soc., India*, 20: 42–49.

Chapter 2

Studies on the Phytoplankton of Olero Creek and Parts of Benin River, Nigeria

T.A. Adesalu and D.I. Nwankwo***

**Department of Botany and Microbiology, University of Lagos, Akoka, Lagos, Nigeria*
***Department of Marine Sciences, University of Lagos, Akoka, Lagos*

ABSTRACT

Studies on the phytoplankton of Olero creek and parts of Benin river were undertaken on four occasions (February, March, May, June 1995). Of the 36 phytoplankton taxa recorded, 31 were diatoms, four blue-green algae and one yellowish green alga. The paucity of phytoplankton populations has been related to high level of suspended solids (> 30.20 mgl^{-1}) which limited light penetration and reduced photosynthetic depth. The low species diversity ($H < 1.0$) and low evenness ($1 < 0.5$) may be due to effect of dredging and re-suspension of sediments as well as occasional oil contamination of the environment.

Keywords: *Oil mining activity, Phytoplankton, Pollution, Species diversity, Niger Delta.*

Introduction

The Niger Delta is a remarkable landmark on the West African coast and one of the world's major hydrocarbon provinces. From its apex at Aboh (Delta State) until it fans out through numerous rivers, creeks and creek-lets into the seas, covering a distance of more than 300km, it provides various environments. F or instance, Powell and Onwuteka (1980) and Ibiebele *et al.* (1984) grouped the rivers and creeks of the Niger Delta based on "white", 'clear"" and "black' water types and conductivity relationships as proposed by Sioli (1973) for tropical waters of the Amazonia.

The Niger Delta is rich in both oil and gas. Oil in the Niger Delta is of light waxy type typical of deltas. According to Chevron Nigeria Limited (CNL) report (1993) Nigerian crude oil enters the environment through seeps as a result of man's activities, through accidental spills or long term low-level discharges. In the Niger Delta, accidental oil spills are serious environmental problems (Bott, *et al.*, 1978; Pudo, 1985). Oil related activity such as dredging introduces enormous quantities of derived solids from adjoining mangroves into the water column, deplete mangrove vegetation and open up wild life sanctuaries for exploitation (EAU report, 1992). The effects of crude oil on aquatic biota have been reported by several workers (McCauley, 1966; Ho and Karim, 1978; Imevbore and Odu, 1980). Laboratory results also reveal specific effects of crude oil on algae (O'Briell and Dixon, 1976; Vandermeulell and Ahern, 1976). In the field, algal communities have been known to be affected by oil related contamination (Minter, 1964; McCauley, 1966; Gaur and Kumar 1985; Singh and Gaur, 1989; Nwankwo, 2000).

In Nigeria, Pudo (1985) reported a correlation between species richness and the level of oil contamination in the Bonny River. Murday *et al.* (1988) observed post oil spill blooms of cyanophyceae on the Escravos mud flats while Powell (1987) stated that generally the Bostrychian algal community on mangrove pneumatophores may not recover from the effect of oil spills for up to a year. In the Challomi creek system, Niger Delta, Nwadiaro (1990) observed a reduction in the diversity and size of algal population in some areas three years after being impacted by crude oil.

Phytoplankton organisms are the basis of aquatic productivity and any serious alteration in their constitution may have detrimental consequences on the food chain and concurrently entire community structure. The aim of the present study was to document the phytoplankton of Olero creek and parts of the Benin river, are prone to stress; particularly increased suspended matter and occasional spills resulting from oil mining activity.

Materials and Methods

Study Area

Olero creek and the Benin river area (Figure 2.1) are located on the western flank of the lower Niger Delta between longitudes 5 20' and 5 27'E and latitudes 5 43' and 5 55'N. The area falls within the hot, humid tropics that experiences two seasons, the rainy season (April–November) and dry season (December–March). The annual rainfall is usually in excess of 3000mm and the area is covered predominantly by seasonal swamps and extensive mangrove forests. The mangrove species (taller ones dominating the riparian and bank vegetation) are dominated by *Rhizophora harrisonii* Leechman, *R. racemosa* G.R.W. Meyer, *R. Mangle* Linn, *Avicennia Africana* P. Beau V., *Acrotiscum aureum* Linn and *Paspalum orbiculare* Forst. A number of floating macrophytes *Eichlornia crassipes* (Mart.) Solms, *Pistia stratiotes* Linn, *Azolla pinnata* R.Br.Var. *Africana* (Desv.) Bak., *Lemna paucicostata* Heglem, *Salvinia, nymphella* Desv. *Ceratophyllum demersum* Linn and *Vossia cuspidata* enter Olero creek and the Benin river in the rainy season.

Owing to the proximity of the site to the sea, there is intensive marine influence and because of extensive upstream activity, lots of dredging and channelization take place introducing much allochthonous materials and acidifying immediate environments. Occasional accidental oil spills are reported either in the open water or surrounding swamps (Powell, 1987).

Collection of Samples

Duplicate water samples were collected in February, March, May and June. 0.5 m below the water surface from various sites along Olero creek and parts of Benin river, with a Ruthner 2.5L non-metallic

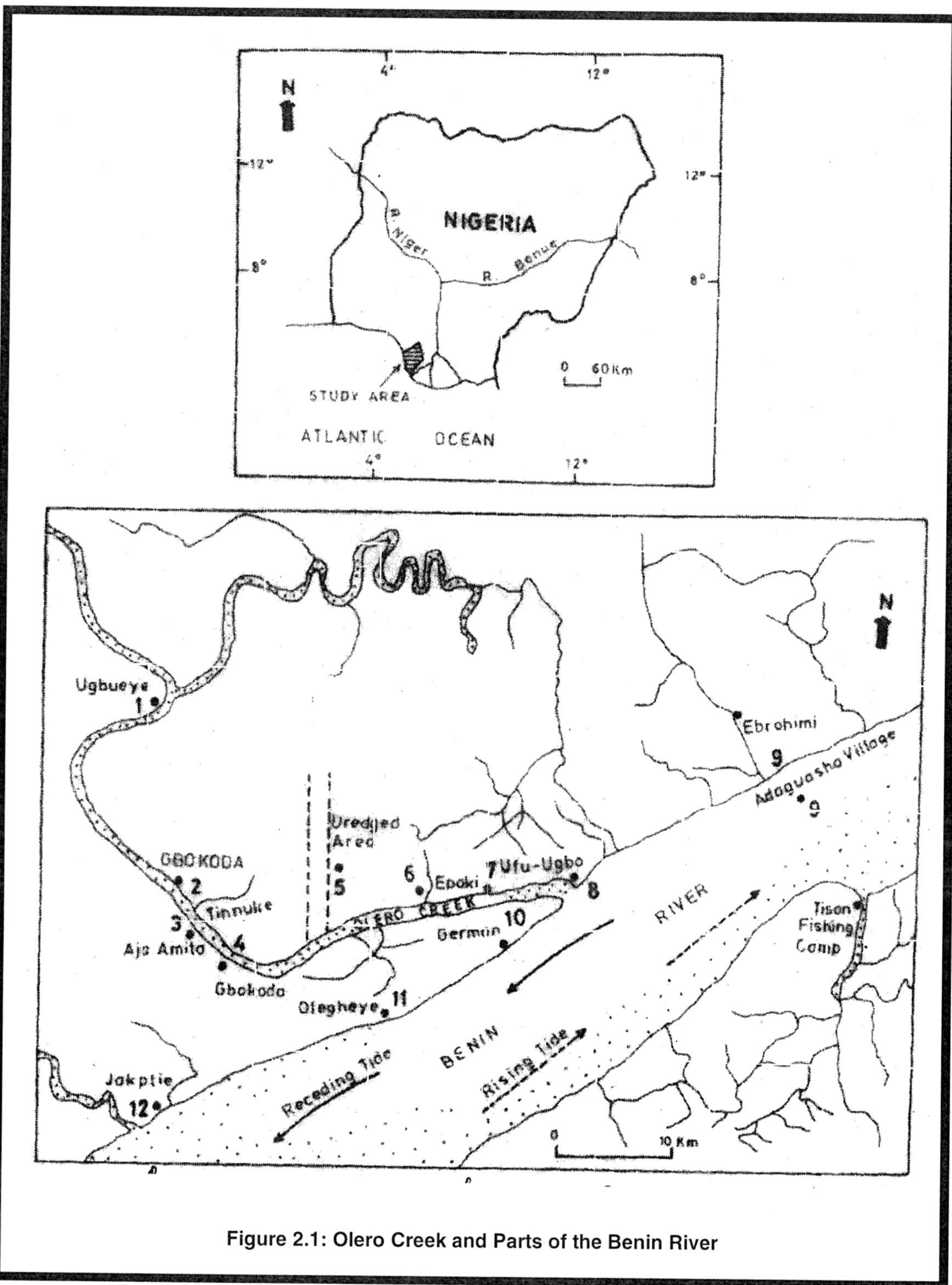

Figure 2.1: Olero Creek and Parts of the Benin River

water sampler and stored in 500ml glass bottles with screw caps, Site 12 (Jackpatie) is near the estuary while site 1 (Ugbueye) is farthest from the estuary. Horizontal phytoplankton hauls were made using a 55 m mesh-size plankton net tied onto a motorized boat and towed at low speed (< 24 knots) for 10min. All biological samples were stored in 500ml plastic bottles and preserved in 4 per cent unbuffered formalin.

Physical and Chemical Analysis

Surface water temperature was measured in the field using a mercury thermometer accurate to 0.1 C while pH, conductivity, total dissolved solids (TDS) and dissolved oxygen (DO) were determined using the Cibacorning meter by attaching appropriate sensor probes. Total suspended solids (TSS) were estimated by the evaporation of 100ml aliquot of sample and transparency measured with a 20cm diameter white and black Secchi disc. Surface water salinity was determined using the silver nitrate-chromate method as described by Barnes (1980).

Phytoplankton Analysis

Biological samples collected on four occasions (February, March, May and June) were investigated using a light microscope (Wild MII Heerbrugg) with a calibrated eye piece. Aliquots of original samples were thoroughly investigated and counted using the drop-count microtransect method as described by Laekey (1938). Sub-samples of the original sample were acid-cleaned (Barber and Harworth, 1981) and empty diatom frustules as well as other phytoplankton were identified at approximate magnifications using relevant texts (Subramanyan, 1946; Desikachary, 1959; Hendery, 1964; Patrick and Reimer, 1966, 1975) Phytoplankton counts were expressed as umber of cells per ml.

Community Structure

Three community composition parameters, species richness d (Margalef, 1951) diversity index H (Shanon and Weaver, 1963) and evenness J (Pielou, 1966) were used to estimate phytoplankton response to environmental stress. Species richness d given by the equation $d = S-1/InN$ and diversity index H given by the equation:

$$H = -\Sigma ni/N \log^2 ni/N$$

where,

S: Number of species in a population;

N: Number of individuals;

N: Total number of individuals in S species;

ni: Number of individuals in the species.

Species evenness was determined by the equation:

$$J = H/H_{max}$$

$$H_{max} = \log S$$

The index of similarity S (Odum, 1971) between two samples were estimated using the equation:

$$S = 2C/A+B$$

A, B: Number of species in compared samples;

C: Number of species common in both samples.

Results

Physical and Chemical Characteristics

The surface water temperature during the sampling period ranged between 29.50 C and 30.20 C in the dry months and between 27.20 C and 29.50 C in the wet months. Transparency in the wet months (0.10–0.20cm) was lower than in the dry months (0.20–0.35cm). Transparency was relatively lower in the creek than in the Benin river. Total dissolved solids in the dry months ranged between 2.04 and 3.84mg^{-1} while in the wet months it ranged between 3.04 and 6.94mgl^{-1}. Total suspended solid values were relatively higher in wet months (30.9–71.2mgl^{-1}) than in the dry season (20.10–32.8mgl^{-1}). In Olero creek T.S.S. values were lower at both dry and wet months (21.4 dry, 71.26 wet) than in the river (20.10 dry, 30.91 wet). Conductivity values ranged between 21.30 and 30.7 µScm^{-1} in the dry months while in the wet months it ranged between 8.12 and 15.9 µScm^{-1}. Salinity values varied between 14.30 and 21.30 per cent in the dry months while in the wet months it ranged between 7.20 and 165 per cent. The pH in the dry months ranged between 5.40 and 7.00 while in the wet months it ranged between 7.00 and 7.80. Dissolved oxygen values varied between 5.52 and 6.01mgl-1 in the dry month and between 8.81 and 15.91mgl^{-1} in the wet months (Table 2.1).

The Phytoplankton Spectrum

The species composition arid abundance of phytoplankton at Olero creek and parts of the Benin river are presented in Table 2.2. In Olero creek and parts of the Benin river investigated, three classes of phytoplankton, bacillariophyceae, cyanophyceae and xanthophyceae were identified. Phytoplankton biomass was low during the four sampling occasions. However, greater number of algal taxa were recorded at sites 3, 6 and 7 representing the boundary between creek and river water (Table 2.2).

The baccillariophyceae (17 genera, 31 species) was the most abundant algal class. Among the bacillariophyceae, 18 species belonging to 10 genera were pennate forms while the centric forms made up of 7 genera comprising 13 species. *Navicula expansa, N. clavata, N. mutic, Nitzchia abruensis, N. irresolute, Pinnularia debesiii* and *Gyrosigma balticum, G. spenceri* and *Tropdoneis semistrata* were the more important pennate diatoms in terms of frequency and abundance. *Biddulphia sinensis, B. monilensis, B. regia, Coscinodiscus centralis, C. eccentricus, C. oculus-iridis, Cyclotella meneghinialna, C. boldanica* and *Podosira tenera* were the most prevalent centric forms. The cyanophyceae (2 genera, 4 species) were all heterocytous forms. The cyanophyceae were more abundant in sites 1 and 3. Prominent species included *Oscillatoria magarittfera* and *Lyngbya martensiana.* The Vaucheriaceae was represented by a soil alga *Botrydium granulatum.*

Community Structure

Species richness d was high (d > 1.8) during the four sampling occasions in sites, 3, 4, 7 and 8 (Figure 2.2). Apart from sites 8 and 10 where evenness was relatively high in the dry months, the other sites experienced low evenness. Shannon and weaver diversity index (H) though generally low (≤1.0) was comparatively high in sites 3. 4 and 10 in March and June (Figure 2.2).

Discussion

Two hydrographic factors, flood water of the rainy months and tidal sea water incursion, naturally influence the Benin river and associated creeks. Flood waters diluted ion concentration and introduced large quantities of derived materials from the benthos and wetlands. According to Hart and Fuller (1972), the concentration of derived suspended materials is increased through channelisation and

Table 2.1: Selected Physico-chemical Characteristics of Olero Creek and Benin River in 1995

Parameters	*Months*	*1*	*2*	*3*	*4*	*5*	*6*	*7*	*8*	*9*	*10*	*11*	*12*
Surface Water Temperature (°C)													
	February	30.20	30.10	29.70	29.60	29.50	29.90	30.10	30.10	29.80	29.80	30.20	30.00
	March	30.00	22.75	29.50	30.10	29.80	30.00	29.80	29.50	30.00	29.60	30.10	30.10
	May	27.30	27.50	27.70	29.60	28.20	29.00	28.50	29.00	29.50	28.50	28.50	28.50
	June	27.30	27.20	27.05	27.90	27.90	27.50	27.50	27.60	27.80	27.20	27.70	27.70
Transparency (1 m)													
	February	0.25	0.20	0.30	0.20	0.24	0.25	0.25	0.35	0.25	0.25	0.25	0.25
	March	0.20	0.20	0.30	0.20	0.20	0.25	0.20	0.30	0.25	0.30	0.30	0.30
	May	0.15	0.15	0.15	0.16	0.15	0.16	0.15	0.10	0.20	0.15	0.20	0.20
	June	0.12	0.11	0.15	0.12	0.10	0.12	0.14	0.10	0.16	0.14	0.16	016
Total Dissolved Solids (mg/l)													
	February	3.27	3.79	3.83	3.25	3.83	3.96	3.84	3.17	3.92	2.74	2.92	2.92
	March	3.82	3.19	3.94	3.10	3.94	3.84	3.72	3.10	2.99	2.24	2.99	2.99
	May	4.25	3.32	4.85	5.73	4.85	4.02	4.80	4.37	3.56	3.04	3.56	3.56
	June	6.40	4.52	6.95	5.94	6.94	5.62	5.92	5.04	3.84	4.17	3.84	3.84
Total Suspended Solids (mg/l)													
	February	41.10	32.80	32.60	32.40	35.50	30.20	32.80	30.90	21.70	22.10	22.90	22.90
	March	43.30	31.70	27.40	29.40	29.50	28.40	33.20	32.90	20.10	26.70	22.70	22.70
	May	60.20	56.30	47.90	50.40	64.50	49.80	52.70	52.20	30.20	34.30	31.10	31.10
	June	64.30	62.15	52.50	59.60	71.20	55.90	60.30	68.40	52.70	57.65	30.91	30.91

Contd...

Table 2.1–Contd...

Parameters	Months	1	2	3	4	5	6	7	8	9	10	11	12
Conductivity (mS dm^{-1})													
	February	27.90	29.30	29.10	29.40	26.40	29.30	21.60	21.60	23.20	24.30	24.70	24.70
	March	28.70	29.90	30.70	30.20	29.80	30.30	26.50	26.50	27.40	28.60	31.30	31.30
	May	11.59	11.03	15.60	15.91	13.90	10.79	12.00.	12.08	10.62	15.40	15.62	15.62
	June	10.67	8.12	10.11	10.58	10.90	8.94	9.00	9.01	8.81	9.94	10.81	10.81
Salinity (%)													
	February	18.30	19.20	17.30	16.20	19.40	15.60	15.60	19.60	16.40	19.20	19.70	19.70
	March	19.70	18.60	18.70	16.11	17.30	17.40	18.40	21.30	18.30	18.10	19.80	19.80
	May	13.10	12.90	14.10	15.50	13.20	13.00	13.30	14.30	13.80	13.95	16.50	16.50
	June	8.00	8.50	8.00	8.70	7.25	7.20	8.10	7.40	7.24	8.20	9.65	9.65
pH													
	February	7.00	6.80	5.69	6.98	6.42	6.20	6.50	6.80	6.80	5.40	5.90	5.90
	March	6.80	6.90	5.69	6.74	6.93	6.40	6.30	6.50	6.60	5.90	6.40	6.40
	May	7.59	7.71	6.46	7.35	7.51	6.80	5.82	7.85	7.24	5.78	6.82	6.82
	June	7.04	7.06	7.00	7.00	7.00	6.90	7.00	7.04	7.00	7.00	7.00	7.00
Dissolved Oxygen (mg/l)													
	February	5.85	5.87	5.71	5.52	5.91	6.01	5.91	5.65	5.93	5.91	5.82	5.82
	March	5.95	5.94	5.80	5.30	5.85	5.92	5.94	5.92	5.73	5.83	5.69	5.69
	May	7.00	6.90	6.90	6.80	7.15	7.00	5.80	7.10	7.50	7.00	7.50	7.50
	June	7.20	7.10	7.04	7.02	7.25	7.20	7.01	7.40	7.80	7.20	7.80	7.80

Table 2.2: The Composition and Abundance of Phytoplankton at the Olero Creek and Part of Benin River (February, March, May, June 1995). No. of Organisms ml^{-1}

Taxa	*UGBUEYE*				*GBOKODA*				*TINNUKE*				*AJA AMITA*				*DREDGED ARE*				*WELLS 4/5*			
	1				*2*				*3*				*4*				*5*				*6*			
	F	*M*	*MY*	*J*	*F*	*M*	*MY*	*J*	*F*	*M*	*MY*	*J*	*F*	*M*	*MY*	*J*	*F*	*M*	*MY*	*J*	*F*	*M*	*MY*	*J*
Class Bacillariophyceae																								
Order 1: Centrales																								
Bacteriastrum hyalinum Lauder	–	–	–	–	4	4	–	–	–	–	–	–	–	–	–	–	–	–	–	–	–	–	–	–
Biddulplria longicruous Greville	–	–	–	–	–	–	–	–	3	3	–	–	–	–	–	–	–	–	–	–	–	–	–	–
B. regia (Schutt) Ostenfeld	–	–	–	–	–	–	–	–	–	–	–	–	–	–	–	–	–	–	–	–	–	–	–	–
B. rhombus (Her.) Wm. Smith	–	1	–	–	–	–	–	–	1	1	–	–	–	–	–	–	–	–	–	–	2	–	–	–
B. sinensis Greville	–	–	–	–	–	–	–	–	1	1	1	2	–	–	–	–	3	4	–	–	–	–	–	–
Coscinodiscus centalis Ehrenberg	–	–	1	–	–	2	–	–	2	2	1	3	–	–	–	–	7	2	–	–	2	4	–	–
C. eccentricus Ehrenberg	–	–	1	–	2	1	–	–	1	1	1	2	–	–	–	–	5	2	–	–	2	–	–	–
C. oculus-tridsi (Ehr.)	–	–	–	–	1	–	–	–	4	3	–	–	–	–	–	–	2	2	–	–	3	3	–	–
Cyclotella boldamica Eulenst	–	1	–	–	–	1	–	–	2	2	–	–	1	–	–	–	4	–	–	–	–	–	–	–
C. meneghiana Kutz.	–	–	1	–	–	1	–	–	1	2	–	–	1	–	–	–	6	4	–	–	2	4	–	–
Leptocylindrus danicus	–	–	–	–	–	–	–	–	2	4	–	–	–	–	–	–	–	–	–	–	–	–	–	–
Podosira tenebro Leudug	–	4	–	–	–	–	–	–	4	6	–	–	–	–	–	–	2	1	–	–	–	–	–	–
Triaceratium flavus Ehr.	–	–	–	–	–	–	–	–	2	3	–	–	–	–	–	–	–	–	–	–	–	–	–	–
Order 2: Pennales																								
Amphiprora augustata	–	–	–	–	–	–	–	–	–	–	–	2	–	–	–	2	2	–	–	–	–	–	–	–
Cocconeis scutellum Ehr.	–	–	–	–	–	–	–	–	1	2	–	–	–	–	–	2	–	–	–	–	–	–	1	2
Diploneis ovalis (Hilse) Cl.	–	–	–	–	–	–	–	–	–	–	–	2	–	–	–	–	–	–	–	–	–	–	–	1
Eunotia glacialis Meist	–	–	2	1	–	–	–	–	–	–	–	–	–	–	3	2	–	–	–	–	–	–	1	1
E. pyramidata Hudst.	–	–	1	1	–	–	–	–	–	–	–	–	–	–	2	2	–	–	–	–	–	–	1	2

Contd...

Table 2.2–Contd...

Taxa	UGBUEYE 1				GBOKODA 2				TINNUKE 3				AJA AMITA 4				DREDGED ARE 5				WELLS 4/5 6			
	F	M	MY	J	F	M	MY	J	F	M	MY	J	F	M	MY	J	F	M	MY	J	F	M	MY	J
Fragilaria construens Ehr.	–	–	–	–	–	–	–	–	–	–	–	–	–	–	–	–	–	–	–	–	–	–	–	2
F. oceanica Cleve	–	6	–	–	–	6	–	–	–	–	–	–	–	–	–	–	–	–	–	–	–	–	–	2
Gyrosigma balticum (Ehr.) Rabenhorst	2	3	–	–	–	–	–	–	2	2	–	–	–	–	4	–	–	–	–	–	–	–	4	3
G. Spenceri (Wm Sm) Cleve	2	3	–	–	–	–	–	–	–	–	–	–	1	2	–	–	2	2	–	–	–	–	13	6
G. scalproides (Rab.) Cl.	1	5	–	–	–	–	–	–	–	–	1	2	1	–	–	–	2	–	–	–	–	–	6	4
Mastoglora exilis Grevile	–	–	1	2	–	–	–	–	–	–	–	–	–	–	–	–	–	–	–	2	–	–	–	10
Navicula clavata Gregory	–	4	2	–	1	–	–	–	1	–	2	1	1	–	2	2	2	2	–	–	2	–	12	10
N. cryptocephala Kutzing	4	6	–	–	2	2	–	–	2	–	–	–	–	–	–	2	1	1	–	–	1	–	13	10
N. expansa	2	–	–	–	–	–	–	–	–	–	–	–	–	–	–	–	–	–	–	–	–	–	–	8
N. mutica	–	–	3	–	–	–	–	–	–	–	–	3	–	–	2	–	–	6	–	–	–	–	–	10
Nitzschia abruensis	–	–	–	–	3	–	–	–	3	–	–	–	–	2	–	–	–	–	–	–	–	–	–	–
N. irresoluta	1	–	–	–	–	–	–	–	–	–	1	–	–	–	–	–	–	–	–	–	–	–	–	–
Pinnularia debesii	–	2	–	–	–	–	–	4	–	–	–	–	–	–	–	2	–	–	–	–	–	–	3	–
Algal Genera	–	–	–	–	–	–	–	–	–	–	–	–	–	–	–	–	–	–	–	–	–	–	–	–
Cyanophyceae	–	–	–	–	–	–	–	–	–	–	–	–	–	–	–	–	–	–	–	–	–	–	–	–
Lyngbya martensiana Meneghini	6	–	4	–	–	–	–	6	–	3	–	8	–	–	–	–	–	–	–	–	–	–	–	–
Oscillatoria agardii Gomont	–	–	–	–	–	–	–	–	–	–	–	5	–	–	–	–	–	–	–	–	–	–	–	–
O. formosa	–	5	2	–	–	1	–	–	4	–	6	–	–	–	–	–	–	–	–	–	–	–	–	–
O. magaritifera	–	9	2	4	–	–	2	–	6	–	3	8	–	–	–	2	–	–	2	4	–	2	2	–
Xanophyceae																								
Vaucheriales																								
Botrydium granulatum	–	–	–	–	–	–	–	–	–	–	–	–	–	2	2	–	–	–	–	–	–	–	–	–

Contd...

Table 2.2–Contd...

Taxa	EPOKI 7				UFU UGBO 8				EBROHIMI 9				GERMAN 10				OLEGHEYE 11				JAKPATIE 12			
	F	*M*	*MY*	*J*	*F*	*M*	*MY*	*J*	*F*	*M*	*MY*	*J*	*F*	*M*	*MY*	*J*	*F*	*M*	*MY*	*J*	*F*	*M*	*MY*	*J*
Class Bacillariophyceae																								
Order 1: Centrales																								
Bacteriastrum hyalinum Lauder	–	–	–	–	–	–	–	–	–	–	–	–	–	–	–	–	–	–	–	–	–	–	–	–
Biddulplria longicruous Greville	–	–	–	–	–	–	–	–	–	–	–	–	–	–	–	–	–	–	–	–	–	–	–	–
B. regia (Schutt) Ostenfeld	–	–	2	1	–	–	–	–	–	–	–	–	–	–	–	–	–	–	–	–	–	–	–	–
B. rhombus (Her.) Wm. Smith	2	2	–	–	–	2	1	2	1	–	–	–	–	–	2	2	–	–	–	–	–	–	2	4
B. sinensis Greville	–	–	1	2	–	–	1	2		–	–	–	–	–	3	6	–	–	–	–	3	3	–	–
Coscinodiscus centalis Ehrenberg	2	4	2	2	–	–	–	–	–	–	–	–	–	–	3	4	–	–	–	–	4	4	–	–
C. eccentricus Ehrenberg	1	2	1	2	–	–	–	–	–	–	2	2	–	–	3	4	2	4	–	–	4	2	–	–
C. oculus-tridsi (Ehr.)	2	2	3	2	1	2	–	–	–	–	–	–	–	–	6	4	–	–	–	–	4	7	–	–
Cyclotella boldamica Eulenst	–	–	2	4	4	2	–	–	–	–	–	–	–	–	6	6	1	2	–	–	2	2	–	–
C. meneghiana Kutz.	3	2	2	3	3	1	–	–	–	2	–	–	–	–	8	6	2	4	–	–	2	–	–	–
Leptocylindrus danicus	5	6	–	–	–	–	–	–	–	–	–	–	–	–	–	–	–	–	–	–	4	2	–	–
Podosira tenebro Leudug	–	–	–	–	4	1	–	–	–	–	–	–	–	–	–	–	–	–	–	–	–	1	–	–
Triaceratium flavus Ehr.	–	–	–	–	–	–	–	–	–	–	–	–	4	2	–	–	–	–	–	–	–	–	–	–
Order 2: Pennales																								
Amphiprora augustata	–	–	–	–	–	–	–	–	–	–	–	–	–	–	–	–	–	–	–	–	–	–	–	–
Cocconeis scutellum Ehr.	1	2	–	–	–	–	1	2	1	2	–	–	–	–	–	–	–	–	–	–	–	–	2	2
Diploneis ovalis (Hilse) Cl.	–	–	–	–	–	–	–	–	–	–	–	–	–	–	–	–	–	–	–	–	2	1	–	–
Eunotia glacialis Meist	–	–	8	2	–	–	–	–	–	–	–	–	–	–	–	–	–	–	–	–	–	–	2	2
E. pyramidata Hudst.	–	–	1	2	–	–	–	–	–	–	–	–	–	–	1	2	–	–	–	–	–	–	1	2

Contd...

Table 2.2–Contd...

Taxa	*EPOKI*				*UFU UGBO*				*EBROHIMI*				*GERMAN*				*OLEGHEYE*				*JAKPATIE*			
	7				*8*				*9*				*10*				*11*				*12*			
	F	*M*	*MY*	*J*	*F*	*M*	*MY*	*J*	*F*	*M*	*MY*	*J*	*F*	*M*	*MY*	*J*	*F*	*M*	*MY*	*J*	*F*	*M*	*MY*	*J*
Fragilaria construens Ehr.	–	–	–	2	–	–	–	–	–	–	–	–	–	–	1	2	–	–	–	–	–	–	–	–
F. oceanica Cleve	–	–	–	1	–	–	–	–	–	–	–	–	–	–	2	2	4	2	–	–	2	1	–	–
Gyrosigma balticum (Ehr.) Rabenhorst	1	1	1	2	–	–	2	2	1	2	3	–	–	–	–	–	–	–	–	–	–	–	–	–
G. Spenceri (Wm Sm) Cleve	4	2	–	–	–	1	1	2	–	–	4	4	–	–	–	–	–	–	–	–	–	–	–	–
G. scalproides (Rab.) Cl.	4	1	–	–	–	2	1	–	–	–	–	–	–	–	–	–	–	–	–	–	–	–	–	–
Mastoglora exilis Grevile	–	–	–	8	–	1	–	–	–	–	5	2	–	–	–	–	–	–	–	–	–	–	–	–
Navicula clavata Gregory	1	5	6	10	–	1	25	–	1	–	4	4	–	–	2	6	–	–	–	–	–	–	–	–
N. cryptocephala Kutzing	1	4	2	4	–	1	15	–	3	–	3	3	–	–	1	2	–	–	–	–	–	–	–	–
N. expansa	–	–	–	4	–	–	–	–	–	–	2	–	–	–	–	–	2	–	–	–	–	–	–	–
N. mutica	–	–	–	–	4	–	–	–	–	–	–	–	–	2	–	–	–	1	–	–	–	–	–	–
Nitzschia abruensis	–	–	–	–	–	–	–	–	–	–	–	–	–	–	–	–	–	–	–	–	–	–	–	–
N. irresoluta	–	–	–	–	2	–	–	–	–	–	–	–	–	–	–	–	–	–	–	–	–	–	–	–
Pinnularia debesii	–	–	–	–	–	–	2	–	–	–	–	–	–	–	–	–	1	–	–	–	–	–	–	–
Algal Genera	–	–	–	–	–	–	–	–	–	–	–	–	–	–	–	–	–	–	–	–	–	–	–	–
Cyanophyceae	–	–	–	–	–	–	–	–	–	–	–	–	–	–	–	–	–	–	–	–	–	–	–	–
Lyngbya martensiana Meneghini	–	–	–	–	–	–	–	–	–	–	3	–	–	–	–	–	–	–	–	–	–	–	–	–
Oscillatoria agardii Gomont	–	–	–	–	–	–	–	–	–	–	–	–	–	–	–	–	–	–	–	–	–	–	–	–
O. formosa	–	–	–	–	–	–	–	4	–	–	–	–	–	–	–	–	–	–	–	–	–	–	–	–
O. magaritifera	–	–	–	–	–	–	1	4	–	–	–	2	–	–	4	–	3	4	–	22	–	–	4	2
Xanophyceae																								
Vaucheriales																								
Botrydium granulatum	–	–	–	–	–	–	–	–	–	–	–	–	–	–	–	–	–	–	–	–	–	–	–	–

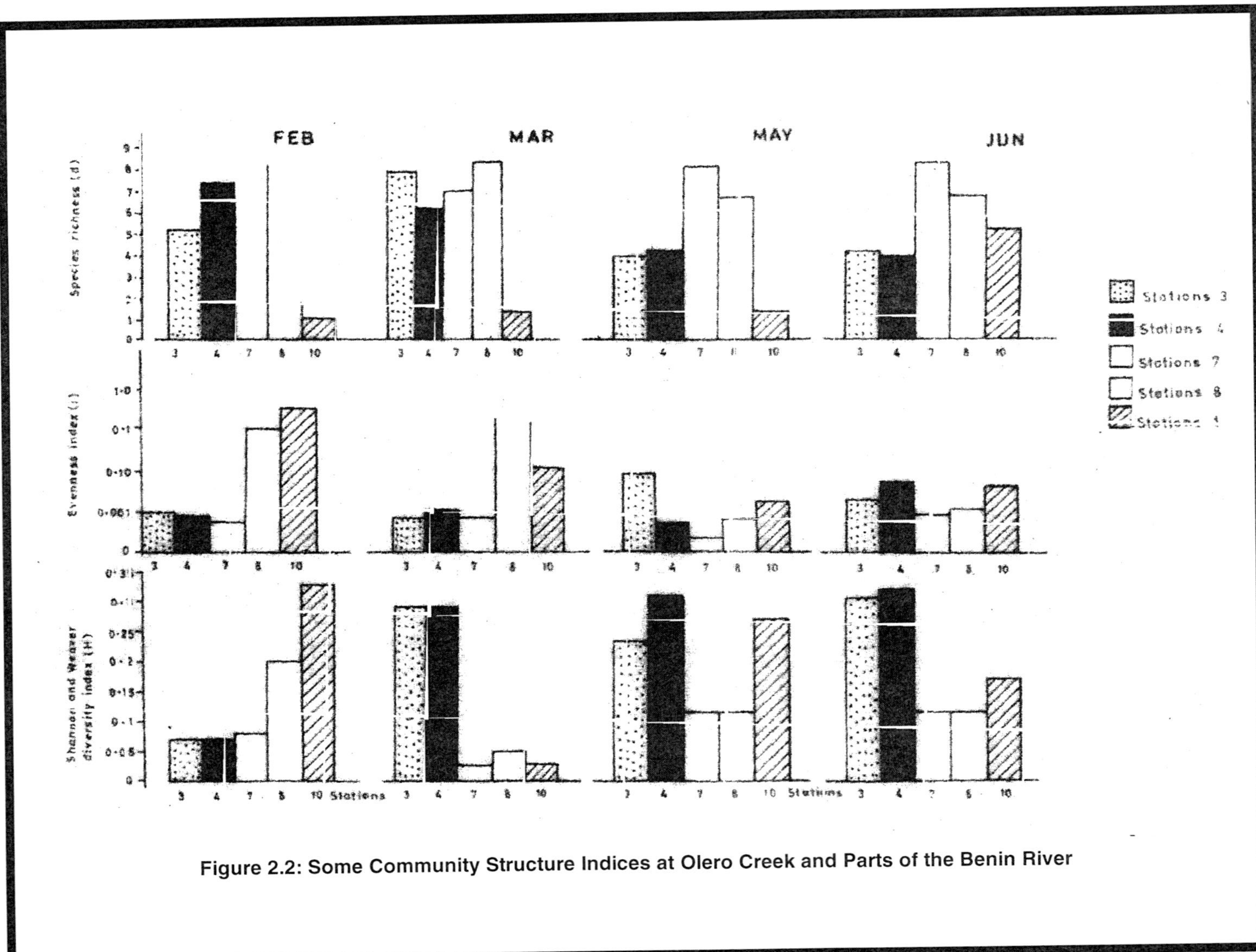

Figure 2.2: Some Community Structure Indices at Olero Creek and Parts of the Benin River

dredging. In the Olero creek and Benin river, oil mining activities may have greatly increased total suspended solids which decreased transparency and lowered photosynthetic depth. Furthermore, dredge spoils could have acidified the environment and possibly contributed to the low pH values (< 7.0) occasionally recorded. According to Karlman (1982) transparency in the Kainji Lake. Nigeria regulates primary production according to flood conditions. Similar observations have been reported elsewhere in the Niger Delta (Erondu, 1984; Erondu and Chindah, 1991; Nwankwo, 1996a, 1996b).

The consequences of channelisation and dredging may account for the paucity of phytoplankton, the dominance of pennate diatoms and the appearance in the phytoplankton of *Botrydium granulatum,* a littoral alga. Secondly, occasional spills may have resulted in the loss of biodiversity. According to Nelson-Smith (–) crude oil pollution as low as 0.01 mgl^{-1} prevented the development of algae.

The low diversity ($H < 1.0$) and low evenness ($j < 0.5$) may be due to the strong hydrographic influence and stress imposed by upstream activity. According to Patrick (1973) diatom communities affected by pollution typically have low diversity and low number of species. In Chanomi creek Nwadiaro (1990) reported a reduction in species richness and abundance due to crude oil mining related activities. Similarly, Pudo (1985) attributed the decline in phytoplankton in parts of the Bonny river to crude oil contamination. The relatively higher diversity in the wet months may be related to the introduction of numerous littoral and bottom dwelling forms into the water columns, a process probably enhanced by high scourging rate, high silt load and high freshwater discharge. Similarly, the increased diversity towards Benin river from Olero creek may be due to the mixture of fresh and brackish water forms. Margalef (1960) and Nwankwo (1996c) reported that diversity trends increase towards the boundary of two water boundaries. It is noteworthy that all the cyanophyta recorded were heterocytous suggesting that nitrates may be limiting. In the tropics because nitrates salts easily go into solution, they are easily leached out of the environment. According to Lee (1999) heterocysts of the blue-green algae endow them with nitrogen fixing capability.

Acknowledgements

The authors are indebted to Triple E and Josef Ross Joint Venture for providing logistics. We also thank the Federal Meteorological Department Oshodi, Lagos, for providing climatic data.

References

Barber, H.G. and Harworth, E.Y. (1981). *A Guide to the Morphology of the Diatom Frustule.* Freshwater Biology Association, Scientific Publication, 44: 122.

Barnes, R.S.K. (1980). *Coastal Lagoons.* Cambridge University Press, London, pp. 106.

Bott, T.L., Rogenmuser, K. and Thorne, P. (1978). Effects of No. 2 Fuel Oil, Nigerian Crude Oil, and used crankcase oil on benthic algal communities. *J. Env. Sci.,* A13: 751–779.

Chevron Nigeria Limited (CNL) Report (1993). Chevron Oil Spill Contingency Plan Revised.

Chindah, A.C. and Pudo, J.K. (1991). A Preliminary Check-list of algae found in plankton of Bonny river in Niger Delta. *Fragon. Flor Geobol.,* 36: 117–126.

Desikachary, T.V. (1959). *Cyanophyta.* Indian Council Agriculture Research, New Delhi, pp. 686.

Erondu, E.S. (1984). Tidal and seasonal influence on the physico-chemical conditions of the New Calabar River at Aluu. *M.Tech. Thesis,* Rivers State University of Science and Technology, Port Harcourt, Nigeria.

Erondu, E.S. and Chindah, A.C. (1991). Physico-chemical and phytoplankton changes in a tidal freshwater station of the New Calabar River, South eastern Nigeria. *Env. & Ecol.*, 9: 561–570.

European Advisory Unit (EAU) Report (1992). Environmental impact studies on the Proposed Nigerian Liquefied natural gas project, pp. 102.

Gaur, J.P. and Kumar, H.D. (1985). The influence of oil refinery effluents on the structure of algal communities. *Arch. Hydrobiol.*, 103: 305–323.

Hart, C.W. Jnr. and Fuller, S.L.H. (1972). Environmental degradation in the Patuscent River Estuary, Maryland. *Acad. Nat. Sci. Phila, Dept, Limnol. Contr.*, pp. 14.

Hendey, N.I. (1964). *An introductory account of the smaller algae of British coastal waters 5. Bacillariophyceae* (Diatoms). Ministry of Agriculture, Fisheries and Food. *Fishery Investigation Series*, 4: 317.

Ho, C.A. and Karim, H. (1978). Impact of adsorbed petroleum hydrocarbons on marine organisms. *Marine Pollution Bull.*, 9: 156–162.

Ibiebele, D.D., Powell, C.B., Shon, P.M., Isou, M., Murday, M. and Sclema, M.D. (1984). Establishment of baseline data for complete monitoring of petroleum-related Aquatic pollution in Nigeria. Proc. Sem. Petroleum Industry and the Nigerian Environment. Port Harcourt 1983. The inspectorate division, NNPC, and Federal Ministry of Housing, Lagos, pp. 296.

Imevbore, A.M.A. and Odu, E.A. (1980). Environmental pollution in the Nigeria Delta. Proc. Sem. "Niger Delta Mangrove Ecosystem". Port Harcourt, 1980.

Karlman, S. (1982). The annual flood regime as a regulatory mechanism for phytoplankton Production in Kainji Lake. Nigeria. *Hydrobiologia*, 80: 93–97.

Lackey, J.B. (1938). The manipulation and counting of river plankton and changes in some organisms due to formalin preservation. *U.S. Public Health Report*, 63: 2080–2093.

Lee, R.E. (1999). *Phycology*. Cambridge University Press, Cambridge, pp. 582.

Margalef, R. (1951). Diversidad de especies en less communides naturals. *Publ. Inst. Bid. Appl. Barcelona*, 9: 5–27.

Margalef, R. (1960). Temporal succession and spatial heterogeneity in phytoplankton. In: *Perspectives in Marine Biology*, (Ed.) Buzzati, Travers, A.H.

McCauley, R. (1966). The biological effects of oil pollution in a river. *Limol. Oceangr.*, 11: 473–486.

Minter, K.W. (1964). Standing crop and community structure of plankton in oil Refinery effluent holding pond. *Ph.D. Thesis*. Oklahoma State University.

Murday, M., Murday, J.H., Sexton, C.B. Powell (1988). Atlas of environmental sensitivity to spilled oil for operational areas of Gulf Oil Company (Nig.) Ltd. Chevron Overseas Inc. San Ramon USA.

Nelso-Smith, A. Biological consequences of oil spills. In: *The Marine Environment*, (Eds.) Lenihan J. and Fletcher, W.W. Environment and Man. Blackie Glasgow and London, p. 46–69.

Nwadiaro, C. (1990). A hydrobiological survey of the Chanomi creek system, lower Niger Delta, Nigeria. *Limnologica* (Berlin), 21: 263–274.

Nwadiaro, C. and Ezefili, E.O. (1986). A preliminary check-list of the phytoplankton of New Calabar River, Lower Niger Delta, Nigeria. *Hydrobiol. Bull.*, 19: 133–138.

Nwankwo, D.I. (1996a). Freshwater Swamp Desmids from in South East Niger Delta. Nigeria. *Pol. Arch. Hydrobiol.*, 43(4): 411–420.

Nwankwo, D.I. (1996b). Notes on the effect of Human Induced Stressors in Parts of the Niger Delta, Nigeria. *Pol. Ecol. Stud.,* 22(1&2): 71–78.

Nwankwo, D.I. (1996c). Phytoplankton diversity and succession in Lagos Lagoon, Nigeria. *Arch. Hydrobiol.,* 135(4): 529–542.

Nwankwo, D.I. (2000). The algae of a crude oil impacted mangrove soil in the Niger Delta, Nigeria. *Tropical Ecology,* 41(2): 1–3.

O'Brien, P.Y. and Dixon, P.S. (1976). The effects of oils and oil components on algae: A review. *Br. Phycol. J.,* 11: 115–142.

Odum, H.T. (1963). Productivity measurements in Texas turtle grass and the effects of dredging on the intercoastal channel. *Pub. Inst. Mar. Sci., Univ. Texas,* 9: 48–58.

Odum, E.P. (1971). *Fundamentals of Ecology.* Philadelphia, W.B. Saunders Company, pp. 374.

Patrick, R. (1973). Diatoms as bioassay organism. In: *Bioassay Techniques and Environmental Chemistry,* (Ed.) Glass, G.E. Michighan, p. 139–151.

Patrick, R. and Reimer, C.W. (1966). *The Diatoms of the United States Exclusive of Alaska and Hawaii,* Vol. 2, Part I, Philadelphia, pp. 213.

Pielou, E.C. (1966). The measurement of diversity to different types of biological collections. *J. Theor. Biol.,* 13: 131–144.

Powell, C.B. (1987). Effect of oil spillage on freshwater fish and fisheries. Proc. Sem. "Niger Delta Mangrove Ecosystem", Port Harcourt, 1986.

Pudo, J. (1985). Investigations on algae in the oil spilled region of Bonny River (Nokoe S. ed). Proc. Sem. "The Nigerian environment: Ecological limits of abuse. Ecological Society of Nigeria. Port Harcourt.

Shannon, C.E. and W. Weaver (1963). *The Mathemathical Theory of Communication.* University of Illinois Press, Urbana, pp. 125.

Singh, A.K. and Gaur, J.P. (1989). Algal epilithon and water quality of a stream receiving oil refinery effluent. *Hydrobiologia,* 184: 193–199.

Sioli, H. (1973). Tropical rivers as expression of their terrestrial environment. In: *Tropical Ecological System Trends in Terrestrial and Aquatic Research,* (Eds.) Goley, F.B. and Medina, E. New York, p. 273–295.

Subramanyan, R. (1946). A systematic account of the marine plankton diatoms of the Madras Coast. In: *Proceedings of the Indian Academy of Science.* 24: 85–197.

Vandermeulen, J.R. and Ahern, T.P. (1976). Effects of petroleum hydrocarbons on algal physiology: Review and progress report. In: *Effects of Pollutants on Aquatic Organisms,* (Ed.) A.P.M. Lockwood. Cambridge University Press, Cambridge, p. 107–125.

Chapter 3

Community Structure and Faunal Assemblages of Zooplankton in an Estuarine Mangrove Ecosystem

N. V. Prasad
Division of Marine Biology, Department of Zoology, Andhra University, Visakhapatnam – 530 003, India

ABSTRACT

Observations were made on the distribution, community structure and species association of zooplankton at 24 stations. A total of 126 species belonging to 25 major groups were encountered in surface collections. Estuarine and estuarine marine forms constitute the bulk of zooplankton population. The neritic species were only straggler in this environment. Some of the species of the same or of different families showed coexistence while others were negatively correlated or associated either over space and time. Based on factor and cluster analysis for Biologically Important Species, four groups of stations concordant with four faunal assemblages was noticed.

Keywords: *Zooplankton, Distribution, Coexistence, Community structure, Backwaters.*

Introduction

The complexity of ecological systems with their numerous species and discordant time course of change in the populations has required fresh approaches. There have been many attempts to define logical frame works for analysing patterns of changes in complex, spatially diverse multispecies assemblages. According to the continua, concept species are believed to be distributed along an environmental continuum independent of most other species (Mills, 1969). It is almost impossible to distinguish boundaries between them and define funcational units classically termed communities. If the continuum concept is rigorously adhered to, each area must be treated as a unique situation and generalisation would be difficult. Fager's (1963) view that associations/assemblages in nature involve both physical and biological factors defined by a set of co-ordinates in time and space is preferred. According to Fager (1963) the co-ordinates will rarely have sharp and impassable limits, but they will often contain broad areas of considerable internal similarity bordered by relatively narrower regions of rapid change. These regions of rapid change grade in turn into other areas with internal consistency. Implicit in this view is that these regions are always open systems and that the decision to recognise recurrent organised systems of organisms with similar structure in terms of species presence and abundance ultimately rests on experience and intuition.

A variety of statistical methods and indices have been in use to delineate associations or assemblages in community analysis. Hubalek (1982) reviewed the properties of the 43 indices that have been used to measure the degree of association between pairs of species. In recent years, multivariate analyses have become standard analytical methods in community ecology due to their efficiency while analysing species/sites matrices for possible pollution and anthropogenic induced effects. The analyses are carried out either through simple calculations (Polar Ordination–Bray and Curtis, 1957; Trellis diagram–Sanders, 1960) or through more complicated methods involving computer aided calculations using specific packages. Multivariate analysis technique has been also used extensively in zooplankton studies, for example in Cochin backwaters (Madhupratap, 1978); Mandovi estuary (Goswami, 1983); Guarau river estuary, Mexico (Lopes, 1994); Bahia Magdalena waters, Mexico (Garcia *et al.,* 1996); Kerala coast (Rao and Jayalakshmy, 1997); Nepean river estuary, Australia (Kobayashi *et al.,* 1998). In the present paper, results of detailed studies on zooplankton community structure and faunal assemblages in estuarine mangrove waters of Kakinada are presented.

Materials and Methods

During the present study (July 1997 to May 1999), an attempt has been made to find out the distribution and community structure of zooplankton at the twenty four selected stations in the estuarine and mangrove waterways of Kakinada based on multivariate analytical procedures using STATISTICA package. Hydrographical parameters such as surface water temperature, salinity, dissolved oxygen and pH were measured by adopting standard methods (APHA, 1995). Zooplankton samples were collected by using a 120 µm mesh size net. For the enumeration of zooplankton, aliquot method was employed (Wickstead, 1965). The classification of zooplankton into clusters was analysed based on Pearson coefficient followed by the hierarchical, Agglomerative method employing group-average linking which is often applied in zooplankton studies (Lopes, 1994). In order to find out the similarity among the constituent species based on their composition and numerical abundance, the BIS were taken into consideration (Table 3.1). Cluster and PCA analyses was carried out after a log transformation.

Results and Discussion

Copepods were the most dominant group in the zooplankton of the estuarine mangrove ecosystem of Kakinada constituting 57 to 78 per cent of the annual zooplankton counts. They formed the bulk of the zooplankton biomass. Altogether 126 species of zooplankton belonging to 27 diverse groups were observed. Density and diversity of zooplankton population varied greatly with seasons. Depending on the environmental parameters various species occur and propagate in the estuary, attain ascendancy

Table 3.1: List and Rank of Biologically Important Species (BIS) of Zooplankton in Coringa Magrove Habitat

Rank	*BIS Species/Group*	*Total Numerical Abundance (no.m^{-3})*	*Percentage by Number*	*Cumulative Percentage*
1	*Acartia erythraea*	1,39,736	13.88	13.88
2	Bivalve veligers	93,589	9.29	23.17
3	*Eucalanus elongatus*	91,500	9.09	32.26
4	*Paracalanus parvus*	78,901	7.84	40.10
5	*Scylla serrata* (zoea)	75,935	7.54	47.64
6	Gastropod veligers	66,021	6.56	54.20
7	*Acartia clausi*	64,073	6.36	60.56
8	*Acrocalanus gibber*	42,069	4.18	64.74
9	*Paracalanus aculeatus*	36,145	3.59	68.33
10	*Evadne* sp.	34,924	3.47	71.80
11	Penaeid mysis	31,534	3.13	74.93
12	Fish eggs and larvae	26,246	2.61	77.54
13	*Macrosetella gracilis*	22,149	20.20	79.74
14	*Sagitta bedoti*	21,774	2.16	81.90
15	*Oikopleura dioica*	16,719	1.66	83.56
16	*Lucifer hanseni*	15,031	1.49	85.05
17	*Acrocalanus gracilis*	14,350	1.43	86.48
18	*Euterpina* sp.	12,534	1.24	87.72
19	*Oncaea venusta*	10,290	1.02	88.74
20	*Oithona rigida*	9,704	0.96	89.71
21	*Pseudodiaptomus aurivilli*	8,653	0.86	90.57
22	*Microsetella* sp.	8,265	0.82	91.39
23	*Euchaeta* sp.	7,035	0.70	92.09
24	*Podan* sp.	6,488	0.64	92.73
25	*Globigerina* sp.	5,960	0.59	93.32
26	*Mesopodopsis orientalis*	5,941	0.59	93.91
27	*Cirripede nauplius*	5,235	0.52	94.43
28	*Acartia* sp.	5,124	0.51	94.94
29	*Tomopteris* sp.	4,702	0.47	95.41
30	Others		4.59	100

and disappear during unfavourable conditions. Species succession which begins in the post monsoon months along with salinity recovery varies which space and time and can be gathered from the numerical dominance exerted by various species. Based on the cluster and factor analysis for Biologically Importance Species (Table 3.1), four groups of stations concordant with four faunal assemblages was observed. From Figure 3.1, it was possible to identify 4 distinct groups of stations.

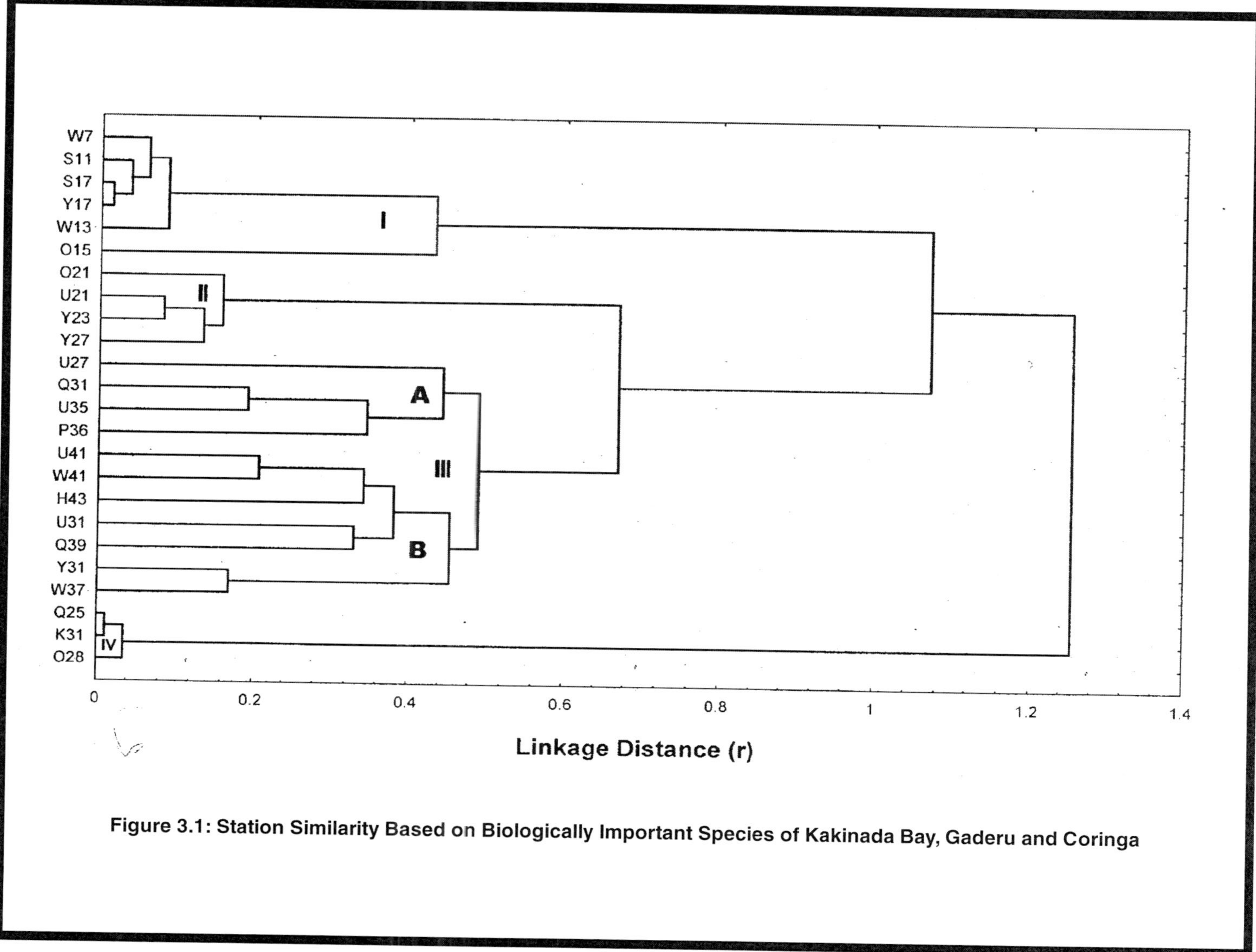

Figure 3.1: Station Similarity Based on Biologically Important Species of Kakinada Bay, Gaderu and Coringa

The first group found in the upper part of the dendrogram constituted of 6 stations (stations W7, S11, S17, Y17, WI3 and O15). Similarity between stations S17, Y17 was particularly remarkable, whereas between stations W7, S11; S11, S17; Y17, W13 and W13, O15, the linkage distance relatively high (r = 0.016–0.074). The second group, comprised stations O21, U21, Y23 and Y27. The third group, being large, consisted of 11 stations. There were 2 sub-clusters represented by stations U27, Q31, U35, P36, U41, W41, H43, U31, Q39, Y31 and W37. The similarity between U4l and W41 was noteworthy. In this cluster, stations Y3l and W37 remained some what independent, and showed highest similarity with station U4l. The fourth group found in the lower half of the dendrogram consisted of station Q25, K31 and O28. The findings revealed a clear demarcation between marine dominated north bay, estuarine waters (south bay), mangrove habitat (Gaderu) and freshwater dominated Coringa stations.

With reference to BIS, four distinct clusters could be identified (Figure 3.2). Of these, the foremost cluster (upper dendrogram), was represented by as many as 14 species of zooplankton of which, there were 7 copepods and 7 belongs to six other different groups. There were 3 sub-clusters represented by *Globigerina* sp., *Paracalanus parvus,* gastropod veligers, *Acrocalanus gracilis, Acartia sewelli* and *Macrosetella gracilis, Lucifer hanseni, Acartia erythraea* and *Microsetella* sp., and *Euterpina* sp., Fish eggs and larvae, *Oikopleura dioica,* bivalve veligers and *Sagitta bedoti.* Among these, the similarity was significant between *Acartia sewelli* and *Macrosetella gracilis,* and *Paracalanus parvus* and gastropod veligers. And, these species were found to be abundant at stations located in south bay.

The 2nd cluster consisted of 7 species (all copepods). These were represented by *Eucalanus elongatus, Euchaeta* sp., *Oithona rigida, Acrocalanus gibber, Oncaea venusta, Paracalanus aculeatus* and *Pseudodiaptomus aurivilli* were numerically abundant at stations located in north bay (Stations W7, S11, S17, Y17, W13 and O15). Among these, the similarity between *Oithona rigida* and *Euchaeta* sp., and *Eucalanus elongatus* and *Euchaeta* sp., was significant. The 3rd cluster comprised of 2 species of cladocerans and one species of mysid. This cluster was represented by *Evadne* sp., *Podon* sp., and *Mesopodopsis orientalis.* These species were found to be dominant at stations located in Coringa (Stations Q25, K31 and O28). *Evadne* sp., and *Mesopodopsis orientalis* showed nearest similarity.

Cluster 4 comprised of five species (2 decapod larvae, 1 copepod and 2 belongs to.other groups). These were represented by *Acartia clausi, Scylla serrata* (zoea), penaeid mysis, *Tomopteris* sp., and cirripede nauplius. Similarity between *Tomopteris* sp., and penaeid mysis was noteworthy. These species were found to be abundant in Gaderu.

In order to search for relationships in species composition between stations, factor analysis was applied. In the R-mode of factor analysis, the first two factors had eigen values representing 58.46 per cent of the total variance (I : 34.08 per cent; II : 24.38 per cent). Using scores obtained from the varimax rotated factor matrix, groups 1, 2, 3 and 4 stations could be established (Figure 3.3). The arrangement of stations revealed a trend similar to the one obtained through cluster analysis (Figure 3.1). The first principal axis explained 34.08 per cent of the total variation. Accordingly, stations O21, Y27, U41, W41, U31, Q31, U35, Q39, Y31 and W37 exhibited highest factor scores (0.954–0.723). Similarly, factor 2 could explain 24.38 per cent of the total variance, the eigen value being 5.851. Stations Y17, S17, W7, S11, W13 and O15 had the highest scores in the order of 0.967, 0.945, 0.943, 0.937, 0.881 and 0.728 respectively.

During the study, factor analysis was also extended to determine the affinity among the BIS (Figure 3.4). In the Q-mode of factor analysis, the first two factors had eigen values representing 73.77 per cent of the total variations (I : 47.10 per cent; II : 26.67 per cent). Using scores of factors I and II for each species, groups I, II, III and IV were established (Figure 3.4). In group I, the most dominant species

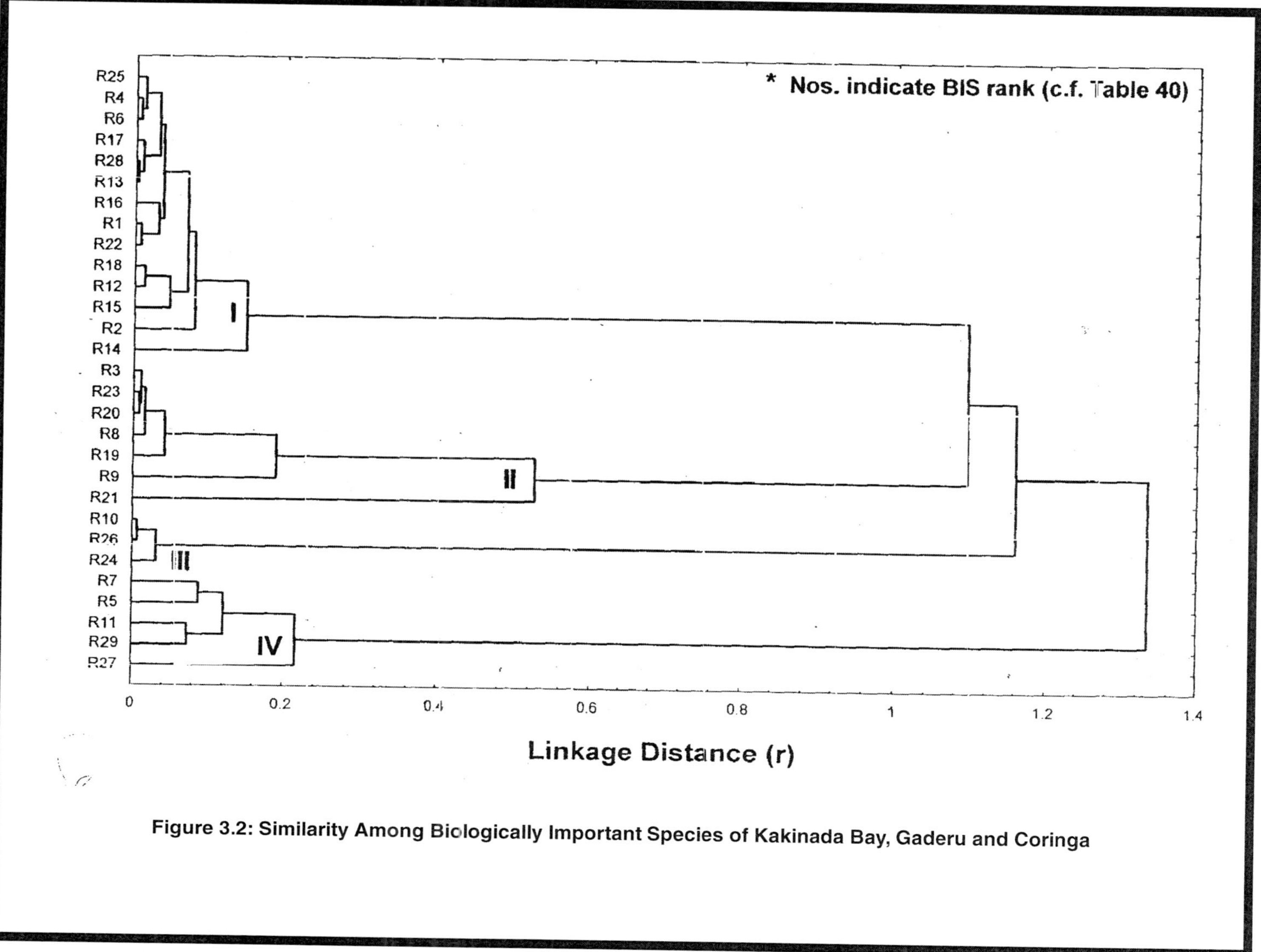

Figure 3.2: Similarity Among Biologically Important Species of Kakinada Bay, Gaderu and Coringa

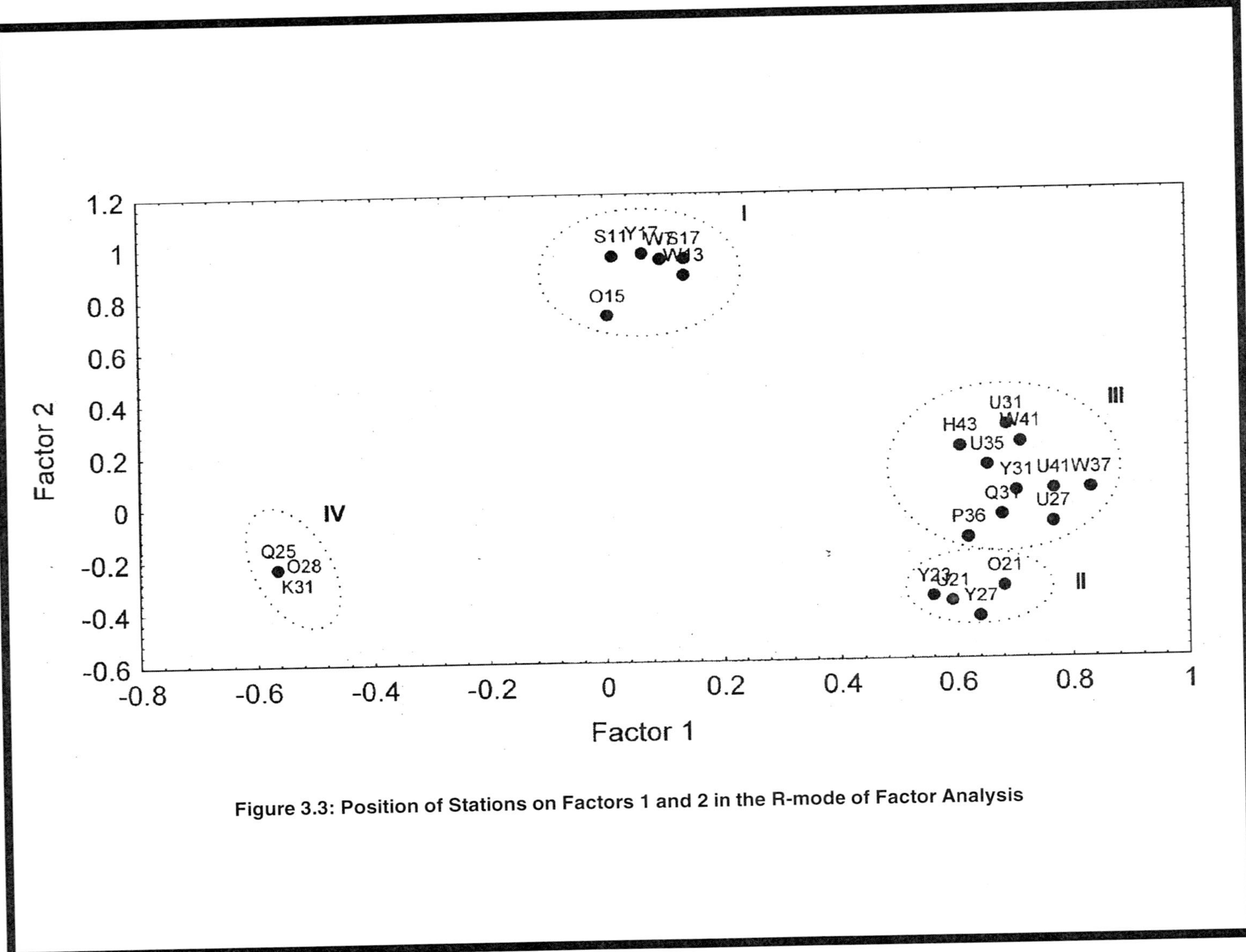

Figure 3.3: Position of Stations on Factors 1 and 2 in the R-mode of Factor Analysis

was the calanoid copepod *Acartia erythraea,* followed by bivalve veligers, *Paracalanus parvus,* gastropod veligers, fish eggs and larvae, *Macrosetella gracilis, Sigitta bedoti, Oikopleura dioica, Lucifer hanseni, Acrocalanus gracilis, Euterpina* sp., *Globigerina* sp., and *Acartia sewelli.* Whereas group II, *Eucalanus elongatus, Paracalanus aculeatus, Acrocalanus gibber, Euchaeta* sp., *Pseudodiaptomus aurivilli, Oithona rigida* and *Oncaea venusta,* were found. In group III, the principal organisms were the cladocerans *Evadne* sp., *Podon* sp., and mysid *Mesopodopsis orientalis.* The fourth group consisted the decapod larvae *Scylla serrata* (zoea) and *Acartia clausi* in large numbers. Penaeid mysis, *Tomopteris* sp., and cirripede nauplius were the other members of this group. Based on station similarities and BIS similarities (Figures 3.1–3.4) certain inferences could be drawn:

Firstly, the zooplankton in group I stations was dominated by *Eucalanus elongatus,* followed by *Paracalanus aculeatus, Acrocalanus gibber, Pseudodiaptomus aurivilli, Euchaeta* sp., and *Oithona rigida.* This faunal group can be considered similar to the faunal assemblage II found in Figure 3.4. On the other hand, group II stations were characterised by the dominance of *Acartia erythraea,* followed by bivalve veligers. *Paracalanus parvus,* gastropod veligers, fish eggs and larvae, *Macrosetella gracilis, Oikopleura dioica* and *Lucifer hanseni.* This faunal group can be considered identical to the faunal assemblage I (Figure 3.4). The third group of stations were inhabited principally by *Scylla serrata,* followed by *Acartia clausi,* penaeid mysis, *Tomopteris* sp. and cirripede nauplius. This faunal group can be considered similar to the faunal assemblage IV. The fourth group of stations was chiefly inhabited by *Mesopodopsis orientalis, Podon* sp., and *Evadne* sp. This faunal group can be considered similar to assemblage III (Figure 3.4).

From the foregoing, it may be concluded that the zooplankton in the estuarine mangrove ecosystem of Kakinada are characterised by the presence of four distinct assemblages (Figures 3.2 and 3.4) designated after the most dominant species (species with the highest coefficients in the factor loadings; Cassie and Michael, 1968) as follows:

1. The zooplankton in group I stations corresponding to faunal group II (Figures 3.1–3.4) may be designated as the *Eucalanus elongates–Acrocalanus gibber* dominated assemblage, characterising areas mostly in north bay locations. This assemblage which exhibited preference to high salinity conditions (mean salinity 28.18 per cent); the numerical density was high (mean 18,276 no.m^{-3}). A total of 63 species represented this assemblage. A strong positive correlation ($r = 0.653–0.741$) between the predominant species namely, *Eucalanus elongatus, Paracalanus aculeatus, Acrocalanus gibber* and salinity suggestive of species preference to high saline conditions.
2. The zooplankton in group II stations, corresponding to faunal group I (Figures 3.1–3.4) may be designated as *Acartia erythraea–Bivalve* veligers dominated assemblage. This assemblage was characteristic of south bay stations. These species preferring estuarine conditions (mean salinity 23.61 per cent). The numerical abundance was high (29,659 no.m^{-3}) relative to other regions. A total of 66 species were noticed in this assemblage. Though positive but insignificant ($r = 0.39–0.49$) correlation was observed between salinity and predominant species namely, *Acartia erythraea, Paracalanu sparvus, Oikopleura dioica* and gastropod veligers.
3. The zooplankton in group III stations corresponding to faunal group IV may be designated as the *Scylla serrata* (zoea)–*Acartia clausi* dominated assemblage, characterising mangrove dominated Gaderu region. The species existing in this assemblage, exhibiting preference to mangrove waterways. The numerical abundance was intermediatary to other locations (13,429 no.m^{-3}), the observed mean salinity was 17.20 per cent. A total of 61 species existing in this assemblage. Majority of species (*Scylla serrata, Acartia clausi* and penaeid mysis)

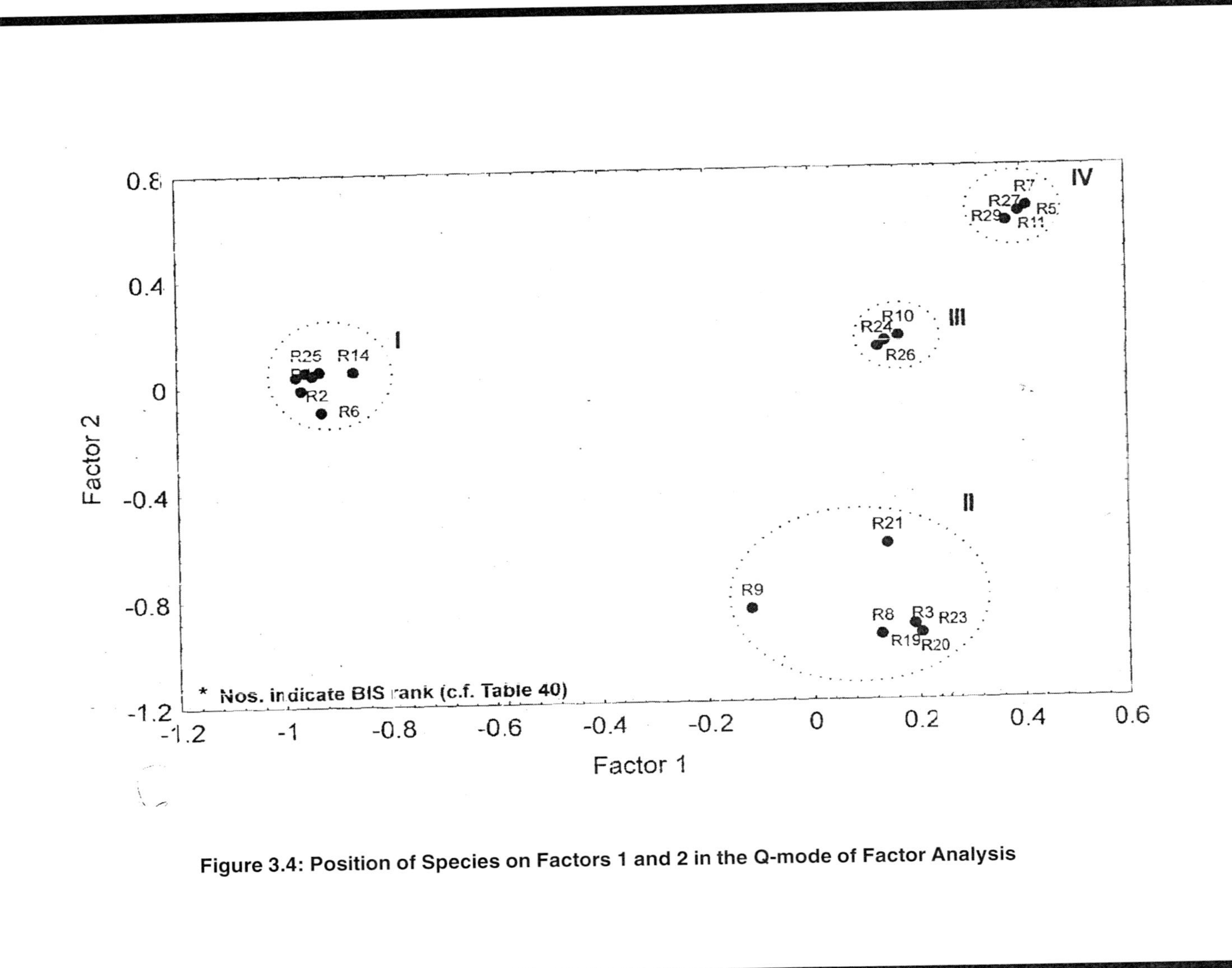

Figure 3.4: Position of Species on Factors 1 and 2 in the Q-mode of Factor Analysis

showed insignificant correlation with salinity (r = 0.28–0.46). In general, majority of these species preferring detritus rich mangrove habitat.

4. The zooplankton in group IV stations corresponding to faunal group III may be designated as the *Evadne* sp.–*Mesopodopsis orientalis* dominated assemblage. These species showed a widespread distribution in Coringa, exhibiting preference to very low saline or freshwater conditions (mean salinity 7.46 per cent). The numerical density was low (9,368 no.m^{-3}). A total of 31 species were existing in this assemblage. A very weak correlation (r = < 0.087) between the predominant species namely, *Evadne* sp., *Podon* sp., and *Mesopodopsis orientalis* and salinity suggestive of species' preference to freshwater conditions.

It is important to recall, that the classical ecological theories usually imply that coexistence of species assemblages requires ecological differentiation between them. Hardin's (1960) competitive exclusion principle stating 'complete competitors cannot coexist' is supported by Slobodkin's (1961) view 'no two species can indefinitely continue to occupy the same ecological niche'.

Goswami (1983), opined that the mechanism responsible for coexistence or association among the related forms is not properly understood. Food may not be a constraint as the rate of primary production in the estuaries far exceeds the rate of consumption by the zooplankton (Qasim, 1970). Peter *et al.* (1977) stated that feeding habitats rather than the quantity of food are responsible for the coexistence among planktonic organisms. Tranter and Abraham (1971) studied the morphological variations in the mandibular structure in the coexisting species of family Acartiidae and reported that the differences which did exist were not sufficient to establish niche separation. The range of salinity tolerance in the congeneric species may lead to niche selection and ecological differences. These species exist as competitors and utilize the available food resources. Jeffries (1967) also observed that congeneric forms efficiently utilize the biotope by competing for common resources. Salinity is one of the important factors for the distribution and abundance of zooplankton in the estuaries (Goswami and Selvakumar, 1977). The structure of associations of the zooplankton are indirectly linked with salinity, because the stability of the niches is controlled by its distribution. The distribution of various species undulates up and down the estuary depending on the salinity variations (Rao *et al.,* 1975). Species having similar distribution can be assumed to invade similar niches. During the present study, the principal component analysis also indicated salinity to be the key factor associated with coexistence of species.

References

APHA (1995). *Standard Methods for the Examination of Water and Wastewater,* 19th Edn. American Public Health Association, Washington D.C., USA, pp. 874.

Bray, J.R. and J.T. Curtis (1957). An ordination of the upland forest communities of southern Wisconsin. *Ecol. Monographs,* 27: 325–349.

Cassie, R.M. and A.D. Michael (1968). Faunal and sediment of an inter tidal mudflat: A multivariate analysis. *J. Expt. Mar. Biol. Ecol.,* 2: 1–23.

Fager, E.W. (1963). Communities of organism. In: *The Sea,* (Ed.) M.N. Hill, Inter-sciences Publisher, New York, 2: 415–437.

Garcia-Soto, G., I. De Madariaga, F. Villate and E. Orive (1996). Day to day variability in the plankton community of a coastal shallow embayment in response to changes in river runoff and water turbulance. *Estuar. Coast. Shelf. Sci.,* 31: 217–229.

Goswami, S.C. and R.A. Selvakumar (1977). Plankton studies in the estuarine system of Goa. Proc. Symp. Warm Water Zoop. Spl. Pub. UNESCO/NIO: 226–241.

Goswami, S.C. (1983). Coexistence and succession of copepod species in the Mandovi and Zuari estuaries, Goa. *Mahasagar-Bull. Nat. Inst. Oceanogr.,* 16: 251–258.

Hardin, G. (1960). The competitive exclusion principle. *Science,* 131: 1292–1297.

Hubalek, Z. (1982). Coefficients of associations and similarity based on binary data: An evolution. *Biol. Rew.,* 57: 669–689.

Jeffries, H.P. (1967). Saturation estuarine zooplankton by congeneric associates. In: *Estuaries* (Ed.) G. Lauff, American. Asso. for the Advancement of Science, Washington, pp. 202–207.

Kabayashi, T., R.J. Shiel and P. Gibbs (1998). Size structure of river zooplankton: Seasonal variation, overall pattern and functional aspect. *Mar. Freshwat. Res.,* 49: 547–552.

Lopes, R.M. (1994). Zooplankton distribution in the Guarau river estuary (South-eastern Brazil). *Estuar. Coast. Shelf. Sci.,* 39: 287–302.

Madhupratap, M. (1978). Studies on the ecology of zooplankton of Cochin backwaters. *Mahasagar-Bull. Nat. Inst. Oceanogr.,* 11: 45–56.

Mills, E.L. (1969). The community concept in marine zoology, with comments on.continua and instability in some marine communities: A Review. *J. Fish. Res. Bd. Can.,* 26: 1415–1428.

Peter, K.J.H., K. Iyer, S. John and E.V.R. Krishnan (1977). Distribution and coexistence of planktonic organisms along southwest coast of India. Proc. Synt. Warm Water Zoop. Spl. Publi., UNESCO/NIO: 80–86.

Quasim, S.Z. (1970). Some problems related to the food chain in a tropical estuary. In: *Marine Food Chain* (Ed.) J. H. Steel, Oliver and Boyd Publications, Edinburgh, p. 45–51.

Rao, K.K. and M. Jayalakshmi (1997). An overview of the planktonic foraminiferal fauna in waters off the Kerala coast, South-west India during summer. *J. Mar. Biol. Ass.,* India, 39: 59–68.

Rao, T.S.S., M. Madhupratap and P. Haridas (1975). Distribution of zooplankton in space and time in tropical estuary. Bull. Dept. Mar. Sci. Univ., Cochin, 7: 695–704.

Sanders, H.L. (1960). Benthic studies in Buzzards bay III, the structure of the soft bottom community. *Limnol. Oceanogr.,* 5: 138–153.

Slobodkin, L.B. (1961). *Growth and Regulation of Animal Populations.* Holt Rinehart and Winston, New York, pp. 184.

Tranter, D.J. and S. Abraham (1971). Coexistence of species of Acartiidae (copepoda) in the cochin backwater, a monsoonal estuarine lagoon. *Mar. Biol.,* 11: 222–241.

Wickstead, J.H. (1965). *An Introduction to the Study of Tropical Plankton.* Hutchinson tropical monograph, Hutchinson and Co. Publication, London, pp. 160.

Chapter 4

Distribution and Ecology of Phytoplanktons in the Dal Lake, Kashmir (India)

M. Jeelani, H. Kaur* & S.G. Sarwar***

**Department of Zoology, Punjabi University, Patiala – 147 002, Punjab*
***Hydrobiology Research Laboratory, S.P. College, Srinagar – 190 001, Kashmir*

ABSTRACT

Phytoplanktons which remain in a variety of aquatic habitats, have been studied with respect to their species diversity and seasonal distribution in the Dal lake. The study was conducted for two years from September 2000 to August 2002 and a total of 127 phytoplankton taxa were recorded. To study the influence of abiotic features on the phytoplankton fauna of this lake, water temperature, dissolved oxygen, pH conductivity, calcium, magnesium, total alkalinity, chloride, nitrate-nitrogen and total phosphorus were also studied.

Keywords: *Phytoplankton, Ecology, Dal lake, Kashmir.*

Introduction

The plankton community is a group of tiny plants and animals floating, drifting or feebly swimming in the water mass. The plant plankters comprise algae and blue green algae (Michael, 1984). The rapidly expanding human population within and the catchment area has brought about a series of changes in the biotic components of the valley lakes (Pandit, 2002). The Dal lake of Kashmir situated at an altitude of 1586m at sea level between 34°5′–34°6′ N latitude and 74°8′–74°9′ E longitude has been attributed to changes in environment due to excessive anthropogenic interferences. To find out its present ecological status, distribution of the phytoplanktons has been recorded and discussed.

Materials and Methods

Selection of Sites

Four sampling sites were selected in the lake which represent different environmental features under the impact of various human activities. The Site–I was selected at the least disturbed region of the lake and was taken as reference site. Site–II was selected in the area infested with floating gardens used for agricultural activity while Site–III and Site–IV were located near the house boats and hotels respectively.

Collection of Samples

The sampling was carried out for a period of two years extending from September 2000 to August 2002. Water samples for physico-chemical analysis were collected from the sampling sites of the Dal lake, once in a month, in one litre polyethylene bottles. The pH and conductivity values were determined electrometrically and dissolved oxygen by unmodified Winkler's method (1888). The detailed chemical analysis of water samples was carried out according to standard methods of Golterman and Clymo (1969) and APHA (1989). Collection of samples for biological study was done using a plankton net (64nm) through which two litres of water from each sampling site was sieved through on each occasion for quantitative analysis. For qualitative analysis, the planktonic net was hauled through vertical and horizontal planes of the lake at each site and the samples were preserved in Lugol's solution. The phytoplanktons were counted using Sedgewick Rafter cell under an inverted microscope (Nikon). Ten replicates were taken from the preserved sample. Identification of different organisms was undertaken with the help of keys given by Heurek (1896), Smith (1950), Randhawa (1959), Ward and Whipple (1959), Round (1964), Philipose (1967), Prescott (1968), and Palmer (1980).

Results and Discussion

The physico-chemical features of water are given in Table 4.1. Water in general ranged from 4–27°C at all the investigated sites of the Dal lake with maximum values in summer and minimum values in winter. The pH values fluctuated between 7.0 and 9.3 indicating the lake to be slightly on alkaline side with no marked difference at different sites. The conductivity ranged from 100–530 µS cm^{-1}. The dissolved oxygen ranged from 0.8 to 12.8 mg/L with maximum at Site–II. Calcium and magnesium concentration varied from 15.2–55.3 mg/L and 2.2 to 30.5 mg/L respectively indicating lake water to be nutrient rich with the cation progression as Ca > Mg. Total alkalinity ranged from 42–460 mg/L. showing water to be moderately hard (Moyle, 1945). The lake water shows low alkalinity values at Site–I and Site–II, higher at Site–III and highest at Site–IV. The lake is rich in Chloride contents ranging from 14.0 to 85 mg/L with a trend as Site IV > Site III > Site II > Site I showing the addition of faecal matter by hotels houseboats. The nitrate-nitrogen contents ranged from 113 to 910 µg/l with highest value at Site–IV. The total phosphorus content fluctuated from 180–621 µg/l with highest value at Site–II as this site receive high concentration of phosphate-phosphorus from the floating gardens.

Phytoplankton population in the Dal lake was composed of six major groups namely Bacillariophyceae, Chlorophyceae, Cyanophyceae, Dinophyceae, Euglenophyceae and Chrysophyceae. Among these, the Bacillariophyceae (62.7 per cent) formed the bulk of phytoplankton followed by Chlorophyceae (17.5 per cent), Cyanophyceae (8.9 per cent), Dinophyceae (2.4 per cent), Euglenophyceae (2 per cent) and Chrysophyceae (0.1 per cent) (Figure 4.1). In all, 127 phytoplankton taxa were encountered (Table 4.2). Seasonal variations in the species diversity of phytoplankton in the studied lake show Bacillariophyceae peak during autumn while Chlorophyceae, Cyanophyceae, Euglenophyceae and Chrysophyceae show their peak during summer. The Dinophyceae do not show any seasonal trend in the present investigation.

Table 4.1: Range of Physico-chemical Characteristics at Four Sites of the Dal Lake

Parameter	*Unit*	*I*	*II*	*III*	*IV*
Water Temperature	°C	4–27	4–27	4–27	4–27
pH	–	7.4–9.3	7.0–7.9	7.0–7.8	7.1–8.9
Conductivity	s Cm^{-1}	100–284	210–265	205–397	112–530
Dissolved oxygen	mg/L	6.0–11.4	8.3–12.8	2.0–10.0	0.8–7.8
Calcium	mg/L	18.6–27.6	19.2–36.8	18.3–35.2	15.2–55.3
Magnesium	mg/L	2.4–16.1	2.2–24.6	2.9–20.6	11.6–30.5
Total alkalinity	mg/L	42–112	38–94	73–195	82–460
Chloride	mg/L	14.0–35.0	8.5–75.0	15–39	15–85
Nitrate-nitrogen	g/L	113–700	152–710	208–570	215–910
Total Phosphorus	g/L	180–360	315–621	210–550	231–490

Table 4.2: Total Number of Phytoplankton Taxa Recorded from Four Investigated Sites

Class	*Sites*				
	Total	*I*	*II*	*III*	*IV*
Bacillariophyceae	75	73	72	68	71
Chlorophyceae	34	33	28	25	25
Cyanophyceae	13	10	11	11	12
Dinophyceae	2	2	2	2	2
Euglenophyceae	2	2	2	2	2
Chrysophyceae	1	–	1	1	1
Total	**127**	**120**	**116**	**109**	**113**

Bacillariophyceae

The Bacillariophyceae taxa recorded from the investigated sites are given in Table 4.3. In all, there are 75 taxa distributed at four sites of the lake with 73 at Site–I, 72 at Site–II, 68 at Site–III and 71 at Site–IV. The population density is highest in autumn at Sites–I, III and IV, but in summer at Site–II. Lund (1965) reported abundance of bacillariophyceae population in colder months and opined that they grow under the condition of weak light and low temperature. Zafar (1967) also observed colder months to be more favourable for multiplication of bacillariophyceae in freshwater bodies. Similar observations have also been made by Venkateshwarlu (1969), Vass *et al.* (1977), Manikya (1984), Sarwar (1985), Singh and Srivastava (1991), Mir (1995), Sarwar *et al.* (1996), Gujarathi and Kanhere (1998), and Sedamkar and Angadi (2003).

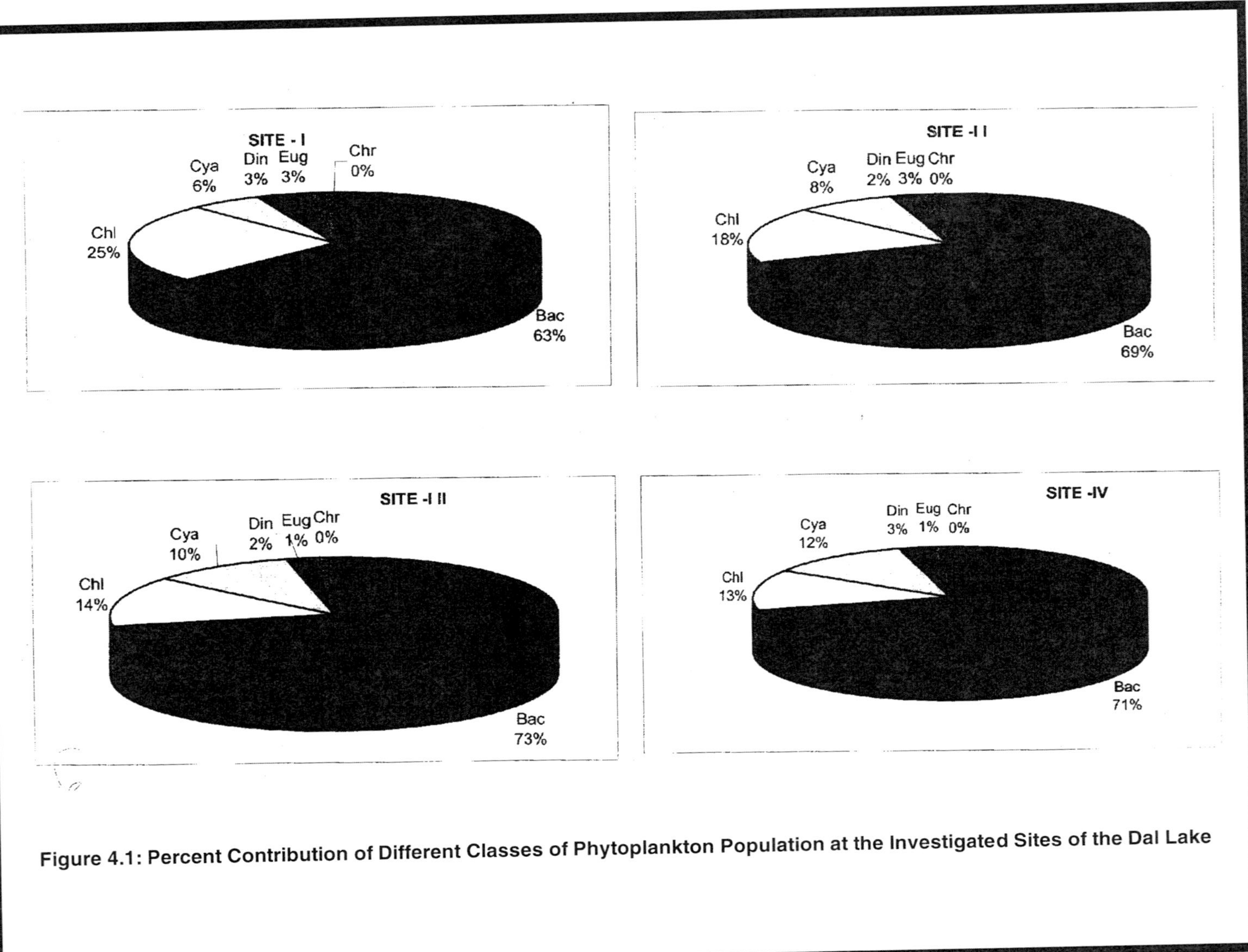

Figure 4.1: Percent Contribution of Different Classes of Phytoplankton Population at the Investigated Sites of the Dal Lake

Table 4.3: Bacillariophyceae Taxa Recorded from the Investigated Sites of the Dal Lake

Taxa	*Site–I*	*Site–II*	*Site–III*	*Site–IV*
Amphora sp. Her	+	+	+	+
Achnanthes sp. Bory	+	+	+	+
Amphora bitumida Prowse	+	+	+	+
Amphora constrictum Donk	+	+	+	+
Amphora falcatus Kutz	+	+	+	+
Amphora normanii Donk	+	+	+	+
Amphora ocellata Kutz	–	+	+	+
Amphora ovalis Kutz	+	+	+	+
Amphora proteus Greg	+	+	–	+
Anomoeoneis sp. Pfitzer	+	+	+	+
Aphanocapsa sp.	+	–	–	–
Asterionella formosa Hassal	+	+	+	+
Caloneis sp. Cleve	+	+	+	+
Ceratoneis arcus (Ehr) Kutz	–	+	+	+
Chaetoceros sp. Her	+	+	+	+
Cocconeis placentula Her	+	+	+	+
Cocconeis sp. (Her) Grun	+	+	+	+
Coscinodiscus sp. Her	+	+	+	+
Cyclotella sp. Kutz	+	+	+	+
Cymbella cistula Hempr	+	+	+	+
Cymbella tumida Breb	+	+	+	+
Cymbella lanceolata Her	+	+	+	+
Cymbella microporum Kutz	+	+	+	+
Cymbella prostrata Ralfs	+	+	–	+
Cymbella sp. Ag	+	+	+	+
Cymbella turgida Greg	+	+	+	+
Cymbella ventricosa Kutz	+	+	+	+
Diatoma elongatum Agardh	+	+	+	+
Diatomella sp. Grev	+	+	–	–
Diploneis sp. Ehr	+	+	–	–
Epithemia sp. Breb	+	+	+	+
Epithemia turgida (Ehr) Kutz	+	+	+	+
Eunotia sorex Kutz	+	+	+	+
Eunotia gracilis (Ehr) Rab	+	+	+	+
Eunotia pectinalis (Kutz) Raben	+	+	+	+
Eunotia sp. Her	+	+	+	+

Contd...

Table 4.3–Contd...

Taxa	Site–I	Site–II	Site–III	Site–IV
Eunotia triodon Ehr	+	+	+	+
Eunotia zebra (Ehr) Kutz	+	+	+	+
Eunotia diodon Ehr	+	+	+	+
Fragilaria acus Desmaz	+	+	+	+
Fragilaria capucina Desmaz	+	+	+	+
Fragilaria construens (Ehr) Grun	+	+	+	+
Fragilaria crotonensis (Ed) Kitt	+	+	+	+
Fragilaria sp. Kutz	+	+	+	+
Fragilaria vaucheriae (Kutz) Peterson	+	+	+	+
G. constrictum Ehr	+	+	+	+
G. subventricosum Ehr	+	+	+	+
Gomphonema augur Ehr	+	+	+	+
Gomphonema geminatum (Lyngb) Ag	+	+	+	+
Gomphonema accuminatum Her	+	+	+	+
Gomphonema olivaceum Kutz	+	–	+	+
Gomphonema sp. Ag	+	+	+	+
Melosira granulata Kutz	+	+	+	+
Melosira sp. (Ehr) Ralfs	+	+	+	+
Meridion sp. (Gruv) Ag	+	+	+	+
Navicula constricta Bory	+	+	+	+
N. rhyncocephala Kutz	+	+	+	+
Navicula radiosa Kutz	+	+	+	+
Navicula sp. Bory	+	+	+	+
Nitzschia longissima (Breb) Ralfs	+	+	+	+
Nitzschia acccicularis W.Sm.	+	+	+	+
Nitzschia epiculata (Greg) Grun	+	+	+	+
Nitzschia plana W.Sm.	+	+	+	+
Nitzschia sp. (Hassal: W.Sm) Grun	+	+	+	+
Pinnularia sp. Her	+	+	+	+
Pleurosigma sp. W.Sm.	+	+	+	+
Rhopalodia gibba (Ehr) O.Mull	+	+	+	+
Rhopalodia sp. O.Mull	+	+	–	+
Stauronies sp. Her	+	+	+	+
Suriella sp. Turpin	+	+	+	+
Synedra lingus Her	+	+	+	+
Synedra sp. Her	+	+	+	+
Synedra ulna (Nitz) Her	+	+	+	+
Tabellaria fenestrata (Lyng)	+	+	+	+
Tabellaria sp. Lyng	+	–	–	–

+: Present; –: Absent.

The distribution pattern of bacillariophyceae indicates that species composition is not much variable at different sites. Out of total of 75, 65 have been found to be present at all the sites. The most frequently encountered are *Amphora ovalis, A. bitumida, Anomoeneis* sp., *Cocconeis placentula, Coscinodiscus* sp., *Cymbella cistula, C. ventricosa, Eunotia gracilis, E. diodon, Eunotia* sp., *Fragilaria construens, F. vaucheria, Navicula rhyncocephala, N. radiosa, Nitzschia acicularis* and *Rhopalodia gibba.* Presence of these at all the sites indicate that lake water is favourable for their growth at all the sites. Syal (1996) found *Cocconies* sp. to be present at polluted and pollution free sites of the river Satluj. However, Kolkwitz and Marsson (1908), Jolly and Chapman (1966) and Zutshi *et al.* (1980) found only it in clean waters. *Cymbella cistula* was also found to be present in clean water as well as water contaminated with fertilizers by Kaur *et al.*, 2002.

The population density of Bacillariophyceae in the lake has been found to be quite high fluctuating between 272 to 1016 ind/mL (Table 4.4). Site–IV shows the maximum with 1016 ind/mL. Here water receives sewage from hotels and houseboats and also high amounts of nutrients added at the preceding sites. Water, therefore, gets loaded with high concentration of nitrate and phosphate as is also revealed by physico-chemical studies. Pearsall and Lind (1942), Munawar (1970), Zafar (1984), and Sedamkar and Angadi (2003) also recorded high density of Bacillariophyceae in the phosphate and nitrate rich waters.

Table 4.4: Fluctuations in the Total Population Density of Bacillariophyceae at the Investigated Sites in the Dal Lake

Site	*Maximum (Ind/mL)*	*Minimum (Ind/mL)*
I	944	272
II	976	332
III	904	440
IV	1016	488

The bacillariophyceae taxon recorded exclusively from the Site–I is *Aphanocapsa* sp. The *Ceratoneis arcus* and *Amphora ocellata* were found to be absent at Site–I but present at rest of the three sites showing their preference for nutrient rich water. *Navicula rhyncocephala* has been found to be more frequently at Sites–II, III, and IV and also its density was found to be higher at these sites. High levels of nutrients seem to be conducive for its growth. Rawson (1956), Round (1964), Richardson (1968), Lowe (1974) and Hickman (1976) have referred this species to be the pollution indicator. *Achnanthes* sp. has been most frequently encountered at Site–II with its density, too, to be highest at this site. This is attributed to high level of phosphate coming from the floating gardens. Sarwar (1991) has related the dominance of this species to overall nutrient enrichment of the lake. *Synedra ulna* is found only once at Site–I whereas it was encountered frequently at sites receiving the effluents from floating gardens, house-boats and hotels thereby indicating its preference for nutrient rich water.

Chlorophyceae

The chlorophyceae constituted the second largest group in terms of population density and species diversity after diatoms. Biswas and Konar (2000) and Yousuf *et al.* (2002) also reported chlorophyceae to be the second largest group of Phytoplankton in the river Ganga and the Anchar lake respectively. Seasonal trend in population density is indicated at Sites–I, II and IV only with higher values in summer. Similar trend is also observed by Kant and Kachroo (1974, 1977), Zutshi and Vass (1982), Gujarathi and Kanhere (1998) and Bhatt *et al.* (1999).

Chlorophyceae is comprised of 34 taxa in all, with 33 at Site I, 28 at Site–II and 25 each at Site–III and Site–IV (Table 4.5). Maximum diversity is recorded at Site–I which is located where water is least polluted and the transparency is highest of all the sites. According to Pearsall (1923), Kaloo *et al.* (1955), Munawar (1970) and Kaul *et al.* (1978), higher transparency is conducive for the growth of chlorophyceae which is in agreement with the present findings.

Table 4.5: Chlorophyceae Taxa Recorded from the Investigated Sites of the Dal Lake

Taxa	*Site–I*	*Site–II*	*Site–III*	*Site–IV*
Ankistrodesmus spiralis	+	+	–	–
Ankistrodesmus sp.	+	–	+	+
Characium sp.	+	+	+	+
Chlorella sp.	+	+	+	+
Chlorococcum sp.	+	+	+	+
Cladophora sp.	+	+	+	+
Clostridium sp.	+	+	+	+
Coelastrum sp.	+	–	–	–
Cosmarium sp.	+	+	+	+
Desmidium sp.	+	+	+	+
Euastrum sp.	+	+	–	–
Gonatozygon sp.	+	+	+	+
Mougeotia sp.	+	+	+	+
Oedogonium sp.	+	+	+	+
Oocystis sp.	+	+	+	+
Pediastrum boryanum	+	+	+	+
P. duplex	+	+	+	+
P. ovatum	+	–	–	–
P. spinosum	+	–	–	–
P. tetras	+	+	–	–
P. simplex	+	+	+	+
Pediastrum sp.	+	+	–	–
Synedra bijugates	+	+	+	+
Scenedesmus dimorphus	+	+	+	+
Scendesmus sp.	+	+	+	+
Selenastrum gracile	–	–	+	+
Selenastrum sp.	+	+	–	–
Sphaerocystis	+	+	+	+
Spirogyra sp.	+	+	+	+
Staurastrum sp.	+	+	+	+
Tetraedon	+	–	–	–
Ulothrix sp.	+	+	+	+
Volvox sp.	+	+	+	+
Zygnema sp.	+	+	+	+

+: Present; –: Absent.

Out of 34, 23 taxa are recovered from all the sites which include *Characium* sp., *Chlorella* sp., *Chlorococcum* sp., *Cladophora* sp., *Clostridium* sp., *Cosmarium* sp., *Desmidium* sp., *Gonatozygon* sp., *Mougeotia* sp., *Oedogonium* sp., *Oocystis* sp., *Pediastrum boryanum, P. duplex, P. simplex, Synedra bijugates, Scenedesmus dimorphus, Scenedesmus* sp., *Sphaerocystis* sp., *Spirogyra* sp., *Staurastrum* sp., *Ulothrix* sp., *Volvox* sp. and *Zygnema* sp. Their presence at all the sites indicate wide distribution of these taxa. Venkateshwarlu *et al.* (1990) and Reddy *et al.* (1991) recorded *Chlorella* sp. in oxygen deficient waters. The taxa restricted to the least polluted and open water Site–I include *Coelastrum* sp.,.*Pediastrum ovatum, P. spinosum,* and *Tetraedon* sp. Presence of *Pediastrum ovatum* in open water is also reported by Sarwar *et al.* (1996). Absence of these species at polluted sites indicate their aversion to organic pollution. *Selenastrum gracile* has been found only at Sites–III and IV showing this to be tolerant to sewage pollution. No species specific to Site–II is recovered. At Sites–III and IV, both species diversity and population density have been found to be lowest. These sites receive sewage from hotels and houseboats which seem to be deterrent for the growth of some chlorophyceae.

Cyanophyceae

In all, 13 species have been recovered with 10 at Sites–I, II each at Sites–II and IV and 12 at Site–IV (Table 4.6). Population density has been found to be higher at Sites–III and IV where water is contaminated with sewage and faecal matter. Pearsall (1932) and Sedamkar and Angadi (2003) had correlated abundance of cyanophyceae with high concentration of organic matter and nutrients. Population density is recorded to be higher in summer as is also observed by Gonzalves and Joshi (1946), Zafar (1967), Munawar (1970), Nandan and Aher (2002) and Angadi (2003). Sabater and Isabel (1990) and Somashekar (2000) have reported abundance of cyanophyceae from late spring to summer while Biswas and Konar (2000) have recorded increased population in rainy season.

Table 4.6: Cyanophyceae Taxa Recorded from the Investigated Sites of the Dal Lake

Taxa	*Site–I*	*Site–II*	*Site–III*	*Site–IV*
Anabaena sp.	+	+	+	+
Anacystis sp.	+	+	+	+
Gleotrichia sp.	+	+	I	+
Gomphosphaera sp.	+	+	+	+
Lyngbia sp.	+	+	+	+
Microcystis sp.	+	+	+	+
M. aeroginosa	+	+	+	+
Merismopedia formosa	–	+	+	+
M. glauca	+	–	–	–
Nostoc sp.	+	+	+	+
Oscillatoria sp.	+	+	+	+
O. princeps	–	+	+	+
Spirulina sp.	–	–	–	+

+: Present; –: Absent.

Out of total of 13, 9 cyanophycean taxa have been found to be present at all the sites. These include *Anabaena* sp., *Anacystes* sp., *Gleotrichia* sp., *Gomphosphaera* sp., *Lyngbia* sp., *Microcystes* sp., *M.*

aerogenosa, Nostoc sp., and *Oscillatoria* sp. Presence of these at all the sites indicate that they can tolerate wide range of pollution. *Merismopedia formosa* and *Oscillatona princeps* were found absent from Site–I indicating their preference for eutrophic water. On the other hand *Merismopedia glauca* was recorded from Site–I only which is the least polluted site. Sarwar *et al.* (1996) also recorded this species in less polluted and open water area of the lake.

Dinophyceae

Only two taxa belonging to Dinophyceae have been recorded (Table 4.7) with total dinoflagellates not exceeding 10 per cent of the total composition of the phytoplankton. Vareethiah and Haniffa (1998) also recorded low contribution of this group towards phytoplanktons. The Dinophyceae taxa include *Gymnodium* sp. and *Peridinium* sp., the former being more common. Tilzer (1973) also found *Gymnodium* sp. dominant throughout the year in the Vorderer Finstertaler lake while Yousuf *et al.* (2002) recorded *Peridinium* sp. to be dominant taxon in this study. No marked seasonal trend was observed.

Table 4.7: Dinophyceae Taxa Recorded from the Investigated Sites of the Dal Lake

Taxa	*Site–I*	*Site–II*	*Site–III*	*Site–IV*
Gymnodium sp.	+	+	+	+
Peridinium sp.	+	+	+	+

+: Present.

In terms of population density, *Peridinium* sp. was recorded highest at Site–I which shows very low human interference and lower levels of phosphorus and nitrogen as compared to other sites. Rodhe (1948) and McMurry and Olive (1975) have opined that nitrate and phosphate have inhibitory effects on *Peridinium* sp. On the other hand *Gymnodium* sp. has been recovered at all the sites showing its tolerance to organic pollution.

Euglenophyceae

Only two euglenophycean taxa have been recorded which include *Euglena* sp. and *Phacus* sp. (Table 4.8). The population density of both showed following trend.

Site–II > Site–IV > Site–III > Site–I

Table 4.8: Euglenophyceae Taxa Recorded from the Investigated Sites of the Dal Lake

Taxa	*Site–I*	*Site–II*	*Site–III*	*Site–IV*
Euglena sp.	+	+	+	+
Phacus sp.	+	+	+	+

+: Present.

Highest population density is found at Site–II where the lake water harbour floating gardens and receive fertilizers and thus is rich in nutrients. The high population density in nutriently rich water has also been recorded by Rao (1953), Zafar (1959), Singh (1960) and Munawar (1970). The population density of euglenoids has been recorded highest in Summer. Jana and Sarkar (1971), Ahmad and Siddiqui (1991) and Syal (1996) also recorded that they flourish well in summer.

The *Euglena* sp. and *Phacus* sp. were present at all the investigated sites of the lake showing that these can thrive well in both clean and polluted waters. David and Ray (1996), Rai (1978) also recorded *Euglena* sp. in both the waters while Palmer (1969), and Visisht and Kapoor (1981) have described it as pollution tolerant species.

Chrysophyceae

Table 4.9: Chrysophyceae Taxa Recorded from the Investigated Sites of the Dal Lake

Taxa	Site–I	Site–II	Site–III	Site–IV
Dinobryon sp.	–	+	+	+

+: Present; –: Absent.

Only one taxon, *Dinobryon* sp. has been recorded. The population density of this taxon is extremely low and its contribution towards algal population is almost negligible. Its presence has been recorded at Sites–II, III and IV (Table 4.9) only. These sites receive good amounts of nitrogen and phosphorus levels. Zutshi and Vass (1982) have also attributed presence its to high levels of nitrate and phosphate in the Dal lake.

Acknowledgement

Kind acknowledgements are made to the Head, Department of Zoology, Punjabi University, Patiala and Principal, S.P. College for providing laboratory facilities.

References

APHA (1989). *Standard Methods for the Examination of Water and Wastewater*, 17th Edition Washington.

Ahmad, M.S. and E.N. Siddiqui (1991). Limnological studies of two ponds of Darbhanga. *Proc. Nat. Planning for Environ. Sus. Dev.*, New Delhi.

Bhatt, L.R., P. Lacoul, H.D. Lekhak and P.K. Jha (1999). Physico-chemical characteristics and Phytoplanktons of Taudaha Lake, Kathmandu. *Poll. Res.*, 18(4): 353–358.

Biswas, B.K. and S.K. Konar (2000). Impact of waste disposal on plankton abundance and diversity in the river Ganga at Hathidah (Bihar). *Poll. Res.*, 19(4): 633–640.

David, A. and P. Ray (1966). Studies on the pollution of the river Daha (N-Bihar) by sugar and distillery wastes. *Environ. Hlth.*, 8: 6–35.

Golterman, H.L. and R.S. Clymo (1969). *Methods for Chemical Analysis for Freshwater.* IBP Handbook No. 8, Blackwell Scientific Publications, Oxford.

Gonzalves, E.A. and D.B. Joshi (1946). The seasonal succession of algae in a tank at Bandra, *Bombay Nat. Hist. Soc.*, 46(1): 154–176.

Gujarathi, S.A. and R.R. Kanhere (1998). Seasonal dynamics of phytoplankton population in relation to abiotic factors of a freshwater pond at Barwani (M.P.). *Poll Res.*, 17(2): 133–136.

Heurck, H.V. (1896). *Diatomaceae.* William Wesley and Sons, W.C.

Hickman, M. (1976). Studies on the epipelic diatom flora of some lakes in the southern Yukon territory Canada. *Arch. Hydrobiol.*, 4: 420–448.

Jana, B.B. and H.L. Sarkar (1971). The limnology of Swetganga. A thermal spring of Bakreswar, W.B. *Hydrobiologia,* 37: 33–47.

Jolly, V.H. and V.A. Chapman (1966). A preliminary biological study of the effects of pollution on Farmers Creek and Cox's river, New South Wales. *Hydrobiologia,* 27: 160–187.

Kaloo, Z.A., A.K. Pandit and D.P. Zutshi (1995). Nutrient status and Phytoplankton Dynamics of Dal lake under *Salvinia natans,* an obnoxious weed growth. *Oriental Science,* 1: 73–85.

Kant, S. and P. Kachroo (1974). Limnological studies of Kashmir lakes. IV. Seasonal dynamics of Phytoplankton in the Dal and Nagin. *Proc. Indian. Natn. Sci. Acad.,* B. 40: 77–99.

Kant, S. and P. Kachroo (1977). Limnological studies of Kashmir lakes. I. Hydrological features, composition and periodicity of Phytoplankton in the Dal and Nagin lakes. *Phykos.,* 16(1–2): 77–97.

Kaul, V., D.N. Fotedar, A.K. Pandit and C.L. Trisal (1978). A Comparative study and plankton populations of some typical freshwater bodies of Jammu and Kashmir State. In: *Environmental Physiology and Ecology of Plants,* (Eds.) D.N. Sen and R.P. Bansal, Bishen Singh Mahendra Pal Singh, Dehra Dun, India, pp. 249–269.

Kaur, H., J. Syal and S.S. Dhillon (2002). Impact of fertilizer factory wastes on physico-chemical and biological features of Satluj River. In: *Ecology of Polluted Waters,* Vol. II., (Ed.) Arvind Kumar, A.P.H. Publishing Corporation, New Delhi.

Kolkwitz, R. and M. Marsson (1908). Oekologie der pflanzlichen Sprolien. *Ber. Bot. Ges.,* 26: 505–519.

Lowe, R.L. (1974). Environmental requirements and pollution tolerance of freshwater diatoms. *Environmental Protection Agency,* Cineinnati, Ohio.

Lund, J.W.G. (1965). The ecosystem of freshwater phytoplankton. *Bioi. Rev.,* 40: 281–293.

Manikya Reddy, P. (1984). Ecological studies in the River Tungabhadra (AP) with special reference to the effect of effluents on the river. Ph.D. Thesis, Osmania University, Hyderabad (AP), India.

McMurry, G. and J.H. Olive (1975). Summer phytoplankton photosynthesis in a north eastern Ohio glacial lake. *Ohio J. Sci.,* 75: 238–250.

Michael, P. (1984). *Ecological Methods for Field and Laboratory Investigations.* Tata MacGraw Hill. Comp. Ltd., New Delhi.

Mir, G.R. (1995). Impact of floating gardens on the limnological features of Dal lake. Unpublished M.Phil dissertation of University of Kashmir.

Moyle, J.B. (1945). Some chemical factors influencing the distribution of aquatic plants in Minnesota. *Amer. Midland Nat.,* 34: 402–420.

Munawar, M. (1970). Limnological studies on freshwater ponds of Hyderabad. I. Biotope. *Hydrobiologia,* 35: 137–162.

Munawar, M. (1970b). Limnological studies on freshwater ponds of Hyderabad. II. Biocenose. *Hydrobiologia,* 36(1): 105–128.

Nandan, S.N. and N.H. Aher (2002). Limnological study of algae of the river Mausam in Maharashtra. *Indian J. Environ. and Ecoplan.,* 6(3): 605–608.

Oslen, S. (1950). Aquatic plants and hydrospheric factors. *Svensk Bot. Tidskr.,* pp. 44.

Palmer, C.M. (1969). A composite rating of algae tolerating organic pollution. *J. Phycol.*, 5: 78–82.

Palmer, C.M. (1980). *Algae and Water Pollution.* Castle House Publication. London, pp. 123.

Pandit, A.K. (2002). Conservation of Lakes in Kashmir Himalaya. In: *Natural Resources of Western Himalaya*, (Ed.) Ashok K. Pandit, pp. 291–328.

Pearsall, W.H. (1923). A theory of diatom periodicity. *J Ecol.*, 11–22: 165–183.

Pearsall, W.H. (1932). Phytoplankton in the English lakes 2. The composition of the phytoplankton in relation to dissolved substances. *J. Ecol.*, 20: 241–262.

Pearsall, W.H. and E.M. Lind (1942). The distribution of phytoplankton in some North-West Irish Loughs. *Proc. Roy Insh Acad.*, 78B: 1–24.

Philipose, M.T. (1967). *Chlorococcales.* Indian Council of Agricultural Research, New Delhi.

Prescott, G.W. (1978). *How to Know Fresh Water Algae*, 3rd edition. Wm. C. Brown Company, USA, p. 17–280.

Rai, L.C. (1978). Ecological studies of algal communities of Ganges river at Varanasi. *Indian J. Ecol.*, 5(1): 1–6.

Randhawa, M.S. (1959). *Zygnemales.* Indian Council of Agricultural Research, New Delhi.

Rao, C.B. (1953). On the distribution of algae in a group of six small ponds. *J. Ecol.*, 41: 62–71.

Rawson, D.S. (1956). Algal indicators of lake types of Saskatchewan. *Limnol. Oceanogr.*, 1: 18–25.

Reddy, P.M., V. Venkateshwarlu and R. Raj Kumar (1991). Phytoplankton in the river Tungabhadra (A.P.) and the impact of paper mill effluent on the flora. *Indian J. Microbiol. Ecol.*, 2: 71–77.

Richardson, J.L. (1968). Diatoms and lake topology in East and Central Africa. *Int. Revue. Ges. Hydrobiologia Hydrogr.*, 53: 299–338.

Rodhe, W. (1948). Environmental requirements of fresh water plankton algae. Experimental studies in the ecology of phytoplankton. *Symb. Bot. Upsal.*, 10: 1–149.

Round, F.E. (1964). The Ecology of Benthic Algae. In: *Algae and Man*, (Ed.) D.F. Jackson, Plenum Press, New York, p. 138–184.

Sabater, B. and Isabel, M. (1990). Successional dynamics of the phytoplankton in the lower part of the river *Ebro. J. Plant. Res.*, 12: 573–592.

Sarwar, S.G. (1991). Trophic status of Dal lake (Kashmir). Proceedings National Conference on Aquatic Sciences in India, New Delhi.

Sarwar, S.G. (1985). Community, species distribution and seasonal variations in Aufwuchs (periphyton) population on natural and artificial substrates. *Ph.D. Thesis*, University of Kashmir (unpublished).

Sarwar, S.G., A.R. Naqshi and G.R. Mir (1996). Impact of Floating gardens on the limnological features of Dal lake. *Poll. Res.*, 15(3): 217–221.

Sedamkar, E. and S.B. Angadi (2003). Physico-chemical parameters of two fresh waterbodies of Gulbarga–India, with special reference to phytoplankton. *Poll Res.*, 22(3): 411–422.

Singh, S.R. and V.K. Srivastava (1991). Investigation on the periodicity of phytoplankton in relation to certain hydrological conditions in the stretch of Ganga river between Buxer and Baffia. *Poll. Res.*, 10(2): 93–101.

Singh, V.P. (1960). Phytoplankton ecology of the inland waters of Uttar Pradesh. *Proc. Symp Algal ICAR,* New Delhi, p. 243–271.

Smith, G.M. (1950). *Freshwater Algae of United States.* McGraw-Hill, N.Y.

Syal, J. (1996). Present Ecological Status of River Satluj in the Region of Punjab. *Ph.D. Thesis,* Punjabi University, Patiala (unpublished).

Tilzer, M.M. (1973). Diurnal periodicity in the phytoplankton assemblage of a high mountain lake. *Limnol. Oceanogr.,* 18: 15–30.

Vareethiah, K. and M.A. Haniffa (1998). Phytoplankton community organisation and species succession in a Bar-Built Estuary. *J. Env. Poll.,* 5(3): 209–214.

Vasisht, H.S. and N.K. Kapoor (1981). The ecology of polluted waters of Yamunanagar (Haryana). *Indian J. Ecol.,* 8(1): 134–135.

Vass, K.K., H.S. Raina, D.P. Zutshi and M.A. Khan (1977). Hydrobiological studies on river Jhelum. *Geobios,* 4: 238–242.

Venkateshwarlu, V. (1969). An ecological study of the algae of the river Mossi, Hyderabad. (India) with special reference to pollution, III The algal periodicity, *Hydrobiologia,* 34(3–4): 533–560.

Venkateshwarlu, V., P.T.S. Kumar and J.N. Kumari (1990). Ecology of algae of the river Moori, Hyderabad: A comprehensive study. *J. Environ. Biol.,* 11: 79–91.

Ward, H.B. and G.C. Whipple (1959). *Freshwater Biology,* 2nd Edition. John Wiley and Sons, New York, USA.

Whipple, G.C., G.M. Fair and M.C. Whipple (1927). *The Microscopy of Drinking Water.* John. Wiley and Sons, New York, N.Y.

Winkler, L.W. (1888). The determination of dissolved oxygen in water. *Berlin. Deut. Chem. Ges.,* 21: 28–43.

Yousuf, A.R., F. Gazala and J.P. Kousar (2002). Ecology and Feeding Biology of Commercially Important Cyprinid Fishes of Anchar lake, Kashmir, with a note on their conservation. In: *Natural Resources of Western Himalaya,* (Ed.) A.K. Pandit, Valley Book House Hazratbal, Srinagar, pp. 243–272.

Zafar, A.R. (1959b). Two years observation on the periodicity of Euglenaceae in two fish breeding ponds. *J. Ind. Bat. Sac.,* 34: 430–450.

Zafar, A.R. (1964). On the ecology of algae in certain fish ponds of Hyderabad, India I. Physico-chemical complex. *Hydrobiologia,* 23: 176–196.

Zafar, A.R. (1967). On the ecology of algae in certain fish ponds of Hyderabad, India III. The periodicity. *Hydrobiologia,* 30(1): 96–112.

Zutshi, D.P. and K.K. Vass (1982). Limnological studies on Dal lake, Srinagar III. Biological features. *Proc. Indian Natn. Sci. Acad.,* B. 48(2): 234–241.

Zutshi, D.P., B.A. Subla, M.A. Khan and A. Wanganeo (1980). Comparative limnology of nine lakes of Jammu and Kashmir Himalayas. *Hydrobiologia,* 72(1–2): 101–112.

Chapter 5

Population Dynamics of Rotifers in the Anchar Lake, Kashmir (India)

M. Jeelani, H. Kaur* & S.G. Sarwar***

**Department of Zoology, Punjabi University, Patiala – 147 002, Punjab*
***Hydrobiology Research Laboratory, S.P. College, Srinagar – 190 001, Kashmir*

ABSTRACT

Qualitative and quantitative analysis of rotifer community of Anchar lake was done on monthly basis from September 2000 to August 2002. The study under taken revealed that rotifer population in the lake had decreased at the site of hospital and sewage outfalls.

Keywords: *Population dynamics, Rotifers, Anchar lake.*

Introduction

As a result of heavy anthropogenic pressures, the lake systems are not only shrinking in their surface area, but the water quality is also deteriorating posing serious health hazard to the people. The rapidly expanding human population and fast urbanization in addition to natural siltation are threatening these ecosystems as well as their natural resources. Since the valley lakes are situated in the low lying areas and floodplains of river Jhelum and Sind, deposits of silt is a continuous process, as seen in the lake Anchar. The lake, a shallow basined valley lake, is situated 12 kms to the northwest of Srinagar city within the geographical coordinates of 34°20′–34 26′ N latitude and 74 82′–74 85′ E longitude.

A network of channels from the river Sind feed the lake. A thick canopy of trees, willows and poplars surround the littorals of the lake. Towards the northeast of the lake is situated the hospital complex of S.K. Institute of Medical Sciences (SKIMS) draining much of its effluents into the lake.

The present study was undertaken to study the detailed impact of various anthropogenic activities on the rotifer community of the Anchar Lake.

Materials and Methods

Selection of Sites

Three sampling sites were selected in the lake that represent different environmental features. The Site–I is located at the point where a channel from the Sind river enters the Anchar lake. It is characterized by flowing water. Site–II is located in the centre of the lake where water is standing while Site–III is located close to the point where hospital (SKIMS) wastes are thrown into lake water. Several sewerage outlets from the catchment area also discharge their contents into water at this site.

Collection of Samples

The sampling was carried out for a period of two years extending from September 2000 to August 2002. Water samples for physico-chemical analysis were collected from the sampling sites of lake once in a month in one-litre polyethylene bottles. The pH and conductivity values were determined electrometrically and dissolved oxygen by unmodified Winklers (1888) method. The detailed chemical analysis of water samples was carried out according to standard methods of Golterman and Clymo (1969) and APHA (1989).

Collection of samples for biological study was done using plankton net (64 nm). Five litres of water was sieved through from each sampling site on each occasion. The samples were preserved in 5 per cent formalin and concentrated through centrifugation. Samples were studied under inverted microscope in a Sedgwick-Rafter cell. Identification was done by using standard monographs of Edmondson (1959), Mellanby (1963), Pennak (1978) and Tonapi (1981).

Observation and Discussion

The physico-chemical features of water are summarized in Table 5.1. The water temperature in general ranged from 3.0–28.0°C showing usual seasonal trend with maximum values in summer and minimum values in winter. The pH values fluctuated from 6.9–8.6 indicating that water is well buffered. Dissolved oxygen ranged from 0.4–9.5 mg/l with minimum at Site–III and maximum at Site–I. The conductivity values of Anchar lake indicate high ionic concentration. The range fall between 205 μScm^{-1} to 580 μScm^{-1}. As per Oslen (1950), the water of the lake can be categorised as of B-mesotrophic type. Calcium concentration varied from 9–43 mg/l and that of magnesium from 4.3–28.5 mg/l indicating the lake water to be nutriently rich with cation progression as Ca > Mg. Total alkalinity ranged from 59–315 mg/l in the investigated lake. Philipose (1960) suggested that a water body with alkalinity values > 100 mg/l is nutriently rich. On this basis, the lake water too seems to be rich in its nutritive contents. The chloride content of water ranged from 9.0–9.5 mg/l with maximum at Site–III showing that water here is polluted. Krenkel (1974) also observed increase in chloride contents in water receiving wastes via sewage outfalls. The content of nitrate- nitrogen ranged from 245–611 µg/l and that of phosphorus from 200–505 µg/l showing that lake is nutriently rich. This is evident by the profuse growth of macrophytes at Site–III and Site–II. Various physico-chemical characteristics reveal that lake is rich in Ca, Mg, Cl, No_3–N and PO_4–P showing it to be nutriently rich and polluted.

Table 5.1: Range in Physico-chemical Characteristics in Anchar Lake

Parameter	*Unit*	*Range*
Temperature	°C	3.0-28
pH	–	6.9–8.6
Conductivity	Scm^{-1}	205–580
Dissolved Oxygen	mg/l	0.4–9.5
Calcium	mg/l	9.0–43.0
Magnesium	mg/l	4.3–28.5
Total alkalinity	mg/l	59–315
Chloride	mg/l	9.0–95.0
Nitrate-nitrogen	g/l	245–611
Total phosphorus	g/l	200–505

The rotifer fauna collected at three sites from September 2000 to August 2002 have been enlisted in the Table 5.2. A total of 13 rotifer species were identified during the whole study period. Seasonal succession of rotifer in the lake reveal the peak in species diversity as well as population density during the summer (Figure 5.1) when they are represented by 12 taxa which include *Brachionus bidentata, B. calyciflorus, B. quadridentata, Bdelloid* sp., *Cephalodella* sp., *Chromogaster ovalis, Epiphanes* sp., *Filinia longiseta, Gastropus* sp., *Keratella valga, K. cochlearis,* and *Synchaeta* sp. In winters, the population density and diversity were found to be the lowest. Only 4 taxa were recorded in waters at the three sites of the lake. These include *Brachionus calyciflorus, Bdelloid* sp., *Filinia longiseta* and *Polyarthra vulgaris.* Species like *Brachionus bidentata, B. quadridentata, Chromogaster ovalis, Cephalodella* sp. and *Synchaeta* sp. were found in warm weather conditions of spring and summer and thus can be regarded as thermophilic as also suggested by Carlin (1943); Beach (1960) and Chakravarty and Kumar (1991).

Table 5.2: Showing Species Composition and Population Density of Rotifers at Different Sites of Anchar Lake

	Site–I	*Site–II*	*Site–III*
Bdelloid sp.	–	+	+
Brachionus bidentata	–	–	+
Brachionus calyciflorus	–	+	+
Brachionus quadridentata	+	–	–
Cephalodella sp.	+	–	+
Chromogaster ovalis	+	–	–
Epiphanes sp.	–	+	+
Filinia longiseta	–	+	+
Gastropus sp.	+	+	–
Keratella cochlearis	+	++	–
Keratella valga	+	+	+
Polyarthra vulgaris	–	+	–
Synchaeta sp.	–	+	–

+: 1–15 ind/ml; ++: 16 and above.

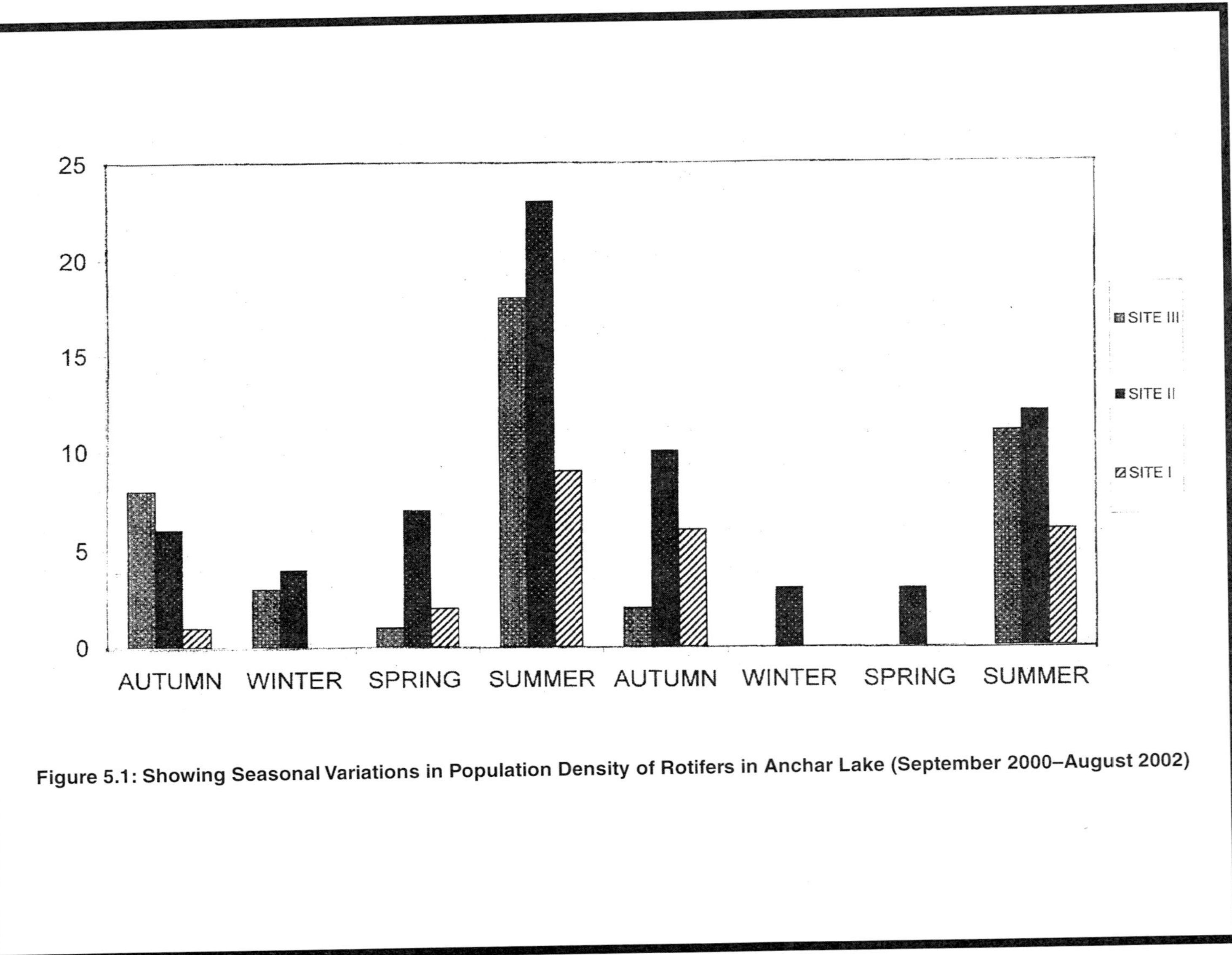

Figure 5.1: Showing Seasonal Variations in Population Density of Rotifers in Anchar Lake (September 2000–August 2002)

From the data it is revealed that majority of rotifer species exhibited wide occurrence throughout the ;ake. Sitewise trend in terms of species diversity and density is Site–II > Site–III > Site–I.

Site–I located in the feeding channel of the lake shows the minimum rotifer fauna in terms of species diversity and population density. Running water conditions and lack of rooted vegetation fail to provide suitable environment for the growth of rotifers at this site. Bath (1996) also observed similarly in Harike wetland of Punjab.

Site–II situated in the centre of the lake, exhibited the maximum population density and species diversity. This site located in standing water with much less anthropogenic interference and having much macrophytic growth, provides the most ideal conditions for rotifer growth among the studied sites. The species *Keratella cochlearis* dominated the rotifer fauna at this site. Yadav *et al.* (2003) also observed *Keratella* sp. to be dominant in a slightly polluted pond of Fatehpur Sikri.

At Site–III, the rotifer fauna appears to be adversely affected as is evident from a fall in population density as well as diversity when compared with Site–II. This site is also located in the standing water area of the lake and is polluted by the sewage and wastes from the hospital. It is concluded that pollution added by the sewage and hospital play a significant role in the dynamics of rotifer community.

Acknowledgement

Kind acknowledgements are made to the Head, Department of Zoology, Punjabi University, Patiala and Principal, S.P. College for providing laboratory facilities.

References

APHA (1989). *Standard Methods for the Examination of Water and Wastewater*, 17th Edition Washington.

Bath, K.S. (1996). Limnological studies on the Harike wetland ecosystem. *Ph.D. Thesis* (Unpublished) Punjabi University, Patiala, Punjab.

Beach, N.W. (1960). A study of the planktonic rotifers of the Ocqueoc river system. Presque. Isk County, Michigan. *Ecol. Monogr.*, 30: 339–357.

Carlin, B. (1943). Die Planktonrotatorien des Motalastrom: zur Taxonomie und Okologie der *Plantorotatorion. Meddn. Lunds. Univ. Limnol. Istn.*, 5: 225.

Chakravarty, T.K. and S. Kumar (1991). Seasonal changes of planktonic Rotifers in Freshwater Ponds of Bhagalpur. In: *Aquatic Sciences in India,* (Eds.) B. Gopal and V. Asthana. Indian Association for Limnology and Oceanography. New Delhi, p. 77–82.

Golterman, H.L., R.S. Clymo and M.A.M. Ohnstad (1978). *Methods for Chemical Analysis of Freshwater.* IBP Handbook No. 8, Oxford, Blackwell-Scientific.

Hutchinson, G.E. (1957). *A Treatise on Limnology Vol. 1.* John Wiley and Sons, New York, p. 10–15.

Krenkel, P.A. (1974). Sources and classification of water pollutants. In: *Industrial Pollution,* (Ed.) N.I. Sax. Van Nostrand Reihold Company, New York.

Mellanby, H. (1963). *Animal Life in Freshwater*, 6th ed. Chapman and Hall Ltd., London.

Oslen, S. (1950). Aquatic plants and hydrospheric factors. *Svensk Bot. Tidskr,* pp. 44.

Pennak, R.W. (1978). *Freshwater Invertebrates of the United States*, 2nd ed. John Wiley and Sons, New York, pp. 803.

Philipose, M.T. (1960). Freshwater of inland fisheries. *Proc. Symp. Algol.,* ICAR, New Delhi.

Ruttner, F. (1968). *Fundamentals of Limnology.* Univ. of Toronto Press, Toronto, pp. 332.

Tonapi, G.T. (1981). *Freshwater Animals of India: An Ecological Approach.* Oxford and IBH, New Delhi, pp. 205.

Ward, H.B. and G.C. Whipple (1959). *Freshwater Biology,* 2nd Edition. John Wiley and Sons, New York, USA.

Winkler, L.W. (1888). The determination of dissolved oxygen in water. *Berlin Deut. Chem. Ges.,* 21: 28–43

Yadav, Gauravi, Pallavi Pundhir and K.S. Rana (2003). Population dynamics of rotifers fauna at Fatehpur Sikri pond, Agra. *Poll. Res.,* 22(4): 541–542.

Chapter 6

Diversity of Phytoplankton and Zooplankton with Respect to Pollution Status of River Tapi in North Maharashtra Region

S.R. Gaikwad, S.R. Thorat* & T.P. Chavan***

**Department of Environmental and Earth Sciences, North Maharashtra University, Jalgaon – 425 001 (M.S.), India*

***Department of Chemistry, Amolakchand Mahavidyalaya, Yavatmal*

ABSTRACT

The quantitative and qualitative analysis of phytoplankton was carried out to assess the impact of waste disposal on plankton community structure in different seasons. A total 29-phytoplankton and 11-zooplankton taxa were identified from four sampling sites *i.e.* S_1 (control 2 km upstream), S_2 (1 km ahead upstream), S_3 (Stagnant zone at Hatnur dam) and S_4 (1 km downstream at thermal power plant towards Bhusawal). Zooplankton contributes 0 to 33.88 per cent of total plankton in community. In contrast planktons were studied and compared with S_1 (control zone). The dominancy of diatom flora at all sites was recorded. Higher abundance of Myxophyceae at all three sites indicated that organic enrichment of these habitats. The role of algae in monitoring river water quality is emphasized. These shifts in the plankton community structure at discharge point indicate nature of pollution in the Tapi river.

Keywords: *Phytoplankton, Zooplankton, Seasonal variation, Physico-chemical parameters and Pollution.*

Introduction

The river Tapi at Bhusawal in Khandesh received effluent from various industrial sectors of multiple nature as well as coal, fly ash effluent from Thermal Power Plant (TPP) and from non-point sources. The addition of these materials not only influences the micro-flora of freshwater but also favours the development of variety of new biota, rendering it unfit for human consumption. A similar case exist for Hatnur dam as its fishes, crab etc. consumed by common people in Bhusawal and Jalgaon city. The water of Tapi river is also used for irrigating crop, fodder and after treatment used for drinking purpose.

Phytoplankton is the fundamental components of aquatic ecosystem as they are major source of biologically important and liable organic carbon, located at the base of food chain. The magnitude and dynamic of phytoplankton population has become an essential tool to assess the general health of an aquatic ecosystem. Phytoplankton's of freshwater river have been studied extensively in India (Bilgrami, 1989; Nandan and Patel, 1992; Kumar, A., 1998 and Thorat *et al.,* 2000).

The importance of river water has been increased now due to the provision of aquatic animals therefore; the present study was carried out to assess the physical and chemical parameters and various forms of bacteria, phytoplankton and zooplankton of Hutnur dam based on river Tapi in different seasons. Such studies have also been conducted by (Anon, 1998; Biswas and Konar, 2000; Thorat and Massrat Sultana, 2000). However, only merge information is available on the ecology of phytoplankton and zooplankton in river Tapi, Bhusawal. Keeping this in view, it was thought important to study the phytoplankton and zooplankton, various forms of bacteria, characteristics of the river water and extend of its pollution.

Materials and Methods

A total 12 water samples were collected from June 2003 to May 2004 from four sites, which comprised four each for rainy, winter and summer June to September was consider as rainy season sample, whereas October to January and February to May were winter and summer collection respectively.

Standard methods (NEERI, 1995) were used during the collection, preservation and estimation of physico-chemical, bacteriological and physiological examinations. Plankton samples were collected from four sites and preserved in small vials with drops of formaline and few drops of glycerin and allow plankton to settle and latter concentration to a known volume. Samples were centrifuge and concentrated to 10 ml in the laboratory for qualitative and quantitative estimation. Plankton was counted as unit per 50 liters instead of conventional unit per liters for avoiding fractional values. Identification of plankton was done with the help of Desikachary, 1959; Sreenivas and Duthie, 1973; Martinez *et al.,* 1975; Battish, 1992; and Agarkar *et al.,* 1994.

Results and Discussion

A total 29 phytoplankton and 11 zooplankton taxa were identified from four sampling sites during the period of investigation out of these 13 genera belong to *Chlorophyceae,* 5 from *Myxophyceae,* 10 from *Bacillariophyceae* and 1 genus from *Euglenophyceae* it indicate in the Table 6.1. Zooplankton was identified at different site are 6 from *Cladocera,* 2 from *Copepoda* and 3 from *Rotifera* indicate in Table 6.2.

Table 6.1: Seasonal Variations in Phytoplankton Abundance and Diversity in River Tapi

Phytoplankton	Rainy				Winter				Summer			
	S_1	S_2	S_3	S_4	S_1	S_2	S_3	S_4	S_1	S_2	S_3	S_4
Cladophora	–	–	–	–	38	–	–	18	–	–	–	–
Microspora	60	174	80	–	18	–	–	60	–	–	–	–
Volvox	–	–	–	–	40	–	72	20	–	64	232	124
Spirogyra	–	–	–	–	14	348	186	350	180	210	904	800
Cylindrocapsa	40	–	–	18	–	–	–	–	–	–	–	–
Monocilia	–	–	–	–	50	–	–	60	–	–	–	–
Bumillaria	–	–	–	–	–	–	–	–	20	20	30	10
Pachycladom	–	–	–	–	–	–	–	–	20	30	–	10
Eudorina	–	–	–	–	–	–	–	–	10	–	–	54
Pseudowella	–	–	–	–	–	–	–	–	40	–	–	70
Staurastrum	–	–	–	–	–	16	30	–	25	44	–	20
Hydrodictyon	–	–	–	–	70	80	90	64	120	350	160	54
Closterium	10	–	–	20	–	–	–	–	–	–	108	164
Total *Chlorophyceae*	**110**	**174**	**80**	**38**	**230**	**444**	**378**	**572**	**415**	**718**	**1434**	**1306**
Oscillatoria	98	108	214	195	10	420	160	220	470	321	441	570
Microcystis	–	–	24	10	–	40	20	10	–	110	60	320
Merismopedia	–	–	–	35	–	70	55	–	–	312	90	142
Anabaena	–	310	260	144	–	65	30	–	210	1850	920	730
Spirulina	–	256	80	30	–	310	210	80	–	80	30	68
Total *Myxophyceae*	**98**	**674**	**578**	**414**	**10**	**905**	**475**	**310**	**680**	**2673**	**1541**	**1830**
Meridion	–	–	96	10	105	130	360	40	334	106	130	220
Anomoeoneis	–	–	–	–	80	–	–	66	–	–	–	10
Navicula	–	10	–	20	65	75	25	–	200	410	165	16
Melosira	–	20	–	800	34	–	–	–	800	40	30	20
Gurosigma	–	40	–	40	–	130	–	140	90	110	40	20
Achnanthes	–	–	60	104	–	90	120	110	80	20	10	94
Tabellaria	–	–	–	–	–	–	110	90	10	105	22	54
Fragilaria	–	42	80	–	–	–	200	80	70	204	12	74
Comphonema	–	10	–	–	10	86	–	60	60	90	28	72
Nitzschia	10	–	90	–	–	76	–	70	90	186	18	32
Total *Bacillario-phyceae*	**10**	**122**	**326**	**974**	**294**	**587**	**815**	**656**	**1734**	**1271**	**325**	**612**
Euglena	–	–	–	–	–	–	–	–	–	10	–	24
Total *Euglenophyta*	**–**	**–**	**–**	**–**	**–**	**–**	**–**	**–**	**–**	**10**	**–**	**24**
Total Phytoplankton	**218**	**796**	**984**	**1426**	**534**	**1936**	**1668**	**1538**	**2829**	**4662**	**3300**	**3772**

Note: All figures in unit/50 litre total of three samples in a year *i.e.* June 2003–May 2004.

Table 6.2: Seasonal Variations in Zooplankton Abundance and Diversity in River Tapi

Zooplankton	*Rainy*				*Winter*				*Summer*			
	S_1	S_2	S_3	S_4	S_1	S_2	S_3	S_4	S_1	S_2	S_3	S_4
Daphnia	40	20	20	50	60	10	20	30	30	–	30	36
Polyphemus	50	40	–	46	–	–	–	–	–	–	–	–
Ceriodaphuia	–	–	–	–	10	50	10	–	–	–	10	–
Bosmina	–	–	–	–	–	–	–	–	20	30	50	70
Macrothrix	–	–	–	–	–	–	–	–	10	10	20	30
Moina	–	–	–	–	–	–	–	–	80	–	16	78
Total *Cladocera*	**90**	**60**	**20**	**96**	**70**	**60**	**30**	**30**	**140**	**40**	**96**	**178**
Diaptomus	80	–	20	40	–	–	–	–	10	–	–	80
Cyclops	–	–	–	–	–	–	–	–	10	30	20	42
Total *Copepoda*	**80**	**–**	**20**	**40**	**–**	**–**	**–**	**–**	**20**	**30**	**20**	**122**
Aspanchna	40	120	90	80	–	–	–	–	–	–	–	–
Brachionus	–	–	–	–	–	–	–	–	–	70	50	30
Keratella	–	–	–	–	10	50	30	70	–	–	–	–
Total *Rotifera*	**40**	**120**	**90**	**80**	**10**	**50**	**30**	**70**	**–**	**70**	**50**	**30**
Total Zooplankton	**210**	**180**	**130**	**216**	**80**	**110**	**60**	**100**	**160**	**140**	**166**	**330**

Note: All figures in unit/50 litre total of three samples in a year *i.e.* June 2003–May 2004.

In phytoplankton all the sites were mostly dominated by diatoms at all the seasons and in Zooplankton *Daphnia* from *Cladocera* and *Aspanchna* from *Rotifera* was dominated in all three stations in a year.

The results of bacteriological analysis are expressed in Table 6.3. The maximum total viable count was observed in rainy season. This may be due to addition of land washing and organic matter by rain. The specific bacteriological examination indicates a notable presence of *Escherichia coli* and *Micrococcus* sp. Sengar *et al.* (1996) reported the presence of similar genera in different sources of water.

Table 6.3: Total Viable Count of Various Bacteria in River Tapi

Species	*Rainy*				*Winter*				*Summer*			
	S_1	S_2	S_3	S_4	S_1	S_2	S_3	S_4	S_1	S_2	S_3	S_4
Escherichia coli–I	468	2300	450	1344	146	80	88	98	340	345	344	90
E.coli–II	220	190	1233	1092	120	94	76	176	234	562	433	198
Staphyllococus aureus	109	1897	1034	234	132	85	59	102	432	536	124	112
Pseudomonas sp.	506	234	1402	654	114	69	78	134	231	452	245	347
Micrococcus sp.	102	654	1230	972	164	90	102	44	341	433	132	454
Aerobactor aerogenes	60	40	96	85	–	10	–	62	687	932	221	416
Others	545	1342	1058	789	372	633	1156	234	1321	2543	1345	2652
Total Count	**2010**	**4657**	**6503**	**5170**	**1048**	**1061**	**1559**	**850**	**3586**	**5803**	**2444**	**4269**

Note: All values are in $x10^{3/1}$.

It was also found that most of the phytoplankton taxas were recorded in considerably higher numbers of different sites; especially during dry months. These abnormalities can be explained as after rainy season a number of small pools were created all along the river course, which were completely isolated except one connection with the main channel. These small pools with their huge storage of organic materials remain until next rainy season; some findings were recorded by Chakraborty and Konar (2000). The water of this pool was greenish and odour is fleshy due to huge proliferation of phytoplankton population *i.e.* indication of eutrophic habitat and occasional release of water from head water dams mixed the phytoplankton of these pools into the main of Tapi river channel.

The second dominant group of S_4 was *Bacillariophyceae,* while at S_2 and S_3 it was *Myxophyceae* and *Chlorophyceae.* Thus, the appearance of blue green algae at S_2 to S_4 was an indication of organic enrichment of these habitats. According to Biswas and Konar (2000), the abundance of blue green algae can be correlated with high concentration of dissolved organic matter in the water and these organic load was obviously supplied from sewage, dead body burning, dumping of animal carcasses and defecation on the river bed. The appearances of tolerant taxas of Chlorophyceae also support the above explanation.

The data of chemical analysis is presented in Table 6.4. The pH ranges from 5.9 to 7.8. Sufficient numbers of total solids has been recorded in all the seasons, maximum being 1642.0 mg/l in rainy season, which may gradual disturbances in sedimentation of solids. The DO value ranges from 6.2 to 9.8 mg/l. While BOD and COD values are quite high and indicates that inorganic and organic pollution load in the river is high. The stretch of river Tapi at Bhusawal is influenced S_4 sites and rest of three sites was not influenced by any major point source of pollutant, it may be due to thermal power plant (TPP) which is situated on bank of the Tapi river released water into it.

Table 6.4: Seasonal Variation in Physico-chemical Parameters of Tapi River at Bhusawal

Parameters	*Rainy*				*Winter*				*Summer*			
	S_1	S_2	S_3	S_4	S_1	S_2	S_3	S_4	S_1	S_2	S_3	S_4
pH	7.2	7.8	7.6	7.2	6.6	6.2	6.5	6.5	6.8	5.9	6.5	6.8
Colour	m	m	m	m	vg	vg	sg	sg	sg	vg	sg	vg
Temperature	28.0	28.0	30.0	32.0	19.0	20.0	22.0	21.0	38.0	35.0	39.0	41.0
Odour	fishy	fishy	fishy	fishy	fishy	fishy	fishy	fishy	fishy	fishy	fishy	fishy
Turbidity, TV	320.0	312.0	432.0	452.0	210.0	222.0	226.0	236.0	190.0	210.0	198.0	232.0
BOD	17.0	19.0	18.0	174.0	68.0	45.0	94.0	160.0	54.0	34.0	88.0	192.0
COD	120.0	210.0	190.0	256.0	200.0	125.0	240.0	310.0	180.0	240.0	216.0	342.0
Total hardness	218.0	210.0	260.0	290.0	216.0	248.0	232.0	264.0.	212.0	238.0	220.0	280.0
Dissolved oxygen	7.2	7.8	8.6	7.8	6.6	9.8	6.2	7.5	8.0	8.4	7.4	9.0
Total solids	1642.0	1430.0	1670.0	1650.0	1220.0	1248.0	1340.0	1320.0	886.0	890.0	730.0	1020.0
Organic Nitrogen	0.024	0.026	0.028	0.022	0.026	0.038	0.034	0.026	0.068	0.082	0.088	0.092
Nitrate	34.0	46.0	44.0	62.0	28.0	34.0	28.46	70.0	56.0	54.0	38.0	74.0

Note: All values as in mg/lit except pH, colour, odour, temperature and turbidity.

sg: Slight green; vg: yellowish green and m: muddy.

In the present study abundance and diversity of phytoplankton and zooplankton was found to be dependent on water quality and co-existing biotic communities at given point of time. Due to organic load disturbances from various non-point sources was clearly reflected by shifting of phytoplankton and zooplankton group at these sites, which enlighten the importance of non-point and point sources to cause pollution in river Tapi.

References

Anon (1998). The river Domodar and its environment. *Bull. Cent.* In Land Fish. Res. Inst. Barrackpore, 79: 32.

Agarkar, M.S., H.K. Goswami, S. Kaushik, S.M. Mishra, A.K. Bajpai and U.S. Sharma (1994). Biology, conservation and management of Bhoj wetland I: Upper lake ecosystem in Bhopal. *Bionature,* 14(2).

Biswas, B.K. and S.K. Konar (2000). Influence of Nunia nallah (canal), discharge on plankton abundance and diversity in the river Domodar at Narankuri (Raniganj) in West Bengal. *Indian J. Environ. and Ecoplan.* 3(2): 209–217.

Battish, S.K. (1992). *Freshwater Zooplankton of India.* Oxford & IBH Publishing Co., New Delhi.

Bilgrami, K.S. (1989). Bio-monitoring of water quality of the Ganga. In: *The Ganga Project Directorate.* North Book Center, New Delhi, p. 101–106.

Chakroborty and S.K. Konar (2000). Studies on the hydrobiology of river Damodar from Raniganj Region (West Bengal) with reference to phytoplankton. *Indian. J. Environ. and Ecoplan.,* 3(3): 485–490.

Desikachary, T.V. (1959). *Cyanophyta: Monographs on Algae.* ICAR, New Delhi.

Kumar, A. (1998). Diversification of phytoplankton and its correlation with certain physico-chemical characteristics of the river Mayurakshi in tribal zone of Bihar. *International J. Ecol. Envi. Sci.,* 24: 215–224.

Martinez, M.R., R.P. Chakroff and J.B. Pautastico (1975). Direct phytoplankton counting techniques using the haemocytometer Phillipine Agri., 59: 43–50.

Nandan, S.N. and R.J. Patel (1992). Ecological studies of algae. In: *Aquatic Ecology.* Ashish Pub. House, Delhi, p. 70–99.

Sreenivas, M.R. and H.C. Duthie (1973). Diatoms flora of the grand river, Ontorio, Canada. *Hydrobiologia,* 42: 161–224.

Sengar, S.G., A.R. Naqshi and G.R. Mir (1996). Impact of floating gardens on the limnological features of Dal lake. *Poll. Res.,* 15(3): 217–221.

Thorat, S.R. and Sultana Masarrat (2000). Pollution status of Salim Ali lake, Aurangabad (M.S.). *Poll. Res.,* 19(2): 307–309.

Chapter 7

Investigations on Blue Green Algae of Satna: A Taxonomical Approach

Rashmi Singh
Department of Botany, Government P.G. (Autonomous) College, Satna (M.P.)

ABSTRACT

Blue green algae are the prokaryotic autotrophic organisms, which represent a link between bacteria and green algae. BGA are prokaryotic characterized by absence of membrane bound organelles like nucleus, chloroplast or mitochondria. BGA are very much identical to photosynthetic bacterial forms, thus BGA are often called as Cyanobacterial forms.

The present communication reveals that BGA of India now consits of about 100 genera and 1500 taxa. The Satna district of M.P. inhabits a very good BGA diversity. There are many beautiful and historical wet lands, which are covered by Cyanophycean blooms.

Our intensive study on taxonomical approach of Cyanophycean flora of Satna revealed that there are 16 genera and 61 species of Blue Green Algae are commonly found in wetlands of Satna. The common species were *Aphanocapsa. Aphanothece, Chroococcus, Gloeocapsa, Oscillatoria, Porphyrosiphon, Phormidium, Lyngbya, Nostoc, Anabaena, Scytonema, Gloeotrichia* and *Rivularia.*

Keywords: *Blue green algae, Taxonomical approach.*

Introduction

Blue green algae (Cyanophyta) are prokaryotic organisms with oxygen evolving photosynthetic process. In cellular organisation they resemble bacteria by the absence of any membrane bound organelles and their photosynthesis is akin to all other Eukaryotic algae and higher plants. Thus they appear to represent a link between bacteria and higher plants.

Blue green algae contain the pigments chlorophyll-a and the phycobillins, the water-soluble billiproteins, like phycocyanin, phycoerythrin and allophycocyanin. They do not possess flagella but some of them show a characteristic gliding movement. Sexual reproduction is absent. Food reserves are glycogen and cyanophycin. The occurrence of blue green algae in a wide range of ecologically extreme habits proves that their capacity of adaptability of high orders.

The studies on blue-greens have gained much importance after the recognition of their role in the natural environment and their ability to provide an alternate source of energy. Several blue green algal members fix, atmospheric nitrogen. This process helps to maintain the nitrogen status of soil (De, 1939). Addition of blue green algae as bio-fertilisers both in lieu of and in combination with chemical fertilisers, has now been proved in our country to yield positive results.

For refinement and review of biofertiliser programmes first step is the comprehensive taxonomic survey of blue green algae. So the present work reports the findings of investigation carried out on Cyanophycean flora of Satna district. The study site is unique for luxuriant growth of various algal forms, of which Cyanophycean members are occurred in majority. Study of algal flora could prove of great biological importance to man.

Only a few workers (Agarkar and Agrakar, 1972; Agarkar, D.S., 1967; Adoni, 1985. Sengar *et al.*, 1984 and Sen *et al.,* 1991) have worked out on algae of the different regions of the state Madhya Pradesh. But no work has been done systematically on the algal vegetation of this region. Since Satna district has rich growth of algae, it has been considered worth to work out about the systematics of Cyanophycean flora of this district undertaken.

Materials and Methods

The study site Satna district is located between 23 58′ and 25 12′ North latitude and 80°21′ and 80 23′ East longitude in the North-Eastern part of Madhya Pradesh. The district occupies all most the central part of Vindhyan region and is surrounded from district Banda of U.P. (North), Rewa (East). Shahdol (South) and Panna (West).

The algal materials were collected from four different sites of Satna district after an interval of a week. These sites are Satna river (Site 1), Uchehra river (Site 2), Maihar Pond (Site 3) and Site 4 belongs to different rice fields of Satna district. After collecting the algal material were preserved in 5 per cent Formaline or in F.A.A. and a portion of the same was brought to the laboratory in living condition for microscopic examination and identification. The systematic identification was done by the help of binocular research microscope at Botany Department of Allahabad. Both living and fixed material were observed and photomicrographs of the different algal forms were taken by using attached Nikon Labo Phot HFX-11A camera at Phycology lab of Botany Department, Allahabad University, Allahabad.

The systematic identification was done with the help of standard works by Fritsch (1935), Smith (1933), Desikachary (1959), Geitler (1932) and Prescott (1951).

Results

In the present investigation the following cyanophycean forms were recorded from the Satna district, based on the monograph like Cyanophyta by Desikachary (1959) and related research papers are also consulted for the same. The classification of Fritsch (1935) has been adopted in the taxonomy of the algae.

CYANOPHYTA Smith

Class–CYANOPHYCEAE Sachs

Order–CHROOCOCCALES Wettstein

Family–CHROOCOCCACEAE Nageli

CHROOCOCCUS Nag.

Chroococcus tenax (Kirchn) Hieron

Geitler, 1932, p. 231, Figure 111a; Desikachary, 1959, p. 103, pl. 26, Figure 7.16 [Plate 7.1, Figure (6)].

Cells generally aggregate in groups of 2–4 to form oval or spherical colonies, colony 15–28 µm broad and 16–28 µm long, sheath colourless, distinct and unlamellated.

Habitat

Planktonic, collected from Sites 2 and 3.

Chroococcus turgidus (Kueetz) Nag.

Geitler, 1932, p. 228. Figure 109b, 110; Desikachary, 1959, p. 101, pl. 26, Figure 6; Starmach, 1966, p. 116, Figure 125 [Plate 7.1, Figure (5)].

Thallus mucilaginous formed by spherical or hemispherical cells, cells do not aggregate but sometimes form small group of 2–4, 9–22 µm broad and 10–18 µm long without sheath, sheath lamellated, 2–3 µm thick. Colony 11–30 µm broad, 11–42 µm long cell contents granular.

Habitat

Planktonic, collected from Sites 1 and 2.

GLOEOCAPSA Kutzing

Gloeocapsa decorticans (A.Br.) Richter

Geitler, 1932, p. 184, Figure 83b; Desikachary, 1959, p. 114, pl. 24, Figure 9 [Plate 7.1, Figure (7)].

Thallus dark blue green; cells spherical or oval; single or 2–4 in a colony; cells covered with colourless, thick and stratified sheath; cell 10–18 µm broad and 10–14 µm long.

Habitat

Presently collected from the moist soil of rice fields.

GLOEOTHECE Nag.

Gloeothece samoensis Wille.

Geitler, 1932, p. 220; Desikachary, 1959. p. 128, pl. 23, Figure 3 [Plate 7.1, Figure (8)].

Cell generally 1 or 2, covered in a single envelope, ellipsoidal, without sheath, 5.5–6.0 µm long; sheath hyaline and slightly lamellated.

Habitat

Presently collected from the moist soil of rice fields.

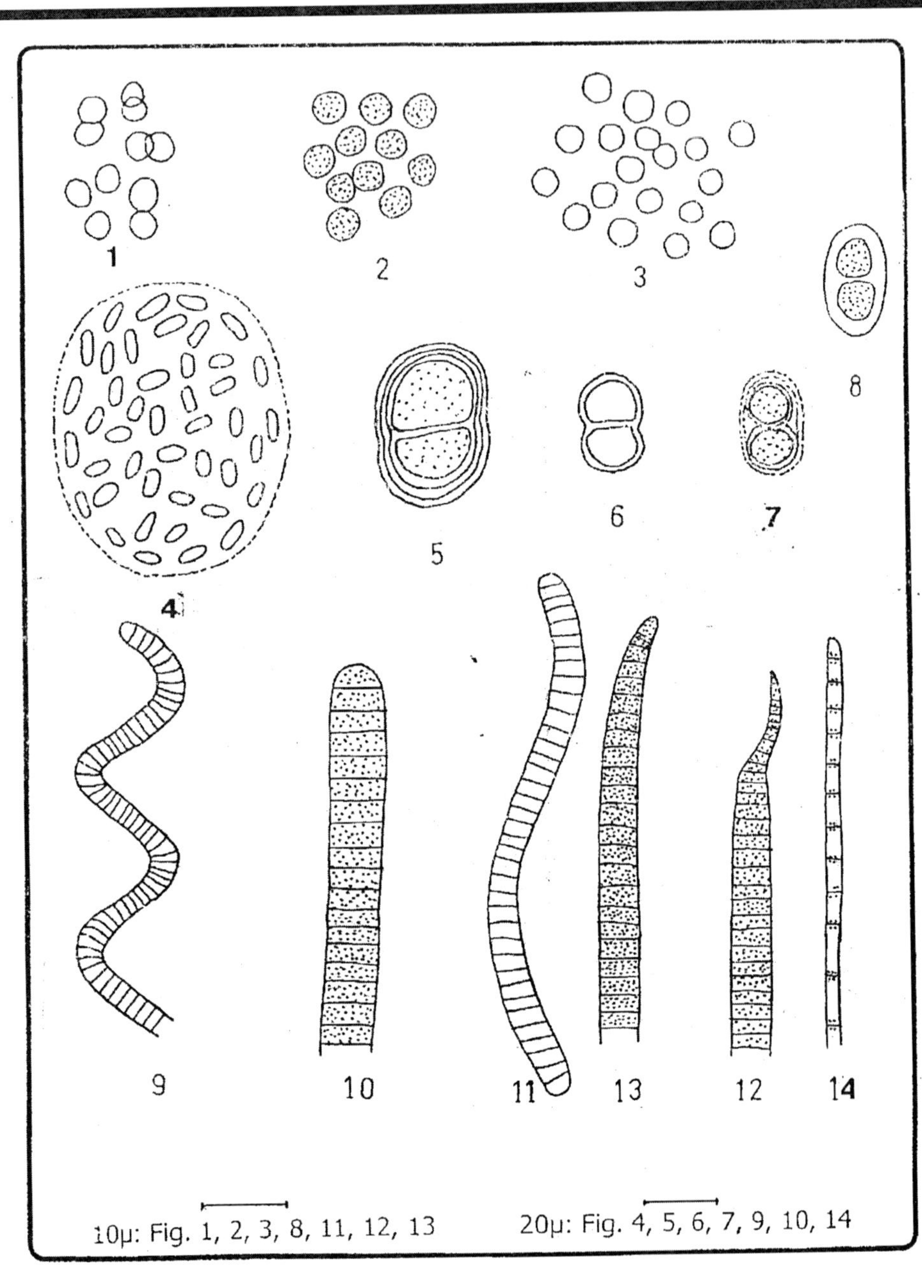

Plate 5.1

(1) ***Aphanocapsa koordersi;*** (2) ***Aphanocapsa pulchra;*** (3) ***Aphanocapsa littoralis;*** (4) ***Aphanothece stagnina;*** (5) ***Chroococcus turgidus***; (6) ***Chroococcus tenax;*** (7) ***Gloeocapsa decorticans;*** (8) ***Gloeothece samoensis;*** (9) ***Arthrospira khannae;*** (10) ***Oscillatoria limosa;*** (11) ***Oscillatoria granulata;*** (12) ***Oscillatoria salina;*** (13) ***Oscillatoria salina*** **f. major;** (14) ***Oscillatoria amphibia.***

APHANOCAPSA Nag.

Aphanocapsa koordersi Storm

Geitler, 1932, p. 155, Figure 68; Desikachary, 1959, p. 132, pl. 23, Figure 1 [Plate 7.1, Figure (1)].

Thallus dull green; cells, sparsely arranged but rarely also seen in-groups of 2 or 4, forming a spherical colonies, cells 2–4 µm in diameter, sheath colourless.

Habitat

Planktonic, collected from Sites 1 and 2.

Aphanocapsa littoralis Hansgirg

Geitler, 1932, p. 153, Figure 66b; Desikachary, 1959, p. 131, pl. 21, Figure 1 [Plate 7.1, Figure (3)].

Thallus of no definite shape; cells loosely arranged without any order; yellowish green in colour; cells spherical to subs-pherical, single or in two forming a formless gelatinous mass, cells 4–5 µm broad.

Habitat

Planktonic, collected from Sites 2 and 3.

Aphanocapsa pulchra (Kuetz).

Geitler, 1932, p. 159, Figure 69g; Prescott, 1951, p. 454, pl. 101, Figure 14; Desikachary, 1959, p. 132, pl. 21, Figure 2; Starmach, 1966, p. 86, Figure 81 [Plate 7.1, Figure (2)].

Thallus dark blue green; attached with substratum or free; mucilaginous, tuberculate; cells spherical; 3–5 µm in diameter; loosely arranged; single or in two; mucilaginous sheath colourless and diffluent.

Habitat

Collected both as free-floating plankton as well as on moist soils.

APHANOTHECE Nag.

Aphanothece stagnina (spreng) A.Br.

Geitler,1932, p.164, Figures 72–75a,b; Prescott, 1951, p. 461, pl.103, Figures 14-16; Desikachary,1959. p. 137, pl. 21. Figure 10; Starmach, 1966, p. 92, Figures 91-93 [Plate 7.1, Figure (4)].

Thallus mucilaginous. spherical, firm, free floating as rounded balls: cells densely aggregated to form oblong or cylindrical in shape; 5-6 µm broad and 5.5-9.5 µm long.

Habitat

Recorded from free floating conditions.

Order: NOSTOCALES Geitler

Family: OSCILLATORIACEAE Kirchner

ARTHROSPIRA Stizenberger.

Arthrospira khannae Drouet et Strickland

Drouet, 1932, p. 141. pl.1, Figure 6b; Desikachary, 1959, p. 189, pl. 29, Figure 12 [Plate 7.1, Figure (9)].

Thallus blue green, multicellular without sheath, trichomes free floating, forming loose spirals, 20–22 µm broad, unconstricted at the joints, little attenuated at the ends, sub-capitate, cells 3–6 µm broad and 3–5 µm long *i.e.* cells as long as broad, cross walls granulated, cells with gas vacuoles.

Habitat

Planktonic form, collected from Site 1.

OSCILLATORIA Vaucher.

Oscillatoria acuta Bruhl et. Biswas orth mut. Geitler.

Bruhl and Biswas, 1922, p. 3, pl. 1, Figures 6a and b; Geitler, 1932, p. 978; Desikachary, 1959, p. 240, pl. 39, Figures 5 and 8 [Plate 7.2, Figure (6)].

Thallus shows variety of sizes, due to the number of parallely aggregated trichomes, trichome straight unconstricted, 4–5 µm broad and 80–140 µm long, and cells acuted, rounded, cells 2–5 µm long, finally granular.

Habitat

It was presently recorded as free-floating from rice fields.

Oscillatoria amphibia Ag. ex. Gomont

Geitler, 1932, p. 966; Prescott, 1951, p. 485, pl. 109, Figure 6; Desikachary, 1959, p. 229, pl. 37, Figure 6 [Plate 7.1, Figure (14)].

Thallus forms thin blue-green masses, trichomes straight, apical cell rounded, unconstricted at the joints, cells 2–5 µm broad and 4–10 µm long, with two granules at the septa, end cell rounded.

Habitat

Occurred as planktonic form.

Oscillatoria chalybea Martens Gomont.

Tilden, 1910, p. 82, pl. 4, Figure 36; Geitler, 1932, p. 956, Figure 808b; Desikachary, 1959, p. 218, pl. 38, Figure 3 [Plate 7.2, Figure (1)].

Thallus blue-green, trichomes straight or sometimes slightly twisted, constricted at the septa, gradually tapering at the apex, 8–12 µm broad, cells 6–8 µm long, end cell generally rounded, septa ungranulated.

Habitat

Presently it was observed as planktonic.

Oscillatoria curviceps Ag. ex. Gomont.

Geitler, 1932, p. 947, Figure 598c; Tiffany and Britton, 1952, p. 344, pl. 94, Figure 1081; Prescott, 1951, p. 487, pl. 108, Figures 17, 18; Desikachary, 1959, p. 209, pl. 38, Figure 2 [Plate 7.2, Figure (7)].

Thallus forming dark blue-green masses, trichomes straight, bent or rarely coiled at the end, slightly attenuated; unconstricted at the cross walls, ungranulated; end cells flat end rounded, 9–16 µm broad and 2–6 µm long.

Habitat

Collected from the moist soil of rice fields.

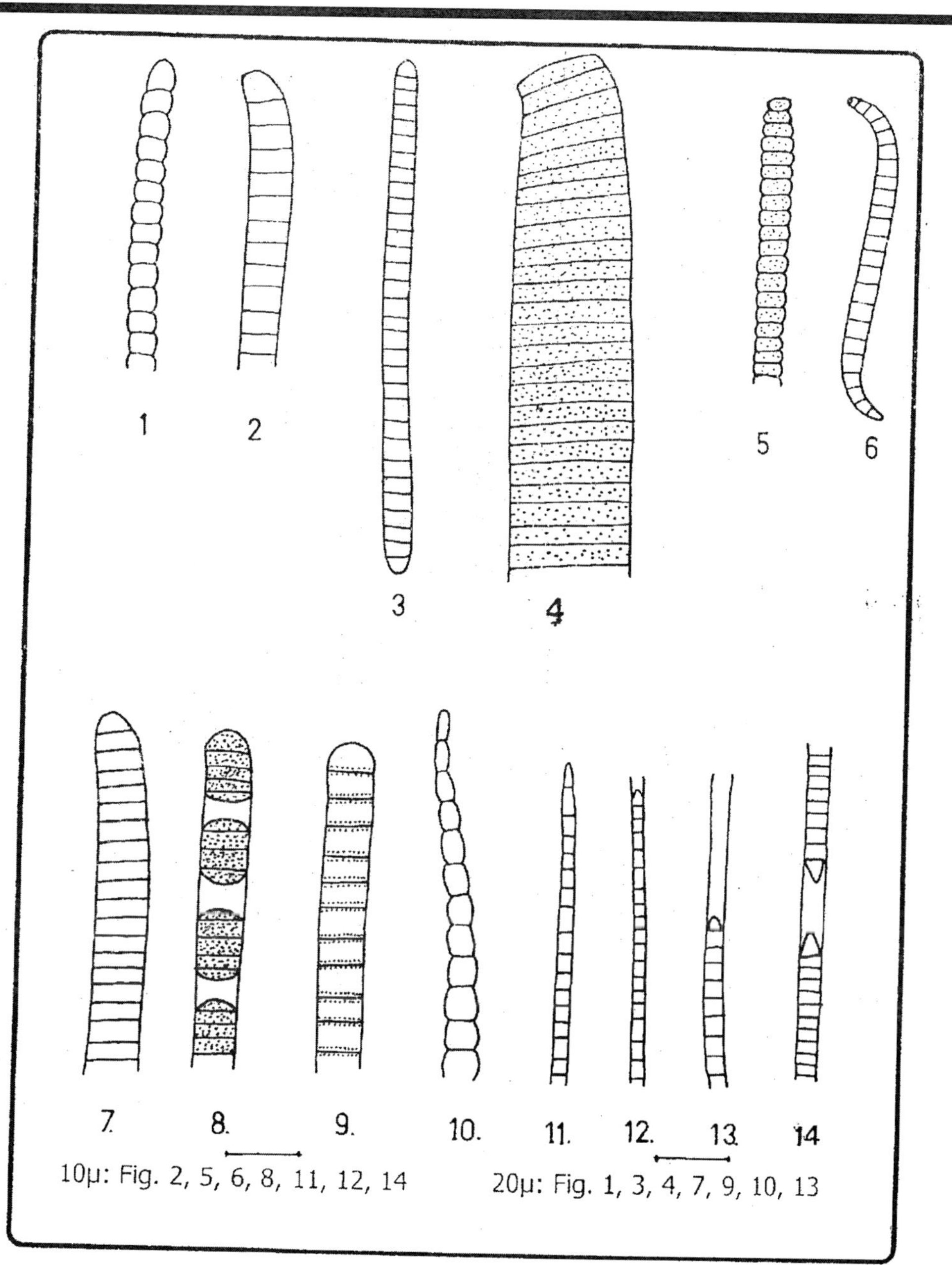

Plate 5.2

(1) *Oscillatoria chalybea*; (2) *Oscillatoria tenuis*; (3) *Oscillatoria tenuis* Ag. *ex.* Gomont Var. *levis* Gardner; (4) *Oscillatoria princeps*; (5) *Oscillatoria sancta* (Kutez) Gomont *f. tenuis* Parakutty; (6) *Oscillatoria acuta*; (7) *Oscillatoria curviceps*; (8) *Oscillatoria subbrevis*; (9) *Oscillatoria irrigua*; (10) *Oscillatoria tanganyikae*; (11) *Phormidium laminosum*; (12) *Phormidium fragile*; (13) *Phormidium corium*; (14) *Phormidium cebennense.*

Oscillatoria granulata Gardner

Geitler, 1932, p. 963, Figure 615b; Prescott, 1951, p. 487, pl. 109, Figures 12, 13 [Plate 7.1, Figure (11)].

Trichomes entangled to form an extensive mass; trichomes straight; unconstricted at the joints, generally granular, cells 3–4 µm broad and 2–3 µm long.

Habitat

Collected as free floating from Site 3.

Oscillatoria irrigua (Kutz.) Gomont.

Geitler, 1932, p. 961, Figures 611a, b; Desikachary, 1959, p. 224, pl. 42, Figures 7–9; Starmach, 1966, p. 336, Figure 472 [Plate 7.2, Figure (9)].

Trichome straight, unconstricted, slightly attenuated at the apex, sub-capitate, cells 8–15 µm broad and 4–10 µm long., apical cell convex with a thick outer wall.

Habitat

Recorded as free floating from Sites 1 and 2.

Oscillatoria princeps Vaucher ex. Gomont.

Geitler, 1932, p. 947, Figures 598a, 601 c–g; Prescott, 1951, p. 289, pl. 110, Figure 1; Desikachary, 1959, p. 210, pl. 37, Figures 1, 10, 11, 13, 14; Starmach,1966, p. 323, Figure 438 [Plate 7.2, Figure (4)].

Thallus blue-green, formed by free floating trichomes, trichomes straight unconstricted, attenuated at the apices, 26–35 µm broad; 6–8 µm long, cells compactly placed, end cell flatly rounded, slightly capitate, with an outer thickend membrane.

Habitat

Occurred as free floating from Site 4.

Oscillatoria salina Biswas

Geitler, 1932, p. 978, Figure 624; Desikachary, 1959, p. 239, pl. 37, Figures 16–17 [Plate 7.1, Figure (13)].

Thallus dark blue green, trichomes straight, briefly tapering, twisted, neither capitate nor calyptrate, cells shorter than broad, 1–2.5 µm long, 5–9 µm broad, unconstricted at the septa.

Habitat

Planktonic form collected from Sites 1 and 4.

Oscillatoria salina Biswas f. *major* Desik.

Geitler, 1932, p. 979, Figure 624; Desikachary, 1959, p. 239, pl. 37, Figures 16–17 [Plate 7.1, Figure (12)].

Thallus dark blue green, membranous mass, trichomes, compactly packed, unconstricted at the joint, single, free floating, 6–8 µm broad, apices hyaline, briefly tapering, ending acuminately into a sharp pointed hook, straight twisted, cells 2–4 µm long *i.e.* shorter than broad.

Habitat

Free floating, collected from Sites 1 and 2.

Oscillatoria sancta (Kutz.) Gomont f. *tenuis* Parakutty

Geitler, 1932, p. 943, Figure 598c; Rao, 1937, p. 366; Parakutty, 1940, p. 120; Tiffany and Britton 1952, p. 490, pl. 110, Figure 4; Desikachary, 1959, p. 203, pl. 42, Figure 10 [Plate 7.2, Figure (5)].

Thallus thin, covered with shining gelatin, trichomes straight or coiled, slightly constricted at the cross walls, ends gradually attenuated, cells shorter than broad, 4–5 μm broad and 2–3 μm long, cross walls granulated, ends cells capitate, with thick outer membrane.

Habitat

Recorded as planktonic as well as free floating from Sites 1, 2 and 3.

Oscillatoria subbrevis Schmidle.

Geitler, 1932, p. 946, Figure 601b; Prescott, 1951, p. 491, pl. 107, Figure 23; Desikachary, 1959, p. 207, pl. 37, Figure 2 and pl. 40, Figure 1 [Plate 7.2, Figure (8)].

Thallus forms a patch of scum on the surface of water; trichomes straight, not attenuated, 5–7 μm broad, cells 1–2.5 μm long, ungranulated at the septa, end cell rounded.

Habitat

Collected from moist soil of rice fields.

Oscillatoria tenuis Ag. ex. Gomont.

Geitler, 1932, p. 959, Figures 611 f–g; Prescott, 1951, p. 491, pl. 110, Figures 10–11; Desikachary, 1959, p. 222, pl. 42, Figures 13–14; Starmach, 1966, p. 338, Figure 476 [Plate 7.2, Figure (2)].

Thallus blue green, thin, trichome straight, slightly constricted at the septa, 4–8 μm broad; briefly bent at the ends, apices neither attenuated nor capitate, cells 3–5 μm long, septa granulated, end cell hemispherical, covered with thick outer membrane.

Habitat

Recorded as free floating from Sites 1 and 2.

Oscillatoria tenuis Ag. ex. Gomont var. *levis* Gardner.

Geitler, 1932, p. 959, Figures 61 f–g; Prescott, 1951, p. 491, pl. 110, Figures 10–11; Desikachary, 1959, p. 222, pl. 42, Figures 13–14; Starmach, 1966, p. 338, Figure 476 [Plate 7.2, Figure (3)].

Thallus blue green, trichome thin, straight, not constricted at the cross walls, 4–8 μm broad, briefly bent at the ends, apices neither attenuated nor capitate, cells 3–5 μm long, septa granulated, end cell more or less hemispherical, covered with thick outer membrane.

Habitat

The alga was recorded both in planktonic as well as in free-floating conditions from Sites 1, 2 and 3.

PHORMIDIUM Kutz.

Phormidium ambiguum Gomont.

Geitler, 1932, p. 1015, Figure 467c; Prescott, 1951, p. 493, pl. 111, Figure 1; Desikachary, 1959, p. 266, pl. 44, Figure 16, pl. 45, Figures 5–8 [Plate 7.3, Figure (2)].

Thallus yellowish blue green, gelatinous and expanded, variously entangled filaments, slightly constricted at the cross wall, not attenuated at the ends, 3.5–5.5 μm broad, sheath thin and firm, cells shorter than broad, 2–2.5 μm long, end cells rounded, without calyptra.

Habitat

Observed on moist soil of rice fields.

Phormidium cebennense Gomont.

Geitler, 1932, p. 1008; Desikachary, 1959, p. 262; Starmach, 1966, p. 294, Figure 394 [Plate 7.2, Figure (14)].

Thallus light brown, thin and expanded, filaments flexuous, straight and entangled, not attenuated, sheath thin and diffluent, not constricted, trichomes 2–4 μm broad; cells of trichome 1.8–2.6 μm broad and 1–3 μm long, end cell not capitates, rounded.

Habitat

Recorded from moist stones and rocks of Sites 1 and 2.

Phormidium corium (Ag.) Gomont.

Geitler, 1932, p. 1018, Figures 649b, c; Prescott, 1951, p. 494; Desikachary, 1959, p. 269, pl. 44, Figures 10–11 [Plate 7.2, Figure (13)].

Thallus dark blackish green, forming a membranous mass, filaments compactly entangled, covered with a thin sheath, trichomes straight, not constricted at cross walls, cells longer than broad, 4–8 μm long, end cells obtuse conical, calyptra absent.

Habitat

Collected from moist soil of rice fields.

Phormidium foveolarum Mont. Gomont.

Geitler, 1932, p. 99, Figure 636b; Tiffany and Britton, 1952, p. 348, pl. 95, Figure 1095; Desikachary, 1959, p. 254 [Plate 7.3, Figure (3)].

Thallus thin and dark blue-green, trichomes flexuous, constricted at the cross walls, pale blue-green, 1–1.5 μm broad, cells quadrated, slightly longer than broad, 1–2.5 μm long, end cells rounded.

Habitat

Observed on moist soil of rice fields.

Phormidium fragile Meneghini Gomont.

Geitler, 1932, p. 999, Figure 636c; Desikachary, 1959, p. 253, pl. 44, Figures 1–3 [Plate 7.2, Figure (12)].

Thallus mucilaginous, light blue-green, trichomes flexuous, entangled in diffluent sheath, distinctly constricted at the joints, cells quadratic 1.5–3 μm broad, 1.5–3.4 μm long, end cell acute conical.

Habitat

Collected from moist soil.

Phormidium laminosum Gomont.

Geitler, 1932, p. 1005, Figure 642c; Desikachary, 1959, p. 259, pl. 44, Figure 6 [Plate 7.2, Figure (11)].

Thallus forming pale blue-green membranous masses, filaments densely entangled, expanded, flexuous, diffluent and covered with a thin mucilaginous sheath, trichomes slightly coiled, not constricted, cross walls granulated, 2–3 µm broad and 1.5–3.5 µm long, cells longer than broad; end cells acute conical.

Habitat

The alga was recorded from moist soil.

Phormidium molle (Kutz) Gomont.

Geitler, 1932, p. 1000; Desikachary, 1959, p. 255, pl. 59, Figure 8; Starmach, 1966, p. 286, Figure 374 [Plate 7.3, Figure (5)].

Thallus thin mucilaginous, trichomes straight, deeply constricted at the cross walls, 2.5–3.5 µm broad sheath thin and diffluent, cells quadratic cylindrical, light blue-green, 2.5–6.5 µm long, end cell rounded.

Habitat

The alga was recorded from moist soil.

Phormidium subfuscum (Kutz) ex. Gomont.

Geitler, 1932, p. 1022, Figures 652d–g; Prescott, 1951, p. 496, pl. 107, Figure 16; Tiffany and Britton, 1952, p. 350, pl. 95, Figures 1096–1097; Desikachary, 1959, p. 273, pl. 44, Figures 22–23; Starmach, 1966, p. 303, Figure 412 [Plate 7.3, Figure (6)].

Thallus forming dark blue green masses, thin, stratified and very much expanded; filaments straight unconstricted and briefly attenuated at the tip, trichomes pale olive-green, 6.5–10.5 µm board and 2.3–4.2 µm long, cross walls granulated, end cells straight, acute conical and capitate, sheath thin and diffluent.

Habitat

The alga was recorded as free-floating conditions from Sites 1, 2.

Phormidium tenue (Menegh.) Gomont.

Geitler, 1932, p. 1004, Figures 642d–e; Desikachary, 1959, p. 259, pl. 43, Figures 13–15, pl. 44, Figures 7–9 [Plate 7.3, Figure (1)].

Thallus yellowish blue-green, expanded to form a membranous mass, trichomes straight densely entangled within mucilaginous sheath, sheath thin, diffluent and pale blue-green, trichomes flexuous and slightly constricted at the region of septa, cells 1–2.5 µm broad and 2–4.5 µm long, end cell acute-conical.

Habitat

The alga was recorded from moist soil.

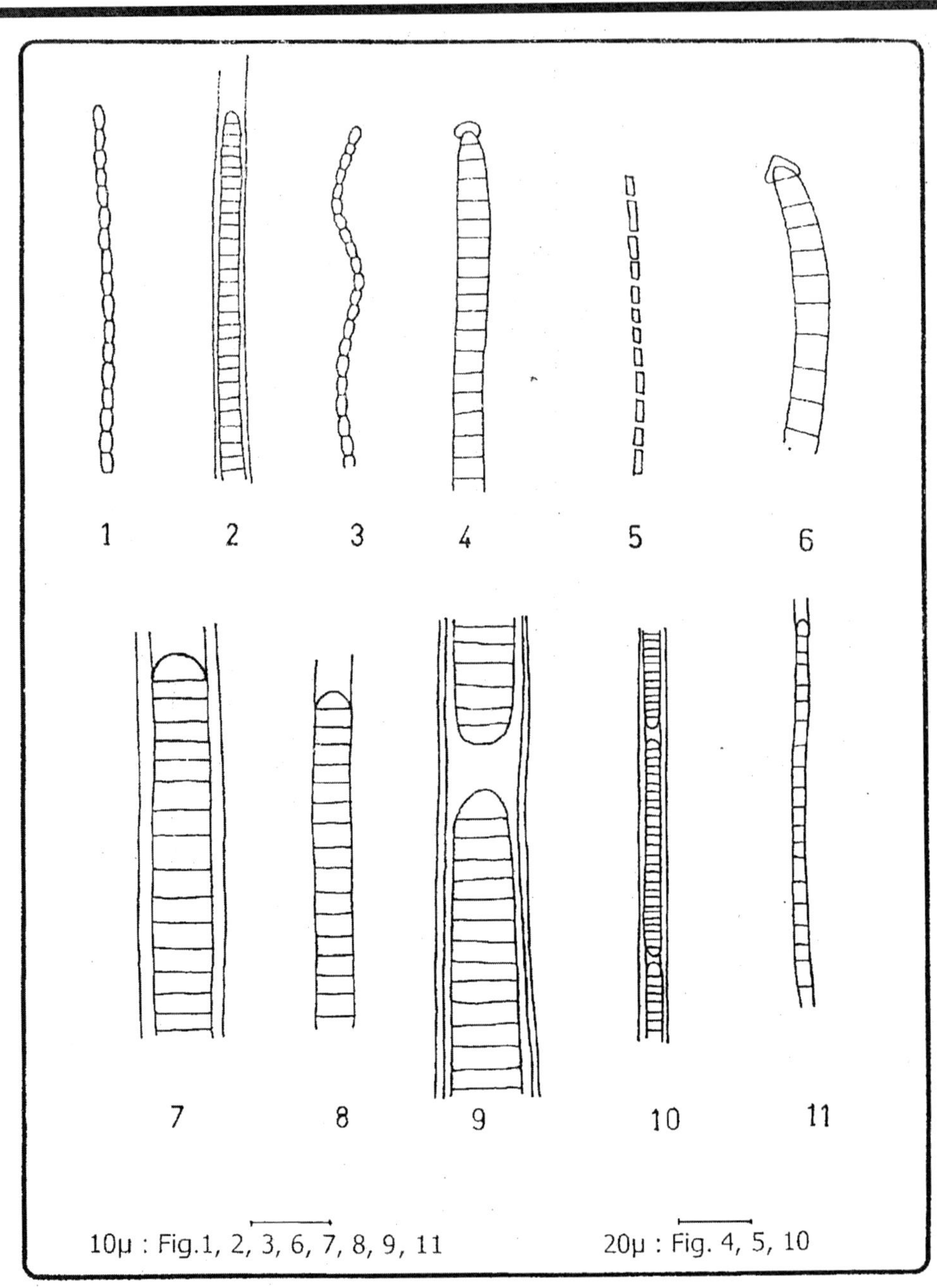

Plate 5.3

(1) ***Phormidium tenue***; (2) ***Phormidium ambiguum***; (3) ***Phormidium foveolarum***; (4) ***Phormidium uncinatum***; (5) ***Phormidium molle***; (6) ***Phormidium subfuscum***; (7) ***Lyngbya ceylanica***; (8) ***Lyngbya allorgei***; (9) ***Lyngbya confervoides***; (10) ***Lyngbya taylorii***; (11) ***Lyngbya limnetica***.

Phormidium uncinatum (Ag.) Gomont.

Geitler, 1932, p. 1025, Figure 652h–i; Desikachary, 1959, p. 276, pl. 43, Figures 1–2, pl. 45, Figures 9–10; Starmach, 1966, p. 304, Figure 414 [Plate 7.3, Figure (4)].

Thallus expanded, formed by straight, free floating and entangled filaments, sheath thin; diffluent in mucilage, trichomes unconstricted at the cross walls, briefly attenuated at the apex, cells granulated, 5–9 µm broad and 3–6 µm long, end cell capitate with a round calyptra.

Habitat

The alga was recorded as free-floating from the Sites 2 and 3.

PORPHYROSIPHON Kutzing.

Porphyrosiphon notarisii (Menegh) Kutz ex. Gomont.

Geitler, 1932, p. 986, Figure 631; Tiffany and Britton, 1952, p. 340, pl. 94, Figures 1093–1094; Desikachary, 1959, p. 248, pl. 47, Figure 9 [Plate 7.4, Figure (5)].

Filaments aggregate to form a compact mass, rarely branched, forming a dull reddish-brown layer, sheath firm, thick lamellated reddish-brown and 4–8 µm thick; trichomes blue-green; constricted at the cross walls; cells 8–16 µm broad and 4–10 µm long; end cell obtuse and attenuated; filaments sometimes shows branching.

Habitat

The alga was collected from tree trunk as well as from moist soil.

LYNGBYA. Ag.

Lyngbya allorgei Fremy.

Geitler, 1932, p. 1059, Figure 671 a; Desikachary, 1959, p. 313, pl. 54, Figure 6 [Plate 7.3, Figure (8)].

Filaments entangled with other algae to form a large thallus, straight solitary and elongate; unconstricted at cross walls; 6–8 µm broad; sheath thin prominent and colourless, 0.5 µm thick; cells of trichome nearly quadrate; 4–6.5 µm broad and 3–6 µm long, cell contents uniformly granular; end cell rounded.

Habitat

The alga was observed on moist soil and stones.

Lyngbya birgei Smith, G.M.

Geitler, 1932, p. 1048, Figure 663; Prescott, 1951, p. 499, pl. 111, Figure 9; Tiffany and Britton, 1952, p. 92, Figures 1064–1065; Desikachary, 1959, p. 296, pl. 50, Figures 7–8; Starmach, 1966, p. 242, Figure 301 [Plate 7.4, Figure (3)].

Filaments straight, free floating, 22–26 µm broad, sheath thick, persistent and unlamellated, 2–3 µm broad, cell contents granular, shorter than broad; 17–22 µm broad and 3–4 µm long, end cell rounded and dome shaped.

Habitat

It was observed as free floating.

Lyngbya ceylanica Wille.

Geitler, 1932, pl. 1055, Figures 668 a, b; Desikachary, 1959, p. 299, pl. 54, Figure 4 [Plate 7.3, Figure (7)].

Thallus comprises straight filaments, 8–14 µm broad, sheath thin and hyaline, trichomes unconstricted at the joints; septa granulated; cells 7–12 µm broad and 2–4 µm long; end cells round.

Habitat

The alga was from moist soil of rice fields.

Lyngbya confervoides C. Ag. ex. Gomont.

Geitler, 1932, p. 1061, Figure 672b; Desikachary, 1959, p. 314, pl. 49, Figure 9 and pl. 52, Figure 13 [Plate 7.3, Figure (9)].

Filaments entangled to form a mass, straight, 8–12 µm broad, blue-green; sheath hyaline and lamellated, trichomes 7–20 µm broad and 2.5–3.5 µm long; not constricted; cross walls granulated; end cell round.

Habitat

The alga was recorded as free floating.

Lyngbya limnetica Lemmermann.

Geitler, 1932, p. 1046, Figures 661 a, b; Desikachary, 1959, p. 294, pl. 50, Figure 11 [Plate 7.3, Figure (11)].

Thallus blue green, filament straight or slightly bent, free floating, 1.5–2.5 µm broad, sheath thin and colourless, cells quadratic, 0.8–1.3 µm broad, and 2–3 µm long, unconstricted, without granule, end cell not attenuated and rounded.

Habitat

The alga was recorded from moist soil of rice fields.

Lyngbya major Menegh ex. Gomont.

Geitler, 1932, p. 1066, Figure 679a; Desikachary, 1959, p. 320, pl. 52, Figure 11; Starmach, 1966, p. 260, Figure 338 [Plate 7.4, Figure (2)].

Filaments entangled to form an expanded thallus; sheath thick and lamellated, trichomes straight, cells 12–16 µm broad, 2.5–3.5 µm long; end cell rounded; covered with a thick outer membrane.

Habitat

The alga was recorded as in free-floating conditions from Sites 2 and 3.

Lyngbya rubida Fremy

Geitler, 1932, p. 1054, Figure 668c; Desikachary, 1959, p. 298, pl. 53, Figure 10 [Plate 7.4, Figure (4)].

Thallus expanded; brownish purple; filaments straight; 6–8 µm broad; sheath firm, unlamellated, unconstricted at the joints; trichomes 4–5 µm broad and 6–12 µm long, end cells round.

Habitat

The alga was collected from Sites 2 and 3.

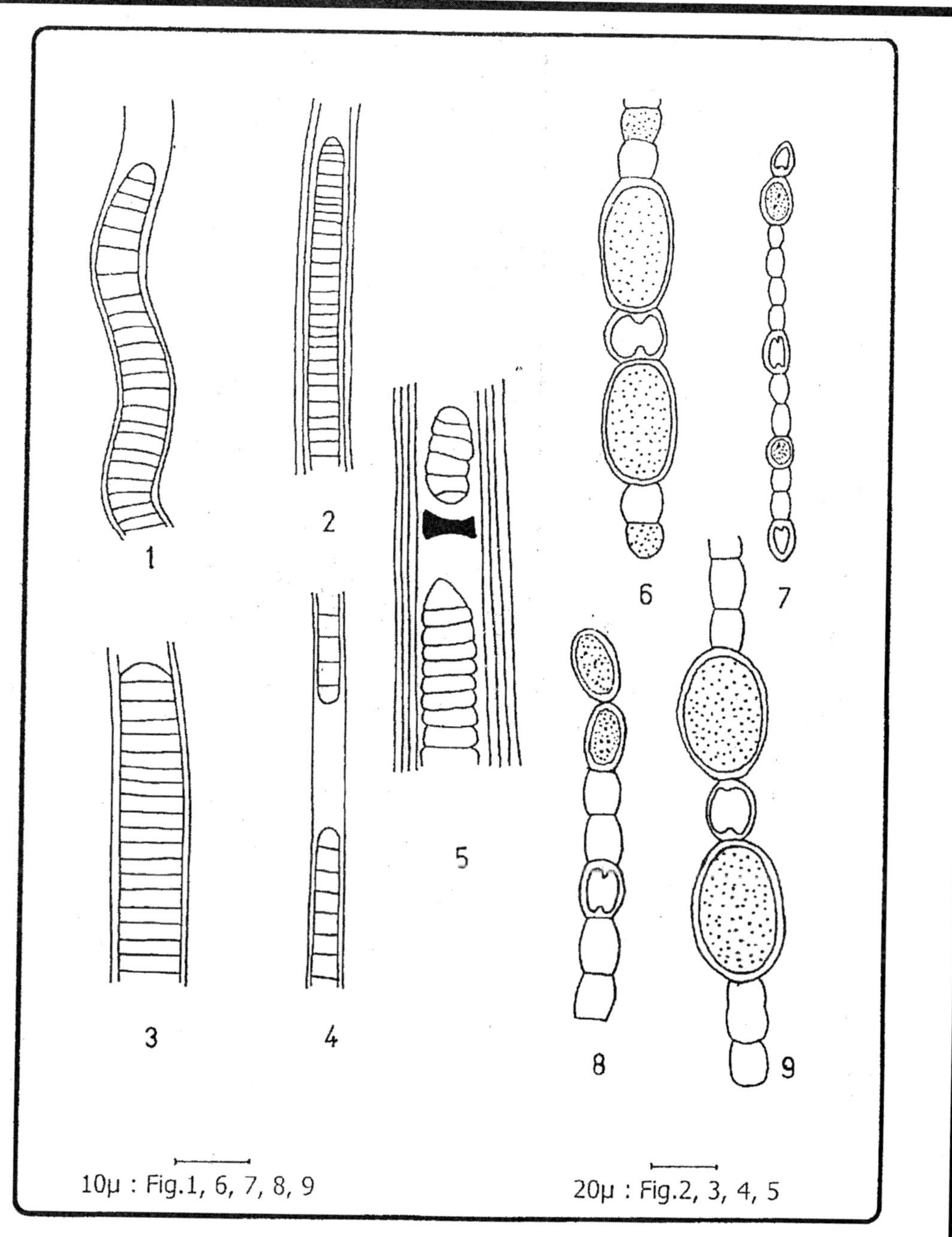

Plate 5.4

(1) *Lyngbya spirallis*; (2) *Lyngbya major*; (3) *Lyngbya birgei*; (4) *Lyngbya rubida*; (5) *Porphyrosiphon notarisii*; (6) *Anabaena ambigua*; (7) *Anabaena oryzae*; (8) *Anabaena variabilis* Kuetzing ex. Born et. Flah. Var. *ellipsospora* fritsch.; (9) *Anabaena sphaerica.*

Lyngbya spirallis Geitler

Geitler, 1932, p. 1042, Figure 659; Desikachary, 1959, p. 289, pl. 48, Figure 1 [Plate 7.4, Figure (1)].

Filaments entangled to form a large leathery thallus; filaments 6–8 µm broad; portions of the trichomes spirally coiled; sheath thin, cells 5–6 µm broad and 1.5–2.5 µm long, end cell rounded.

Habitat

Recorded as planktonic conditions from Sites 1, 2 and 3.

Lyngbya taylorii Drouet and Stickland

Prescott, 1951, p. 503, pl. 113, Figure 3; Tiffany and Britton, 1952, p. 339, pl. 92, Figures 1068–1069 [Plate 7.3, Figure (10)].

Trichomes single, straight, covered with thin, firm and hyaline sheath, olive green in colour, cells 2–6 µm broad and 1.5–3.5 µm long, unconstricted at the cross walls.

Habitat

Reported from moist stones.

Family: NOSTOCACEAE Kuetzing

Sub-family: ANABAENAE

NOSTOC Vaucher

Nostoc commune Vaucher ex. Born *et al.*

Geitler, 1932, p. 845, Figures 536–537; Prescott, 1951, p. 523, pl. 119, Figure 13; Desikachary, 1959, p. 387, pl. 68, Figure 3; Starmach, 1966, p. 533, Figures 817–818 [Plate 7.5, Figure (3)].

Thallus firm, gelatinous, globular and leathery with many perforations; filaments closely packed, expanded; sheath thick and distinct; trichomes more compact in outer dense brown layer of colony and less in central less dense, soft part of colony; cells spherical or barrel shaped; 3–5 µm broad; and 15–40 µm long; heterocyst nearly spherical 6– 7.2 µm in diameter, size of spores equal to the vegetative cells; epispore smooth and hyaline.

Habitat

The alga was observed on moist soil of rice fields.

Nostoc muscorum Ag. ex. Born et. Flah.

Geitler, 1932, p. 844, Figure 535; Prescott, 1951, p. 524, pl. 120, Figure 6; Tiffany and Britton, 1952; p. 364, pl. 100, Figure 1149; Desikachary, 1959, p. 385, pl. 70, Figure 2; Starmach 1966, p. 538, Figure 822 [Plate 7.5, Figure (4)].

Thallus expanded gelatinous, tuberculate, attached by the lower surface; without a firm outer layer; much crowded; irregular, entangled; 3.2–5.5 µm broad; cells barrel shaped; longer than broad, 3–6 µm long; heterocysts spherical 5–7 µm in diameter; spore oblong; many in a series; 3–7 µm broad and 8–10 µm long; epispore smooth and yellowish in colour.

Habitat

The alga was recorded from moist soil of rice fields.

Nostoc paludosum Kuetzing ex. Born et. Flah.

Tilden, 1910, p. 165, pl. 6. Figure 38; Geitler, 1932, p. 826c, Figure 528a; Desikachary, 1959, p. 375, pl. 69, Figure 2; Starmach, 1966, p. 525, Figure 803 [Plate 7.5, Figure (2)].

Thallus microscopic, gelatinous, soft and dark blue-green; sheath smooth hyaline mucilaginous; trichomes compactly packed, less densely coiled: barrel shaped, irregularly arranged, 3.2–4.5 μm broad and 3–6 μm long; heterocysts 3.4–5 μm broad and 4–8 μm long; spore oval, 4–5 μm broad, 7–8 μm long.

Habitat

The alga was recorded from the moist soil of rice fields.

Nostoc spongiaeforme Agardh. ex. Born et. Flah.

Geitler, 1932, p. 839, Figure 531; Prescott, 1951, p. 525, pl. 121, Figure 10; Tiffany and Britton, 1952, p. 366, pl. 100, Figure 1150; Desikachary, 1959, p. 380; Starmach, 1966, p. 542, Figure 829 [Plate 7.5, Figure (5)].

Thallus expanded; mucilaginous; tuberculate; irregularly torn; without a firm outer layer; lightly blue-green; filaments not densely entangled; trichomes longer than broad, cell 3.3–4 μm broad and 3–7 μm long; cells barrel shaped, sheath hyaline and diffluent; heterocyst spherical, oblong, sometimes in pairs 3.5–8.5 μm in diameter, spores oblong 6–8 μm broad and 8–10 μm long, epispore hyaline and smooth at first, and later yellowish.

Habitat

The alga was collected from moist soil of rice fields.

ANABAENA Bory

Anabaena ambigua Rao, C.B.

Desikachary, 1959, p. 400, pl. 76, Figure 2 [Plate 7.4, Figure (6)].

Trichomes free floating, enclosed in a common mucilaginous sheath, straight or bent; slightly tapering at the ends, end cell conical, cells barrel shaped with deep constrictions at the joints, 4–6 μm broad and 3–5.5 μm long, heterocyst spherical sometimes with flattened ends, 6–10 μm in diameter, akinetes on both the side of heterocyst, ellipsoidal, with rounded ends, 8–10 μm broad and 13–16 μm long; exospore thick and hyaline, endospore thin.

Habitat

The alga was reported as free floating or attached to submerged stones on Sites 1 and 2.

Anabaena fertilissima Rao, C.B.

Rao, 1937b, p. 363, Figures 6A–C; Desikachary, 1959, p. 398, pl. 74, Figure 1 [Plate 7.5, Figure (1)].

Thallus amorphous, blue-green, trichomes straight or bent, cells barrel shaped, end cell rounded, 4–7 μm broad and 3–6 μm long, heterocyst single, terminal or intercalary; almost spherical, 4–6 μm broad and 3.5–8 μm long; akinetes in long chain, generally the whole trichome becomes, sporogenous near the heterocyst, spherical, 4.5–8.5 μm broad and 4–8 μm long, outer wall smooth and colourless.

Habitat

The alga was reported from moist soil of the rice fields.

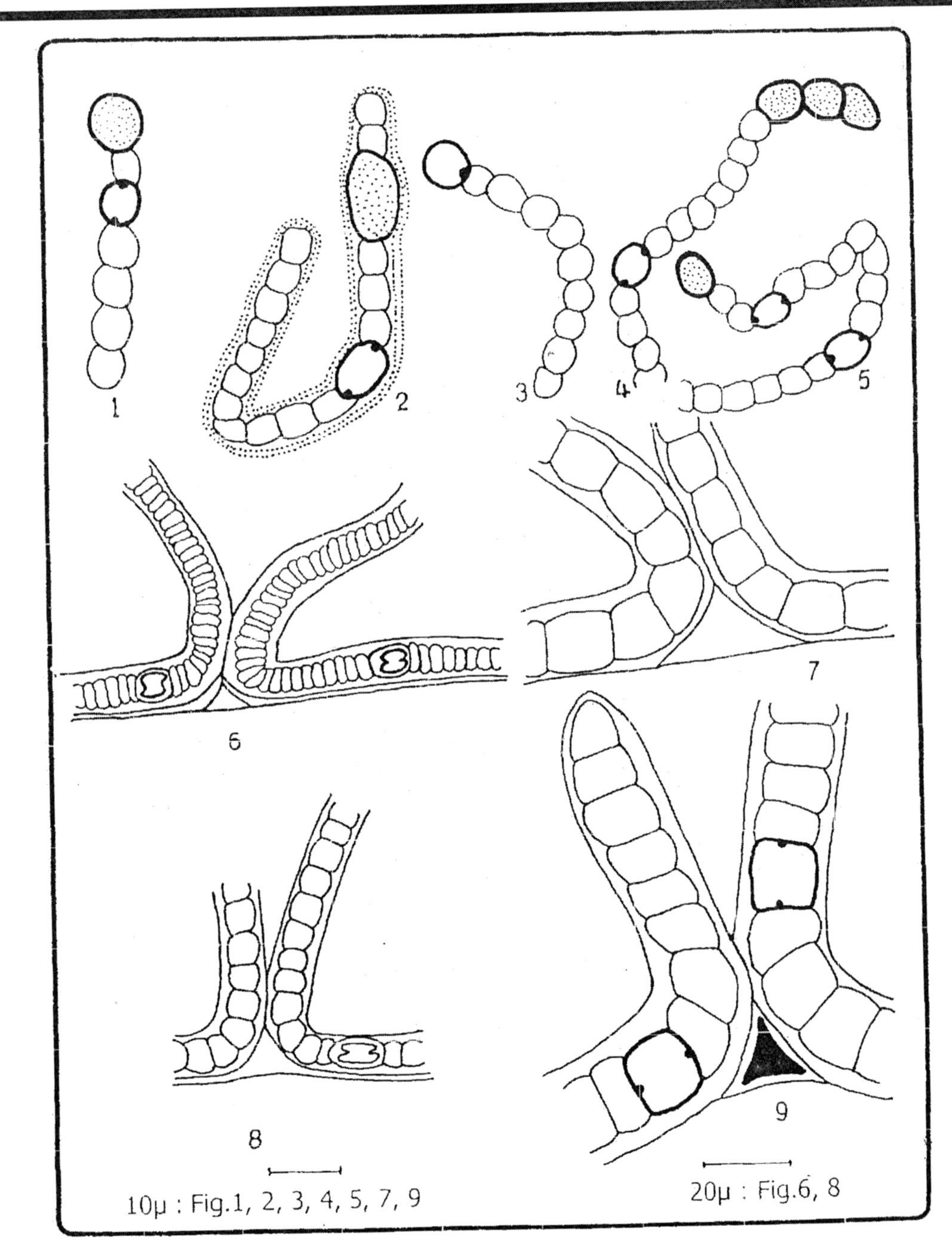

Plate 5.5

(1) ***Anabaena fertilissima***; (2) ***Nostoc paludosum***; (3) ***Nostoc commune***; (4) ***Nostoc muscorum***; (5) ***Nostoc spongiaeforme***; (6) ***Scytonema stuposum***; (7) ***Scytonema bohneri***; (8) ***Scytonema schmidtii***; (9) ***Scytonema hofmanni.***

Anabaena oryzae Fritsch.

Desikachary, 1959, p. 396, pl. 72, Figure 3 [Plate 7.4, Figure (7)].

Thallus gelatinous, trichomes straight, short and parallel, densely aggregated, cells barrel shaped, 3–4 μm broad, 4–8 μm long, heterocysts terminal and intercalary heterocyst broader than vegetative cells, sometimes 2–3 in a series, generally barrel shaped, terminal heterocyst is conical 3–7 μm broad and 3–9.5 μm long, akinetes either next to terminal heterocysts or away from intercalary heterocyst, single or 2–8 in a series, sub-spherical, 5–5.5 μm broad and 5–7 μm long exospore yellowish brown.

Habitat

It was collected from moist soil of the rice fields.

Anabaena sphaerica Born. et. Flah.

Geitler, 1932, p. 878, Figure 560b; Desikachary, 1959, p. 393; Starmach, 1966, p. 482, Figure 717 [Plate 7.4, Figure (9)].

Thallus blue-green, trichomes moniliform, straight, 5–6 μm broad; sheath distinct, mucilaginous; cells barrel shaped; 3.5–4.5 μm broad and 4–7 μm long; end cell conical rounded; heterocyst sub-spherical or spherical; 6–7.5 μm broad; akinetes on one or both sides of heterocyst; 9–12 μm broad and 10–16 μm long; epispore smooth; yellowish brown.

Habitat

Reported as plankton from the Sites 2 and 3.

Anabaena variabilis Kuetzing ex. Born. et. Flah. Var. ellipsospora

Fritsch, 1935, p. 142, Figures 40–50; Desikachary, 1959, p. 411, pl. 72, Figure 1 [Plate 7.4, Figure (8)].

Thallus gelatinous, thin and blue-green; trichomes loosely aggregated; constricted at the cross walls; apical cell rounded; cells cylindrical 3–4 μm broad, 5–10 μm long; heterocyst terminal or intercalary; single; spherical; 4–6 μm broad and 4.5–8.5 μm long; akinetes away from the heterocyst; formed in a series; ellipsoidal; 4.5–8 μm broad and 8–12 μm long; epispore thin; exospore separated from the surface.

Habitat

Reported from moist soil from rice fields.

Family: SCYTONEMATACEAE Rabenhorst

SYCTONEMA Ag.

Scytonema bohneri Schmidle.

Geitler, 1932, p. 753; Desikachary, 1959, p. 457, pl. 87, Figure 1; Starmach, 1966, p. 666, Figure 983 [Plate 7.5, Figure (7)].

Thallus blackish-green, formed by heterotrichous filaments; 10–25 μm broad; false branched; sheath colourless 1.2–1.8 μm thick; cells of trichomes longer than broad; 4–8 μm broad and 5–15 μm long; heterocysts ellipsoidal to rectangular, compressed, 6–9 μm broad and 4–16 μm long.

Habitat

Reported from the moist wall of the pond.

Scytonema cincinnatum Thuret ex. Born. et. Flah.

Geitler, 1932, p. 748, Figure 477; Desikachary, 1959, p. 453, pl. 93, Figure 1 [Plate 7.6, Figure (1)].

Thallus brownish-green, filaments long 18–36 µm broad and 3–4 cm long; false branches mostly geminate; sheath thick and hyaline, 1–2.5 µm broad; trichomes 14–28 µm broad; cell broader than long, 3–9 µm long; heterocyst quadrated, single or many.

Habitat

It was collected from moist soil of the rice fields.

Scytonema hofmanni Ag. ex. Born. et. Flah.

Geitler, 1932, p. 772, Figure 495; Desikachary, 1959, p. 476, pl. 91, Figure 2; Starmach, 1966, p. 666, Figure 982 [Plate 7.5, Figure (9)].

Thallus forming an expanded cushion like stratum; dark blue-green, filaments 8–13 µm broad, false branches aggregated to form vertical fascicles, sheath firm and membranous; trichomes 5.5–1.0 µm broad, cells unequal in length, generally 3–8 µm long, heterocyst oblong.

Habitat

It was reported from tree trunk.

Scytonema schmidtii Gomont.

Geitler, 1932, p. 759, Figure 483b; Desikachary, 1959, p. 459, pl. 92, Figure 1 [Plate 7.5, Figure (8)].

Thallus expanded; blackish-red with wrinkled surface; filaments irregularly undulated; primarily prostrate; abundantly and repeatedly false branched; sheath firm, hyaline and yellowish brown; trichomes 8–12 µm broad cells sub-quadrate; 2–8 µm long; heterocysts quadrated; 4–6 µm broad and 2–3 µm long.

Habitat

It was reported from moist soil of rice fields.

Scytonema stuposum (Kutzing) Bornet ex. Born. et. Flah.

Geitler, 1932, p. 756, Figure 482; Desikachary, 1959, p. 459, pl. 93, Figure 4 [Plate 7.5, Figure (6)].

Thallus expanded, blackish-green, filaments 15–18 µm broad and 4–9.5 µm long, false branches as broad as the main filaments, single, or geminate, sheath thick gelatinous, trichomes slightly constricted at the septa 11–16 µm broad and 4–6 µm long.

Habitat

Presently reported from the moist soil of rice fields.

Family: RIVULARIACEAE Rabenhorst

RIVULARIA (Roth) Ag.

Rivularia aquatica De Wilde

Geitler, 1932, p. 652; Desikachary, 1959, p. 552 [Plate 7.6, Figure (2)].

Thallus spherical, large 2 mm in diameter, filaments slightly pressed together, sheath thin hyaline, unstratified, attenuated at the ends, trichome ending into a long thin hair, cells at the base broader and at apex longer, at base 7–10 µm broad and 6–8 µm long, at apex 5–7 µm broad and 4–5 µm long, heterocyst 7–12 µm broad and 1–15 µm long.

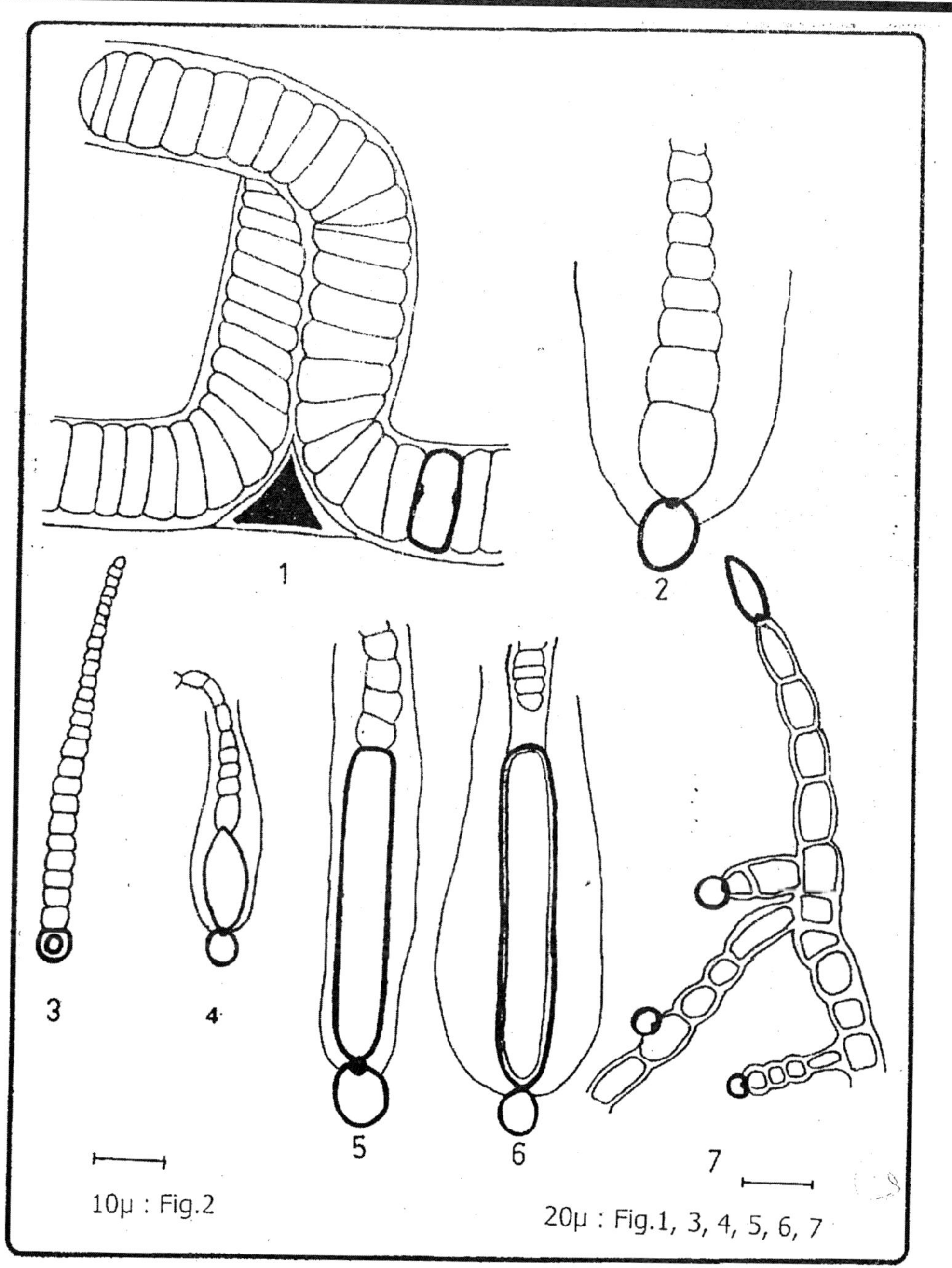

Plate 5.6

(1) ***Scytonema cincinnatum***; (2) ***Rivularia aquatica***; (3) ***Rivularia hansgirgi***; (4) ***Gleotrichia intermedia*** (lemm) ***Geitler*** Var. ***Kanwaensis*** Rao, C.B.; (5) ***Gleotrichia*** ghosei; (6) ***Gleotrichia*** natans; (7) ***Nostochopsis lobatus.***

Habitat

Recorded as planktons from Sites 2 and 4.

Rivularia hansgirgi Schmidle

Geitler, 1932, p. 649; Desikachary, 1959, p. 549, pl. 112, Figure 7 [Plate 7.6, Figure (3)].

Thallus thin gelatinous but solid, expanded to form a Nostoc like ball; spherical blackish brown: filaments loosely arranged and radial in colony; attenuated at the ends; apices of trichomes 2–3.5 µm broad, ending in a hair, middle part of trichome 5–8 µm broad, cells sub-quadrate; 7–9 µm broad and 5–7 µm long; heterocyst basal solitary and spherical, 7–8 µm broad and 8–9 µm long; sheath thick, gelatinous, unstratified hyaline or yellow.

Habitat

It was observed on moist wall.

GLOEOTRICHIA Ag.

Gloeotrichia ghosei, Singh, R.N.

Singh, 1939b, p. 64, Figures A–C, Desikachary, 1959, p. 561, pl. 118, Figures 1–3 [Plate 7.6, Figure (5)].

Thallus free floating, spherical or ellipsoidal, brown; filaments covered with a thin brown sheath; trichome ending in a long hair, constricted at the joints, at base 8–10 µm broad, at apices 3–6 µm broad, at base cells barrel shaped and 4.5 to 6.5 µm long; cells at hair long and cylindrical, 10–16 µm long; heterocyst single and spherical, 8–12 µm broad and 8.5 to 10.5 µm long; akinetes long, ellipsoidal with a hyaline smooth wall, with sheath 24–25 µm broad; without sheath 13–16 µm broad; 40–42 µm long.

Habitat

It was recorded as free floating from Sites 1 and 2.

Gloeotrichia intermedia (Lemm.) Geitler Var *Kanwaensis* Rao, C.B.

Rao, 1936, p. 168, Figures 2A–C; Desikachary, 1959, p. 560, pl. 118, Figures 9–11 [Plate 7.6, Figure (4)].

Thallus small, soft and spherical, blackish green, 2–10 mm in diameter; filaments less densely aggregate; sheath thin, hyaline and closely and pressed; trichomes ending in a hair; generally slight bent; the cells of trichomes 6–10 µm broad at base; at apices 5–7 µm broad; cells at the base barrel shaped 3–7 µm long; at apices quadratic 5–12 µm long; heterocysts terminal, spherical or sub-spherical, single, 6–12 µm broad and 12–15 µm long, akinetes cylindrical 16–18 µm broad with sheath and 12–15 µm broad without sheath, 30–50 µm long.

Habitat

Recorded from the moist soil of rice fields.

Gloeotrichia natans Rabenhorst ex. Born. et. Flah.

Geitler, 1932, p. 639, Figures 406–407; Desikachary, 1959, p. 561, pl. 118, Figures 7–15 [Plate 7.6, Figure (6)].

Thallus blackish brown, soft hollow, spherical, filaments loosely arranged; trichomes 7–10 µm broad and 6–9.5 µm long, attenuated into a long hair; cells at base barrel shaped, at apices cylindrical; heterocysts basal, spherical, 6–12 µm broad; akinetes cylindrical, straight, without sheath, 10–18 µm broad and 80–200 μm long; sheath hyaline or brownish, saccate and transversely constricted; with sheath about 40 μm broad.

Habitat

It was observed on submerged rocks of Sites 1 and 2.

Order: STIGONEMATALES Geitler

Family: NOSTOCHOPSIDACEAE Geitler

NOSTOCHOPSIS Wood em. Geitler

Nostochopsis lobatus Wood em. Geitler

Desikachary, 1959, p. 570, pl. 120, Figures 1–8; Geitler, 1932, p. 475, Figures 285–286; Stramach, 1966, p. 705, Figure 1034 [Plate 7.6, Figure (7)].

Thallus spherical, irregularly lobed, solid or hollow, 3–4 cm in diameter, blue-green or yellowish-green or brownish; trichomes radially arranged, straight or sometimes bent, sparsely branched; branches lateral and stretching upwards, parallel to the main filament; cells of branches, barrel shaped 4–6 μm broad; 6–10 μm long; trichome with slightly tapered end; heterocysts lateral and sessile or terminal and pedicellate; on 2–4 celled lateral branch, ellipsoidal, 6–10 μm broad; rarely intercalary or basal; mucilaginous sheath hyaline or yellowish or brownish, in branch diffluent at apices, confluent at base.

Habitat

It is observed as attached or free floating.

Discussion

In wetlands of Satna the algal vegetation is so varied that there consideration as a whole, afford great difficulties. Therefore in the present investigation blue green flora of Satna (M.P.) were taken.

It was seemed that this locality was adjourned with 16 Cyanophycean genera with 61 species belonging to 6 families and 3 orders (Table 3.1).

Table 3.1 Order and Family-wise Numerical of Cyanophycean Flora (BOA) of Satna (M.P.)

Sl.No.	*Order*	*Family*	*Genera*	*Species*
1.	Chroococcales	Chroococcaceae	5	8
2.	Nostocales	Oscillaoriaceae	5	34
		Nostocaceae	2	9
		Scytonemataceae	1	5
		Rivulariceae	2	4
3.	Stignematales	Nostochopsidaceae	1	1
	3 Orders	6 Families	16	61

From viewing the Table 3.1 it is clear that all the three orders records their presence in wetlands of Satna. The most promising family among these was Ocillatoriaceae, which comprised of 5 genera and 34 species. Family Chroococcaceae also comprised of 5 genera with only 8 species. Family Nostocaceae shows 2 genera with 9 species, while family Scytonemataceae shows 1 genera and 5 species. Family

Rivulariaceae shows 2 genera and 4 species. In order Stigonematales, family Nostochopsidaceae seems with only 1 genera and 1 species.

The author chooses 4 major wetlands as a site for investigation. Blue green algae are ubiquitous in water and occur as blooms in water bodies (Stewart and Pearson, 1970). The algal biodiversity index was highest for Sites 1, 2 and 4, while for 3 was lowest. The Site 4 belongs to rice fields it shows improvement of soil aggregation and organic matter (Singh, R.N., 1939).

References

Adoni, A.D. (1985). *Workbook on Limnology.* Pratibha Publishers, Sagar, India, pp. 216.

Agarkar, D.S. (1967). Myxophyceae of Gwalior, M.P. *Phycos,* 6: 1–6.

Agarker, M.S. and D.S. Agarkar (1972). Zygnemataceae of Madhya Pradesh. *Phycos,* 11(1&2): 71–77.

Bruhl, P. and K. Biswas (1922). The algae of Bengal filter beds. *J. Dept. Sci. Calcutta University,* 4: 1–17.

De, P.K. (1939). The role of blue-green algae in nitrogen fixation in rice fields. *Proc. R. Soc. Lond. Ser. B,* 127: 121–140.

Desikachary, T.V. (1959). *Cyanophyta Monograrh of Algae.* ICAR, New Delhi, India.

Drouet, F. (1932). A list of algae from Missouri. *Bull. Torrey. Bot. Club.,* 59.

Fristch, F.E. (1935). *The Structure and Reproduction of the Algae 1.* Cambridge, pp. 791.

Geitler, L. (1932). Cynophyceae in Rabenhorst's Krypto gamenoflora. *Lerpzig,* 14: 1196.

Parakutty, P.R. (1940). The Myxophyceae of the Travancore State. India. Proc. Indian. *Acad. Sci.,* B11(3): 117.

Prescott, G.W. (1951). *Algae of Western Great Lakes Area.* Gran book Inst. of Science. Bull. 31, Bloom Field Hills, Misch, p. 946, pl. 136.

Prescott, G.W. (1965). Ecology of Alasken fresh water algae V: Limnology and flora or Malikruk lake. *Trans Amer. Micros. Soc.,* 84(4): 427–478.

Rao, C.B. (1936). The Myxophyceae of the United Provinces, India II *Proc. Indian Acad. Sci.,* 38: 165–174.

Rao, C.B. (1937a&b). The Myxophyceae of the United Provinces, India III *Proc. Indian Acad. Sci.,* 6B: 339–375.

Sen, U.S., V.S. Bais and S. Yateesh (1991). The phytoplankton of the strawboard mill effluent mixing zones in comparison with diluted zones of river Dhasan (M.P.) India. *Acta. Hydrochem. Hydrobiol.,* 19(1): 45–55.

Senger, M., S.P. Singh and D.C. Pandey (1984). A systematic account on algal vegetation of Puchmarhi (M.P). University of Allahabad Studies, 16(6).

Singh, R.N. (1939). An investigation into the algal flora of paddy field soils of the United Provinces I. *Indian J. Agri. Sci.,* 9(1): 55–77.

Smith, G.M. (1933). *Freshwater Algae of the United States,* 1st ed. New York, London, p. 716, 449f.

Starmach, K. (1966). Cyanophyta-since Glaucophyta Glaukofity Panstwowe Wydawnitwo naukowe, Waryawa, pp. 807.

Stewart, W.D.P. and H.W. Pearson (1970). Effect of aerobic and anaerobic conditions on growth and metabolism of blue-green algae. *Proc. R. Soc.,* 175B: 293–311.

Tiffany, L.H. and M.E. Briton (1952). *The Algae of the Illinois.* 407 I, 136 F, Chicago and London.

Tilden, J.E. (1910). *Myxophyceae in Minnesets,* Algae I, Minneapolis, pp. 328.

Chapter 8

Physico-chemical Factors Controlling the Growth of Diatoms in Two Lakes of Mysore

J. Mahadev & S.P. Hosmani***

**DOS in Environmental Science, **DOS in Botany, Manasagangotri, Mysore – 570 006*

ABSTRACT

Seventeen species of Diatoms were recorded in Karanji lake and eighteen species in Dalvoi lake during the study period. Higher pH, calcium and oxidizable organic matter coupled with low concentration of nitrates and phosphates were found to favour the growth of Diatoms.

Keywords: *Diatoms, Karanji lake, Dalvoi lake, Langlier's index.*

Introduction

Eutrophication and pollution are bogging up and shrinking the lakes. Several workers have studied the ecology of fresh water algae in general and Diatoms in particular. Hosmani and Bharati, 1982; Kuldeep Kumar, 1990; Pandey *et al.*, 1995; Naganandini and Hosmani, 1998; Hosmani and Vasanth Kumar, 1996. Present study was aimed at monitoring the Diatom population in Karanji and Dalvoi lakes and correlating their occurrence with various physico-chemical parameters including calcium carbonate saturation index.

Materials and Methods

Water samples were collected in 1L plastic carbuoys from Karanji and Dalvoi lakes, during June 1999 to May 2001. Sampling was done between 10.00 am and 11.00 am by selecting suitable areas in the lake to constitute a composite sample. Collection, preservation and enumeration of diatoms and various methods adopted for analysis of physico-chemical parameters were followed as per the procedures described in Trivedy and Goel, 1986, and Standard Method for Examination Water and Wastewater, APHA, 1995, 19th edition.

Results and Discussion

The growth of fresh water diatoms has been dealt by many researchers. Some of the classical contributions are by West and West, 1912; Delf, 1915; Naumann, 1917; Atkins, 1923; Yoshimura, 1932; Patrick, 1948. Periodicity of diatoms has been studied by Singh, 1960; Philipose, 1960; Zafar, 1964 and Munawar, 1970. The important factors considered where Nitrate, Phosphate, Calcium and pH. Diatoms increased with increase in Nitrate but in the present study the concentration of Nitrate quite low, but the diatom population was significantly high (Table 8.1).

Table 8.1: Occurrence of Bacillariophyaceae (or/L)

Organisms	*Karanji Lake*			*Dalvoi Lake*		
	June 1999 to May 2000	*June 2000 to May 2001*	*Average*	*June 1999 to May 2000*	*June 2000 to May 2001*	*Average*
Cocoonies palcentula lineate	7400	9660	8530	5880	4200	5040
Cymbella cymformis	3360	5460	4410	–	–	–
Cymbella simulate	–	6720	3360	–	–	–
Cymbella aspera	–	–	–	4620	–	–
Cymbella turgid	–	–	–	–	420	210
Cyclotella striata	–	–	–	4200	2100	3150
Cyclotella catenata	–	5460	2730	–	–	–
Eunotia monodon	–	7140	3570	–	–	–
Gyrosigma granula	–	6300	3150	–	6720	3360
Gyrosigma elongatum	–	–	–	–	7980	3990
Gyrosigma accuminatum	–	–	–	4200	–	2100
Gomphonema tenelum	–	7560	3780	–	2940	1470
Gomphonema sumatrense	4200	–	2100	5460	–	2730
Gomphonema gracile	3360	5040	4200	–	–	–
Gomphonema subapicatum	–	–	–	–	5040	2520
Navicula rhombo	9960	8500	9230	–	13020	6510
Navicula sphaerophora	7980	5040	6510	7300	11340	9320
Nitzschia palea	–	4620	2310	8400	5460	6930
Pinnularia gibba	5040	2940	3990	–	2100	1050
Pinnularia simplex	–	5040	2520	–	–	–
Stauroneis phoenicentron	4520	4200	4360	2520	2520	2520
Stauroneis angulare	–	6300	3150	–	–	–
Synedra ulna danica	8400	12860	10630	19320	13020	16170
Synedra acus	–	–	–	7140	9240	8190
Rhopalodia gibba	–	–	–	7140	–	3570
Total organisms per year in two lakes	**54220**	**102840**	**78530**	**76180**	**81600**	**81140**

Figures represent organisms L^{-1}.

Panday *et al.,* 1994; Tripathy and Pandy, 1990; Naganandini and Hosmani, 1998, observed that maximum diatoms occurred during winter and summer. While Parvateesam and Maneesh Mishra, 1993, observed that diatom were maximum during summer seasons. The low concentration of calcium may influence moderate number of diatoms in Karanji lake. Sengar and Singh, 1986, and Chowdhary and Amazumbar, 1981, have concluded that pH may influence the growth of diatoms.

In the present study pH in the range of 7 to 8.25 influences the growth of diatoms. Nitrate and phosphate are considered to be important in the growth of Diatoms. Kuldeep Kumar, 1990, observed that high phosphate during winter favour diatoms. Rajendra Nair, 1990, observed that diatoms where directly correlated to phosphate. Hosmani, 1975, reported the relation of diatoms with the phosphate content. In the present study diatoms where directly proportional to phosphate and nitrate. Both the water bodies support the higher number of diatoms but have varied amount of dissolved oxygen. In Karanji lake dissolved oxygen was 5.25 mg/l and the diatoms population was 6539 or/L. In Dalvoi lake dissolved oxygen content was low the diatom population was 8277 or/L (Table 8.1). This indicates that oxygen has a lesser effect on growth of diatoms.

Table 8.2: Seasonal Occurrence of Diatoms

Seasons	*Karanji Lake*			*Dalvoi Lake*		
	1999–2000	*2000–2001*	*Total Two Season 1999–2001*	*1999–2000*	*2000–2001*	*Total two Season 1999–2001*
Rainy	4305	9440	6872	9765	6195	7980
Winter	2499	840	5449	5880	8400	7140
Summer	4200	10395	7297	9660	9765	9712

Figures represent organisms L^{-1}.

Table 8.3: Physico-chemical Parameters and Langlier's Index of Two Lakes

Parameters	*Karanji Lake*	*Dalvoi lake*
Water temperature °C	27.91	28.17
pH	7.96	8.07
Total solid	757.73	961.2
Free carbon dioxide	11.89	5.32
Dissolved oxygen	5.24	4.67
Calcium	46.03	66.52
BOD	8.6	13.06
COD	12.58	20.21
Phosphate	0.84	1.65
Nitrate	0.31	0.66
Nitrite	0.074	0.16
Langlier's index	+ 0.80	+ 1.404
Diatoms organisms or/L^{-1}	6539	8277

All values in mg L^{-1} except, water temperature and pH.

The periodicity of diatoms in Karanji lake was regular except for a few species like *Cymbella Cymbliformis, Staurnoneis phoenicentron, Gyrosigma granulatas* and *Gomphonema gracile*. These species were abundant, while *Gomphonama tennulam, Syndra ulna, Navicula rhomboids* were significantly high. In Dalvol lake *Gomphonama tennulam, Navicula rhomboids* and *Syndra ulna* dominated and occurred as blooms. The Langlier's index which is a measure of calcium carbonate saturation index has no correlation in growth of diatoms in the present study.

It may inferred from the present study that pH in the range 7 to 8 is an important factor along with where Nitrate, Phosphate, Calcium. Diatoms increased with increase Nitrate.

References

APHA, AWWA, WPCF (1995). *Standard Method for the Analysis of Water and Wastewater*, 19th ed. APHA, INC. Washington DC.

Atkins, W.R.G (1923a). The phosphate content of fresh and salt water in its relationship to the growth of algae plankton. *J. Mar. Boil. Assoc U.K.*, 13: 119–150.

Chowdhary, S.H. and Amazumbar (1981). Limnology of lake Kapti, physico-chemical feature. *Bangladesh Jr. Zool.*, 10: 59–72.

Delf, E.M. (1915). The algal vegetation of some ponds on Hampstead Heath. *New Phytol.*, 14: 63–80.

Hosmani, S.P. and L. Vasanth Kumar (1996). Calcium carbonate saturation index and its influence in phytoplankton. *Poll. Res.*, 15(30): 285–288.

Hosmani, S.P. (1975). Limnological Studies in Ponds and Lakes of Dharwad. *Ph.D. Thesis*, Karnataka University, Dharwad.

Hosmani, S.P. and S.G. Bharati (1982). Use of algae in classifying water bodies. *Phykos.*, 21: 48–51.

Kumar, Kuldeep (1990). Limnological reconnaissance and fluctuation of planktonic fauna of anoxbow lake of West Bengal: A case study. *Recent Trends in Limnology*, pp. 301.

Munawar, M. (1970). Limnological studies on fresh water pond of Hyderabad India I: The Biotope. *Hydrobiologia*, 35: 127–162.

Naganandini and S.P. Hosmani (1998). Ecology of certain in land waters of Mysore district, Biological indices of pollution. *J. Env. & Poll.*, (4): 269–272.

Naumann, E. (1917). Beitriage Zur Kenntnis des Teichananoplanktons II Uber das Neuston des Subwassers, Biol. Zbl., 37: 98–106.

Panday, B.N., A.K. Jha and P.K.L. Das (1994). Hydrobiological study of swamp at purine Bihar in relation to its phytoplankton fauna. *J. Eco. Bio.*, 6(1): 13–16.

Parvateesam, N. and Maneesh Mishra (1993). Algal of Pushkar lake including pollution indicating forms. *Phykos.*, 32(1&2): 27–39.

Patrick, R. (1948). Factor effecting of distribution of diatoms. *Bot. Rev.*, 14(8): 473–524.

Philipose, M.T. (1960). Fresh water phytoplankton of inland fisheries. *Proc. Sym. Algology*, ICAR, New Delhi, p. 272–291.

Rajendra Nair, M.S. (1999). Seasonal variations of phytoplankton in relation to physico-chemical factors in a village pond at Imalia (Vidisha). *Indian J. Ecotoxicol. Env. Monit.*, 9(3): 177–182.

Sengar, R.M.S. and S. Singh (1986). Pollution status of Keethan lake (Soor Sarovar) at Agra. *Geobios,* 13: 56–61.

Singh, V.P. (1960). Phytoplankton ecology of the inland waters of Uttar Pradesh, *Proc. Sym. Algol.,* ICAR, New Delhi, p. 243–271.

Tripathy, A.K. and S.N. Pandey (1990). *Water Pollution.* Ashish publishing House, p. 1–336.

Trivedy and P.K. Goel (1986). *Chemical and Biological Methods for Water Pollution Studied.* Environmental Publication, Karad, p. 94–96.

West, W. and G.S. West (1921). On the periodicity of phytoplankton of some British lakes. *J. Linn. Soc. Bot.,* 40: 395–432.

Yoshimura, S. (1932). Seasonal variation in the content of nitrogenous compounds and phosphate in the water of Takasuka ponda, Saitanma, Japan. *Arch. Hydrobiol.,* 24(1): 155–176.

Zafar, A.R. (1964). On the ecology of algae in certain fish pond of Hyderabad, India. *Hydrobiologia,* 24(4): 556–566.

Chapter 9

Tidal and Diurnal Variability of Zooplankton in Mangrove Habitat of Gaderu, East Coast of India

N. V. Prasad

Division of Marine Biology, Department of Zoology, Andhra University, Visakhapatnam – 530 003

ABSTRACT

The ebb and flow of the tides in an estuary affect not only the hydrographical conditions but also to a marked extent the abundance of zooplankton population. To determine hourly changes in the qualitative and quantitative distribution of zooplankton, 24 hours cycles were carried out at a fixed station at Gaderu river. Among zooplankton, copepods and larvae of molluscs and polychaetes contributed to the bulk of zooplankton component. The day time planktonic composition was rich with phytoplanktonic community and detritus, whereas copepods and veligers showed greater densities during the night. The biomass and numerical abundance varied between 0.21–1.62 $ml.m^{-3}$ and 8948–35,082 $no.m^{-3}$ respectively. Higher displacement volumes (biomass) and numerical abundance was observed during night samples (*i.e.*, 18.00, 21.00, 00.00 hrs). The temperature and salinity reflected the conditions of the bay while the distribution and abundance of different groups of zooplankton had direct relationship with the nature of tide, strength of the current and direction of flow. In spite of the fluctuating currents, regular diurnal rhythm was observed for many of the holoplanktonic and meroplanktonic forms.

Keywords: *Zooplankton, Diel variation, Composition, Gaderu.*

Introduction

Mangroves and estuarine water biotopes are rich in nutrients and plankton are influenced mainly by nearshore waters through semidiurnal tides. The ebb and flow of the tides in these waters effect not only the physico-chemical conditions but also to a marked extent the biotic component of the estuarine and mangrove waters. Diurnal differences in zooplankton are mainly caused by vertical migration, the cause of which is chiefly attributed to responses of zooplankton to variations in light intensities (Madhupratap and Rao, 1979). Diurnal vertical migration of zooplankton communities is also an adaptation shown for the procurement of food and to avoid the oxygen poor waters at the bottom during the night time. Tidal studies are considered important for any estuarine environment as the tides exert varying influence on physical, chemical and biological characteristics including the dynamics of zooplankton.

In order to study in greater detail the hourly changes in the hydrographical conditions diversity and distribution of zooplankton, collection of samples were undertaken during one tidal cycle lasting 24 hrs in the month of April 1999 in Gaderu river which is the main, receptacle for all the influx of nutrients from the surrounding mangrove swamps. Similar studies on the diel variation of zooplankton distribution in relation to changing physico-chemical variables have been studied by Chandra Mohan and Rao (1972) from Godavari estuary; Vijaya Lakshmi and Venugopalan (1973) from Vellar estuary; Madhupratap and Rao (1979) from Cochin backwaters; Goswami *et al.* (1979), Padmavati *et al.* (1997) from Mandovi-Zuari estuary; Srinivasan and Santhanam (1991) from Pullavazhi backwaters; Goswami and Goswami (1992) from the mangrove habitat of Goa.

Materials and Methods

Surface zooplankton and water, samples were collected at intervals of three hours at a fixed station in middle of the Gaderu river (82 08′E; 16 46′N). The sampling started 09.00 hrs in the morning (01–04–1999) and continued till 09.00 hrs on succeeding day (02–04–1999). The surface temperature of the water was measured directly with a thermometer as soon as the sample of water was drawn. The transparency was recorded only during daytime with the help of a Secchi disc. Hydrographical parameters such as temperature, salinity, dissolved oxygen and pH were measured by adopting standard methods (APHA, 1995). Zooplankton samples were collected using 40 cm diameter net of 120 μm mesh size. Biomass was estimated by displacement volume method (Wickstead, 1965). For numerical analysis the plankton was sorted out and individual groups were counted and expressed as no.m^{-3} with flow meter readings. The relative percentage composition of different groups of zooplankton were calculated for all the 9 samples collected from 9.00 a.m. of 01–04–1999 to 9.00 a.m. of 02–04–1999.

Results and Discussion

The tides in mangrove waterways of Gaderu are semi-diurnal type with a maximum range of about 2 m with constant flushing and flooding by the semi-diurnal tides makes this estuary a dynamic environment. The sampling started (9.00 hrs, 01–04–1999) corresponding to mid high tide period and the depth was about 2.9 m, and the maximum depth was noticed during high tide period *i.e.* about 3.5 m. Because of the receding water from adjacent small creeks even during low tide there is not much variation in the depth. The surface water turbidity ranged from a minimum of 16 ppm during mid high tide (09.00 hrs) to a maximum of 23 ppm at low tide period (12.00 hrs) (Table 9.1). The turbidity of surface water showed greater variation depending upon the changing tide. In general, high turbidity values were observed during low tides. It may be due to the large quantity of suspended matter carried by the river at the time of the low tide. The high tide lead to the penetration of less turbid inshore waters and so during such periods low turbidity values recorded. During the present tidal cycle waters were more transparent at mid-day hours when the secchi disc was visible up to a depth of 1 m, which coincided with the high tide.

Table 9.1: Zooplankton Biomass and Numerical Abundance along with the Hydrographical Parameters During 24 hours

Time (hrs)	*Nature of Tide*	*Secchi disc (cm)*	*Turbidity (ppm)*	*Temp. (°C)*	*Salinity (%)*	*D.O. (ml/l)*	*Biomass (ml/m⁻³)*	*NA (no.m⁻³)*
09.00	MHT	59	21	29.6	27.72	4.48	0.32	27139
12.00	LT	46	23	31.8	31.21	4.70	0.21	8948
15.00	MLT	66	19	32.2	30.21	5.09	0.22	11342
18.00	HT	74	18	30.4	22.38	5.43	0.70	21508
21.00	MHT	–	21	29.8	24.33	5.06	1.62	35082
00.00	LT	–	22	28.8	26.47	4.64	0.84	21992
03.00	MLT	–	21	09.2	24.49	4.70	0.23	13482
06.00	HT	104	18	30.2	17.09	4.76	0.59	17748
09.00	MHT	121	16	30.4	16.55	4.81	0.59	28859

HT: High Tide; LT: Low Tide; MHT: Mid High Tide; MLT: Mid Low Tide; NA: Numerical Abundance.

The surface water temperature ranged from a minimum of 28.8 C during low tide at 00.00 hrs to a maximum of 32.2 C during mid low tide at 15.00 (01–04–99). Variation in temperature of the surface water appears more closely that of the atmosphere. The day time temperatures were higher and more variable than night temperatures (Table 9.1). The range of diurnal variation of surface water temperature was 3.2 C during a period of 24 hours. The temperature showed a single peak during 12.00 hrs, 15.00 hrs and a gradual decrease in temperature from 18.00 hrs was noticed. The causative factor for the variation of the temperature in the study area seems to be diurnal changes in the atmosphere and not the tides (Chandra Mohan and Rao, 1972). Dissolved oxygen values ranged from a minimum of 4.48 ml/l during mid high tide at 9.00 hrs, to a maximum of 5.43 ml/l during high tide period at 18.00 hrs (Table 9.1). The high water accompanied by a higher dissolved oxygen and similarly with the ebb the oxygen values showed a marked decrease. In the mangrove dominated estuarine waters, oxygen fluctuated in accordance with phytoplankton activity and flow of water from submerged vegetation (Murthy, 1997; Prasad, 2003). During the present study, the salinity fluctuations were interesting. In general, the salinity is high during the high tide and low during low tide. But during the present study at certain hours, the low tide waters also have higher salinity than the high tide waters. In the study, at 12.00 hrs. the salinity was 31.21 per cent corresponding to low tide similarly at 18.00 hrs. the salinity of surface water was 22.38 per cent, corresponding to high tide waters. At 00.00 hrs corresponding to low tide the salinity increased gradually and it was 26.47 per cent, by 06.00 hrs (02–04–99), corresponding to high tide the salinity again decreased to 17.09 per cent (Table 9.1).

The only possible explanation that can be given for higher salinity during the low tide period, was that the water during the high tide period enters into the mangrove swamps and creeks and it is enriched with the nutrients and weathered sediments from the mangrove mud-flats and also receives land drainage and decaying litter. The ground water or interstitial water in the mangrove mud-flats usually recorded higher salinities. The water, which inundates the mud-flats during the high tide period returns into the main canal Gaderu, after some stagnation and evaporation in the shallow creeks during the low tide period, resulting in a situation where the low tide water shows higher salinities than the high tide water. During the high tide period even though the higher saline water coming from the near by bay is well mixed with the stagnant diluted, low saline water in Gaderu and results in less salinity than the low tide water. Similar observations of higher saline low tide waters was reported earlier from the Zuari river, Goa (Desouza, 1977). The rise in the salinity is attributed to the physiographic characteristics of estuary (Dehadri, 1970).

Zooplankton biomass values ranged from a minimum of 0.21 ml.m^{-3} (12.00 hrs, low tide), to a maximum of 1.62 ml.m^{-3} (21.00 hrs, mid high tide), the mean being 0.59 ml.m^{-3} (Table 9.1). Whereas, the numerical abundance of zooplankton varied from a minimum of 8,948 no.m^{-3} (12.00 hrs, low tide), to a maximum of 35,082 no.m^{-3} (21.00 hrs, mid high tide). The zooplankton biomass and numerical abundance showed some relationship with the tide and strength of the current and the direction of the flow. However, samples collected during night had higher values of numerical abundance as well as biomass, irrespective of state of the tide and direction of the tidal flow (Table 9.1). In the early hours of the day the zooplankton biomass was high (overall 22 per cent). During mid-day hours (12.00 hrs and 15.00 hrs) there was reduction in the plankton biomass, again from evening onward there was a rise in the biomass, which reached a peak at high hours (18.00 hrs, 21.00 hrs and 00.00 hrs) (Figure 9.1). Later on the zooplankton biomass started falling down. A distinct rise in biomass during night irrespective of the nature of the tide was observed. During day time the increase in the biomass was due to flow of neritic waters, but at night the increase in biomass seems to be due to other factors operating like, the vertical migration of the planktonic organisms to the surface (Chandra Mohan and Rao, 1972). In the study, zooplankton numerical abundance also showed similar trend to that of biomass, highest numerical abundance was recorded during night samples (18.00 hrs, 21.00 hrs and 00.00 hrs) (Figure 9.1). Sankarankutty *et al.* (1979), from the Potengi river estuary waters, Brazil; reported that the largest number of zooplankton was recorded at 23.00 hrs and also reported that zooplankton in general, was more abundant during low tide, more conspicuously so during midnight (23.00 hrs to 03.00 hrs). During the present study, similar observation of higher biomass and numerical abundance of zooplankton were recorded during the mid-night collected samples *i.e.*, 21.00 hrs and 00.00 hrs. The high biomass as well as numerical abundance during night samples relative to day samples, probably because during night many of the bottom dwelling forms like veligers and other invertebrate larvae come up to the surface along with the copepods, mysids, amphipods etc. This fact can be well supported by the study of density distribution of zooplankton *i.e.*, numerical abundance in relation to biomass (Figure 9.1). Similar observation was noticed by Seshaiya (1959) in Vellar estuary.

Altogether 17 groups of zooplankton were observed. Copepods formed a major component contributed 41.65 per cent of total zooplankton. Among the meroplankton bivalve and gastropod veligers, decapod larvae constitute the bulk of the composition. Occurrence and abundance of different taxa over the study cycle depend on the time of collection of the samples. Decapod larvae (megalopa, brachyuran zoeae), chaetognaths, amphipods, polychaete larvae, herpecticoid copepods were abundant during night samples (18.00 hrs, 21.00 hrs, 00.00 hrs and 03.00 hrs) relative to day samples, while calanoid copepods, ostracods, were more in the day samples. Some groups such as cladocerans, turbularians showed more or less uniform density during day and night collection.

The copepods showed an interesting distribution, their abundance was very low during noon hours. From evening onwards their percentage proportion in the sample increased up to about 63 per cent, recorded late in the night (21.00 and 00.00 hrs). Their dominance in the samples during nights was irrespective of the hydrographical conditions and the state of the tide. This could be due to the

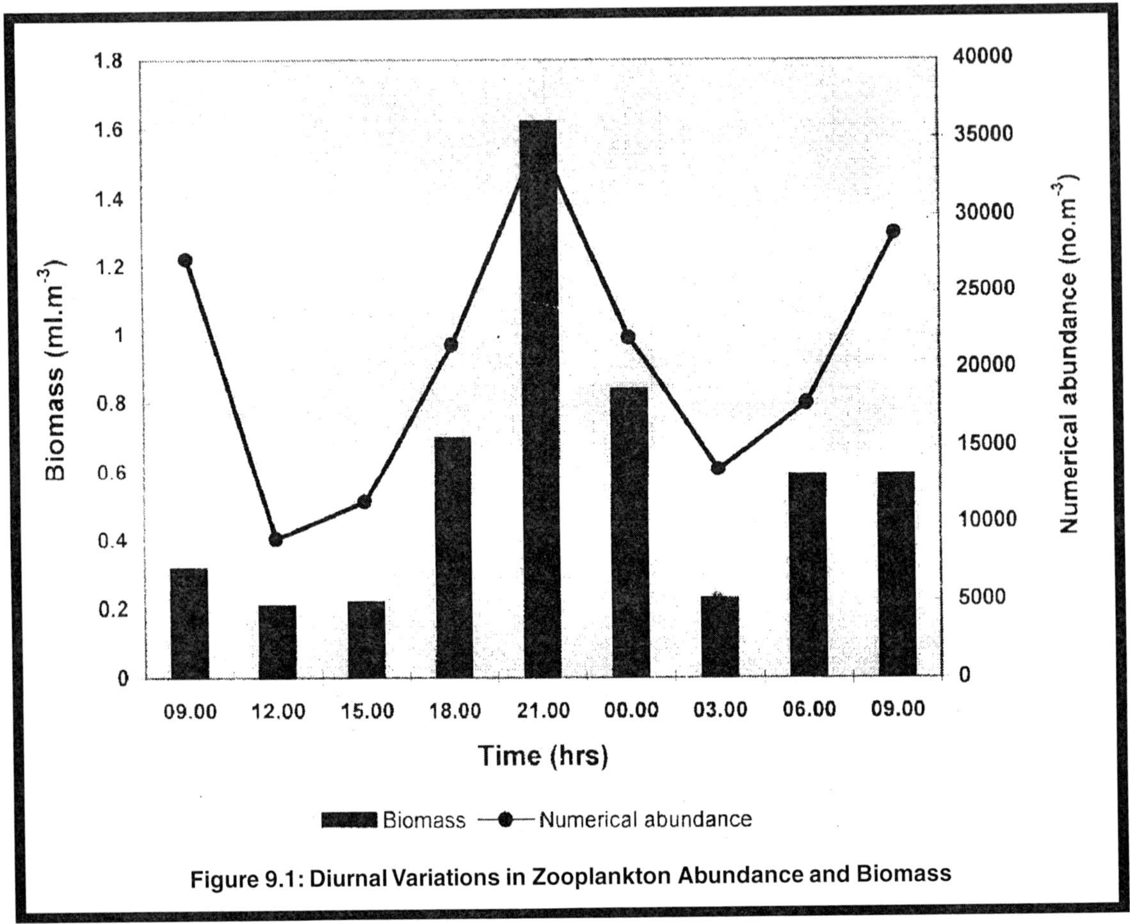

Figure 9.1: Diurnal Variations in Zooplankton Abundance and Biomass

photonegative behaviour of the copepods associated with the phenomena of diurnal vertical migration exhibited by these forms to a marked extent (Chandra Mohan and Rao, 1972). Copepodid stages out numbered the adult copepods during the day and the reverse trend was observed in the night samples. Copepods like *Paracalanus parvaus, Acrocalanus gibber, Eucalanus elongatus, Acartia erythrae, Macrosetella gracilis, Centropages furcatus, Corycaeus speciosus* were seen in good numbers. The day samples were rich of the naupli and zoeae stages as reported from Mandovi estuary (Goswami and Goswami, 1992). It is noteworthy that the post larvae stages of shrimps were fairly abundant during night samples (about 50 to 75 per net haul). The mangroves are the nursery grounds for the shrimps and also other benthic organisms dwelling in the estuary (Osore, 1992). Among the adult crustacean groups lucifers were recorded in large numbers, *Lucifer hanseni* being very dominant in the collections. Their abundance was high during high tide, whereas, the mysids *i.e. Rhopalopthalmus kempi* was observed mostly during night samples. Mysids exhibited significant diel variation irrespective of tides. They appear to come out from the mud into the water column only during night as reported earlier by Padmavati *et al.* (1997) from the Mandovi-Zuari estuaries. This phenomena was observed for amphipods and mysids from Cochin backwaters (Madhupratap and Rao, 1979).

The abundance of coelenterates in the samples was observed during day hours. Among this group the hydromedusae dominated in numbers. Siphonophores and Ctenophores were represented in few numbers. The group chaetognatha was dominated by *Sagitta* species. Their abundance was high during 18.00 hrs corresponding to high tide, when there was large scale incursion of neritic water into the Gaderu. Gastropod and. bivalve veligers were seen in all the samples and their number was fairly high during night and to that of day samples as reported from the Coleroon estuarine complex (Ayyakkannan, 1989). Fish larvae and pelagic tunicates *Oikopleura* sp. were observed more in night samples coincided with high tide. In general, there were more groups were found during flood tide than during ebb tide.

The present study revealed that the fluctuating hydrographical parameters influence the plankton diversity and distribution. The salinity which is the main physical parameters that can be attributed to the plankton diversity act as a limiting factor which influence the plankton community and its distribution (Sri Krishna Das *et al.,* 1993). Slobodkin and Sanders (1969) state that if the changes in an environment are predictable such as the tidal changes in an estuary, the inhabiting animals may develop behavioural mechanisms in response to tidal changes, which keep them in the water of same salinity throughout the tidal cycle by having some kind of biological clock or signal. Madhupratap and Rao (1979), studies on zooplankton diel variation in the Cochin backwaters, concluded that the diurnal differences in zooplankton are usually caused by vertical migration, the cause of which chiefly attributed to response to variation in high intensities of light. However a complete movement of whole plankton population, but rather in tendency of part of the population to spread towards surface at night.

Pillai and Pillai (1973) reported three pattern of zooplankton from the Cochin backwaters:

1. Those forms dominating in the samples during night irrespective of the hydrographical conditions and the state of the tide such as amphipods, cumaceans and fish larvae.
2. Second category of zooplankton showing a primary maximum during night and a secondary one during flood period of the day in the estuary, as in the case in pelagic polychaete, mysids, isopods, lucifers and decapod larvae.
3. The third type of distribution was evinced by typically neritic forms, and their fluctuations were closely associated with the state of tide, and their maximum being recorded during the flood period *e.g.,* coelenterates, cladocerans, stomatopods, ostracods, chaetognaths, pelagic tunicates and fish eggs.

During the present study also these three categories of zooplankton can be distinguished which were mainly dependent on the salinity as well as the phase of tide prevailing during that period.

With regard to tidal cycle studies in tropical riverine estuaries it is to be noted that each area needs to be dealt separately, since the physiographical conditions like riverine discharge, flow characteristics, vegetation etc., are not uniform. The present study area Gaderu, has unique physiographical conditions. The Gaderu canal is open at both ends, one into the Kakinada Bay and other end in the proximity of the opening of the Gautami-Godavari into Bay of Bengal. In view of its two connections into the bay at either end, the Gaderu canal presents unique situation where the mixing processes and flow pattern have to be given paramount importance in determining the salinities and other biological factors. The Gaderu river is subjected to the influence of tides from both ends. On account of this during high tide water enters the canal from both ends and gets piled up in the mid region and recede on either side during subsequent low tide. The region where it gets piled up and the extent of water involved in this phenomena, depends on various factors like intensity of tides, currents and physiography of the area. In addition to it, Gaderu receives drainage from innumerable creeks all along its length. On account of these peculiar physiographical conditions, the mangrove ecosystem here undergoes rapid changes and the populations living in such a situation exhibit wide tolerance to euryhaline conditions. Inspite of these variations certain amount of endemism is noticed in the faunal characteristics. The survival and distribution of these organisms obviously depend on certain behavioral aspects in particular like diurnal vertical migration, which is the universal phenomena after exhibited by the zooplankton community. Inspite of distinct variations noticed from different regions some uniform general observation can be made:

1. Irrespective of the direction of flow, nature of tide etc., always night abundance is observed allover, which definitely needs to be correlated with diurnal vertical migration of the populations which respond to light.
2. Always the flood tide influences to some extent as a major factor in extending the distribution of neritic population to penetrate deeper into the estuary for necessary colonization.
3. Since many of the organisms are euryhaline they can very well adapt themselves to changing salinity and their distribution depends on net movement of water either seaward or towards the interior of the river.
4. Certain extent of endemism cannot be ruled out in such situations which may contribute to the distinct annual distribution pattern, which is more or less regular.

References

Ayyakkannan, K. (1989). Diel variation and relative abundance of plankton larvae in Coleroon estuarine complex, south east coast of India. *J. Mar. Biol. Ass.*, India, 3: 276–286.

APHA (1995). *Standard Methods for the Examination of Water and Wastewater* 19th edn. American Public Health Society Pub., Washington. D.C. pp. 874.

Chandra Mohan, P. and T.S. Satyanarayana Rao (1972). Tidal cycle studies in relation to zooplankton distribution in Godavari estuary. *Proc. Indian Acad. Sci.*, 75: 23–31.

Dehadri, P.V. (1970). Changes in the environmental features of Zuari and Mandovi estuaries in relation to tides. *Proc. Indian Acad. Sci.*, 57: 68–80.

Desouza, S.N. (1977). Monitoring of some environmental parameters at the mouth of the Zuari river, Goa. *Indian J. Mar. Sci.*, 6: 114–117.

Goswami, S.C., R.A. Selvakumar and U. Goswami (1979). Diel and tidal variation in zooplankton populations in the Zuari estuary, Goa. *Mahasagar-Bull. Nat. Inst. Oceanogr.*, 12: 247–258.

Goswami, S.C. and U. Goswami (1992). Lunar, diel and tidal variability in penaeid prawn larval abundance in the Mandovi estuary, Goa. *Indian J. Mar. Sci.*, 21: 21–25.

Madhupratap, M. and T.S.S. Rao (1979). Tidal and diurnal influence on estuarine zooplankton. *Indian J. Mar. Sci.*, 8: 9–11.

Murthy, N.V.V.S. (1997). Hydrography. In: *An Assessment of the Ecological Importance of Mangroves in Kakinada Area, Andhra Pradesh, India.* Final report, European Community, INCO-DC Project, p. 16–25.

Osore, M.K.W. (1992). The ecology of mangrove and related ecosytems. *Hydrobilogia*, 247: 119–120.

Padmavathi, G., S.C. Goswami and P.S. Vindya (1997). Diurnal variations in zooplankton in the Zuari Estuary, west coast of India. *J. Mar. Biol. Ass.,* India, 339: 166–170.

Pillai, P.P. and A. Pillai (1973). Tidal influence of the diel variation of zooplankton with special reference to copepods in the Cochin backwaters. *J. Mar. Biol. Ass.,* India, 15: 411–417.

Prasad, N.V. (2003). Composition and abundance of meroplankton in Coringa mangrove ecosystem with special reference to aquaculture. *J. Aqua. Biol.,* 18: 29–34.

Sankarankutty, C., G.F. Demederios and De. Q. Sentos, N. (1979). Diurnal variation of zooplankton in a tidal estuary. *J. Mar. Biol. Ass.,* India, 21: 187–190.

Seshaiya, R.V. (1959). Some aspects of estuarine hydrology and biology. *Curr. Sci.,* 28: 54–59.

Slobodkin, L.B. and H.L. Sanders (1969). Growth and regulation of animal population. *Brookhaven. Sym. Biol.,* 22: 82–84.

Sri Krishna Das, B., V. Sundara Raj and K. Ramia Moorthy (1993). Zooplankton of Portonovo coastal zone with special reference to invertebrate larvae. *J. Mar. Biol. Ass.,* India, 35: 141–144.

Srinivasan, A. and R. Santhanam (1991). Tidal and seasonal variation in zooplankton of Pullavazhi brakish waters, south east coast of India. *Indian J. Mar. Sci.,* 20: 182–186.

Vijaya Lakshmi, G.S. and V.K. Venugopalan (1973). Diurnal variation in the physical and biological properties in Vellor estuary. *Indian J. Mar. Sci.,* 2: 19–22.

Wickstead, J.H. (1965). *An Introduction to the Study of Tropical Plankton.* (Eds.) J. E. Web and G.E. Newwell, Hutchinson tropical monograph, Hutchinson and Co (Pub.), London, pp. 160.

Chapter 10

Trace Metals in Zooplankton from the Mangrove Waterways of Coringa, East Coast of India with Special Reference to Industrial Effluents

N. V. Prasad

Division of Marine Biology, Department of Zoology, Andhra University, Visakhapatnam – 530 003

ABSTRACT

Industrial effluents and urban waste are major source of pollution in Coringa mangrove waterways. They contain mixture of toxicans such as pesticides, trace metals, industrial products and a variety of other substances. Among these, trace metals are significant in the sense that they act as limiting factors in the growth of planktonic population and productivity in the aquatic ecosystem. In the present study, the trace metal concentration in zooplankton showed the order of Fe > Cu > Zn > Pb. Higher concentrations of trace metals were observed in Gaderu and Coringa relative to Kakinada Bay. During the investigations higher values of metal concentrations were recorded during monsoon season and lower levels during the summer period. The results revealed that the concentration of trace metals in mangrove water and Bay environment of Kakinada are within the limits of FAO/WHO standards.

Keywords: *Trace metals, Zooplankton, Mangrove-Bay environment, Kakinada, Industrial effluents.*

Introduction

Study of the distribution and concentration levels of trace metals in the coastal environment is essential to assess their accumulation in organisms and their possible transfer to man through food chain (Satyanarayana *et al.*, 1985). Build up of metal concentration in coastal environment receiving sewage and industrial effluents affect the growth and development of plankton leading to decrease in productivity of the region (Davies, 1978; Bernhard and Zattera, 1978; Matkar *et al.*, 1981; Kress *et al.*, 1999). The uptake of metals by plankton provides on entry into marine food chains (Gajbhiye *et al.*, 1985; Lau and Lane, 2002). Phytoplankton and zooplankton play an important role in the biogeochemical cycle of trace metals due to bioaccumulation and later transfer to higher levels. Zooplankton organisms are known to accumulate chemicals by direct absorption from water and through food substances intake. Since biological magnification and bioaccumulation are the phenomena in which the trace elements get accumulated at various trophic levels in varying concentrations, a close monitoring of these trace elements needs to be carried out particularly at the secondary producer level *i.e.*, in zooplankton. The present article deals with trace metals concentration in the mixed zooplankton from the mangrove waters and the bay environment of Coringa mangroves.

Materials and Methods

Trace metals in zooplankton from Coringa mangrove waterways including Kakinada Bay, Gaderu and Coringa were studied during July 1999 to June 2000. Altogether three stations, one in Kakinada Bay (Station 1), and 2 in estuarine regions of Gaderu (Station 2) and Coringa (Station 3) surrounded by mangroves (82 14′–82 22′ E and 16 50′–17 00′ N) (Figure 12.1) were selected for the trace metal analysis of zooplankton. Monthly surface zooplankton was collected by oblique haul using 40 cm diameter of 120 mn mesh size net. As soon as the plankton net was hauled the contents were deep frozen on the boat, later brought to the laboratory and washed with distilled water and dried at 60 C until constant weight was obtained. Finely powdered dried material (0.5–2.0 gm) was weighed accurately, later the material was dissolved in 50 per cent 3N Nitric acid followed by concentrated perchloric acid until a clean solution was obtained. The acid digestion was carried out in the water bath. The solution made upto 50 ml by triple distilled water (APHA, 1995). The concentration of 4 metals *viz.* copper, zinc, lead and iron were determined by Atomic Absorption Spectrophotometer and expressed in terms of $mg.gm^{-1}$ dry weight.

Results and Discussion

The trace metals concentration in zooplankton exhibited greater fluctuations both in time and space with a marked variation from station to station. Table 12.1 contain data on the trace metals concentration at selected stations (Range and mean values) and the following is an account of the findings of each metal.

In the present study, the concentration of zinc in the zooplankton from Kakinada Bay varied between 0.11 $mg.g^{-1}$ during March 2000 and 0.76 $mg.g^{-1}$ during December 1999, with a mean of 0.29 $mg.g^{-1}$ (Table 12.1). In the river Gaderu the concentration of zinc varied from a minimum of 0.08 $mg.g^{-1}$ (October 1999) to a maximum of 0.73 $mg.g^{-1}$ (May 1999), with an annual mean of 0.31 $mg.g^{-1}$. While in the Coringa river the zinc content varied between 0.05 $mg.g^{-1}$ (August'99) and 1.54 $mg.g^{-1}$ (January 2000), with an annual mean of 0.47 $mg.g^{-1}$. The observed values of zinc in the zooplankton samples are in agreement with the observations made by George and Kureishy (1977) from the Bay of Bengal and Gajbhiye *et al.* (1985) from the Bombay coastal waters. During the present study, the higher values of metal concentration were associated with large amount of land and river

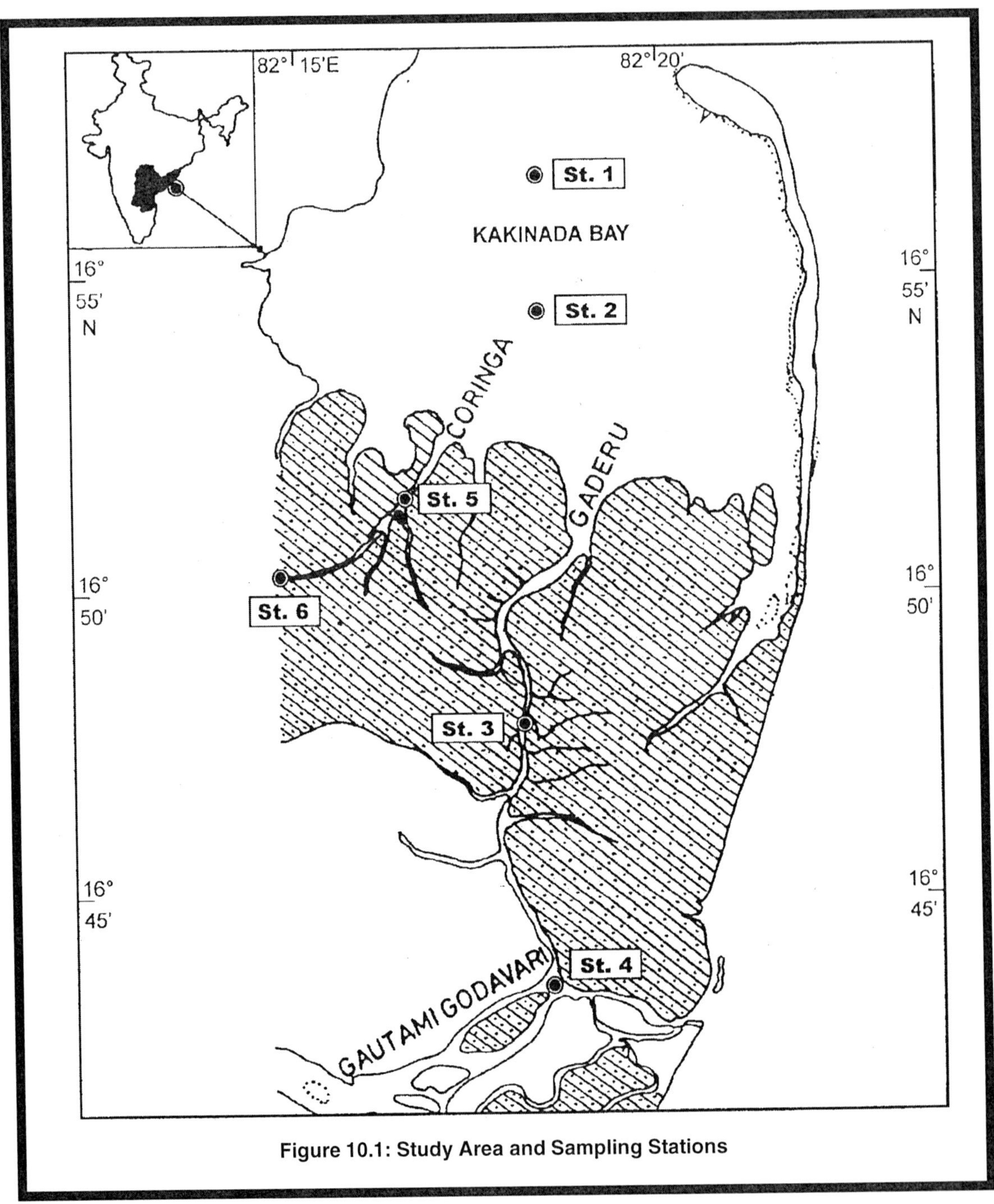

Figure 10.1: Study Area and Sampling Stations

drainage during the monsoon (October–November) and lower levels were observed during the summer period (April–June) due to poor source and active utilization of metal by organisms.

Table 12.1: Range of Trace Metal Conentration (mg.g^{-1}) in Zooplankton at Selected Stations (values in parentheses denote mean)

Trace Metal	*Kakinada Bay (Station 1)*	*Gaderu (Station 2)*	*Coringa (Station 3)*
Zinc	0.11–0.76	0.08–0.73	0.05–1.54
(mg.g^{-1})	(0.29)	(0.31)	(0.47)
Lead	0.15–1.00	0.08–0.44	0.05–0.56
(mg.g^{-1})	(0.35)	(0.26)	(0.38)
Iron	1.65–8.25	1.24–8.55	1.53–9.83
(mg.g^{-1})	(3.98)	(3.83)	(4.53)
Copper	0.12–1.68	0.12–1.27	0.24–1.83
(mg.g^{-1})	(0.85)	(0.55)	(0.91)

The lead concentration in Kakinada Bay samples varied between 0.15 mg.g^{-1} (October 1999) and 1.00 mg.g^{-1} (May,), the observed mean being 0.35 mg.g^{-1} (Table 12.1). In Gaderu river the lead concentration varied between 0.08 mg.g^{-1} (December 1999) and 0.44 mg.g^{-1} (May 2000), with an annual mean of 0.26. Similarly in Coringa river the lead concentration ranged from a minimum of 0.05 mg.g^{-1} (August 1999), to a maximum of 0.56 mg.g^{-1} (January 2000), the mean being 0.38 mg.g^{-1}. Valenta *et al.* (1983) from the Ooster, Westerscheldt and Siersaleone rivers of Germany reported that the lead was found in high concentration in river waters than the estuarine waters. In the present investigation also zooplanktonic organisms in the low saline river water *i.e.* Gaderu and Coringa regions had high lead concentration with a mean of 0.38 and 0.26 mg.g^{-1} respectively. Satyanarayana *et al.* (1985) from the Visakhapatnam harbour waters reported that the dissolved and particulated Pb in coastal waters showed enrichment in the surface, primarily due to its anthropogenic input. During the present study, the station in Gaderu and Coringa rivers receive anthropogenic as well as aquaculture effluents, and so the lead concentration was relatively high than in the bay.

In general, the accumulation of lead in the animals is not correlated with the salinity, but, the lead in the dissolved state is correlated with salinity. It is significant to note that high levels of lead load was observed in the bivalves may be due to the biological availability of lead. The apparent selectivity for lead and other metals largely depends upon its availability in the environment, its chemical and physical properties, the kind of number of ligands available for chelation, its transport, storage and stability for the complex formed (Eisler, 1981).

During the present study, the iron concentration in the zooplankton samples from the Kakinada Bay ranged between 1.65 mg.g^{-1} (January 2000) and 8.25 mg.g^{-1} (February 2000), the observed mean being 3.98 mg.g^{-1} (Table 12.1). In Gaderu the concentration varied between 1.24 (October 1999) and 8.55 (March 1999), with an annual mean of 3.83 mg.g^{-1}. While in Coringa, the concentration of iron ranged from a minimum of 1.53 mg.g^{-1} (August 1999) and 9.83 mg.g^{-1} (January 2000), the mean being 4.53 mg.g^{-1}. In the present study, the zooplankton samples collected from Gaderu (Station 2) and Coringa (Station 3) showed higher Fe metal concentration than Bay zooplankton. This may be due to the removal of the metal in solution, where the inflowing river water meets the saline water (Braganca and Sanzgiry, 1980).

In the present investigations the observed Fe values of zooplankton were comparatively higher because the process of biological degradation of mangrove leaves, litter decomposition, results into the input of nutrients into the aquatic phase. The zooplankton populations dwelling in mangrove areas are mostly detritivores and herbivores which mostly feed on the detritus and phytoplankton, so the Fe concentration was relatively high in the zooplankton samples.

The copper concentration in zooplankton of Kakinada Bay varies between 0.12 $mg.g^{-1}$ (March 1999) and 1.68 $mg.g^{-1}$ (July 1999), with a mean of 0.85 $mg.g^{-1}$ (Table 12.1). In Gaderu, the copper concentration ranged from a minimum of 0.12 $mg.g^{-1}$ (June 1999) to a maximum of 1.27 $mg.g^{-1}$ (January 2000) with an annual mean of 0.55 $mg.g^{-1}$. In Coringa river the copper in zooplankton varied between 0.24 $mg.g^{-1}$ (August'99) and 1.83 $mg.g^{-1}$ (December 1999), with an annual mean of 0.91 $mg.g^{-1}$. From the results it may be seen that the bay zooplankton has relatively higher concentration of copper content than the Gaderu and Coringa. Seasonal variations of the copper concentration in the surface waters of the sea have been ascribed to two important factors, one the phytoplankton activity, and the other contribution the land drainage (Chalapati Rao and Satyanarayana Rao, 1974). Sarma *et al.* (1996) from the sediment of Kakinada Bay, Gaderu and Coringa reported that the concentration of copper was 73 $mg.g^{-1}$, 80 $mg.g^{-1}$, and 65 $mg.g^{-1}$ respectively. The present observations were similar with the reported values.

During the present investigation, the trace metal concentration in zooplankton showed the order of Fe > Cu > Zn > Pb. In the Gaderu (Station 2) and Coringa (Station 3) rivers relatively higher concentrations of trace elements were recorded. This may be due to the land run off along with freshwater inflow from the river Godavari (Figure 12.1). The composition of zooplankton may also have a role in determining the concentrations of trace elements. In the Gaderu and Coringa (Stations 2 and 3) the meroplankton component was high with relatively higher values of copper and zinc than the bay. Where as in the Bay (Station 1) the holoplankton composition was high with iron and lead concentrations markedly high in the samples. Lead may be the element getting accumulated due to the effluents from the surrounding industries like fertiliser, gas and textile factories in the vicinity of Kakinada Bay. In the meroplankton component organisms like the decapod larvae, gastropods and bivalves had relatively higher copper and zinc content in their body tissues, on the contrary the adult forms like copepods, chaetognaths, lucifers, pteropods and ostracods had relatively higher concentration of iron and lead. Among different group of organisms gelatinous forms, copepods and lucifers were known to be very sensitive to copper whereas brachyuran larvae, mysis etc. were tolerant groups among zooplankton (Srinivasa Rao, 1990). In general, the higher copper concentration in the zooplankton coincided with the phytoplankton.

In the present study, the results revealed that the trace metal concentration in the zooplankton was within the admissible limits of FAO/WHO (FAO, 1975) standards. Inspite of the estuarine land drainage into the river on with periodic of tides/currents much of stagnation could not be seen in the study area. Due to proximity of the bay, the waters are carried out with continuous mixing and finally exit out into bay. These factors can explain for the presence of harmless low levels of trace metals in the organisms dwelling in the Mangrove Bay environment of Kakinada.

References

APHA (1995). *Standard Methods for Examination of Water and Wastewater*, 19th ed. American Public Health Assoc., New York.

Bernhard, M. and E. Zattera (1978). Major pollutants in the marine environment. In: *Marine Pollution and Marine Waste Disposal*. Pergamon Pub, Oxford, p. 195–300.

Braganca, A. and S. Sanzgiry (1980). Concentration of few trace metals in coastal and offshore regions of the Bay of Bengal. *Indian J. Mar. Sci.*, 9: 283–286.

Chalapathi Rao, V. and T.S. Satyanarayana Rao (1974). Distribution of the trace elements iron, copper, manganese and cobalt in the Bay of Bengal. *J. Mar. Biol. Ass.,* India, 16(1): 94–112.

Davies, A.G. (1978). Pollution studies with marine plankton part II, Heavy metals. *Adv. Mar. Biol.,* 15: 381–508.

Eisler, R. (1981). *Trace Metal Concentration in Marine Organisms.* Pergman press, New York, pp. 685.

FAO (1975). Manual of methods in aquatic environmental research Part I: Methods of detection measurement and monitoring of water pollution. Food and Agricultural Organization of the United Nations, p. 221–227.

Gajbhiye, S.N., V.R. Nair, P.V. Narvekar and B.N. Desai (1985). Concentration and toxicity of some metals in zooplankton from nearshore waters of Bombay. *Indian J. Mar. Sci.,* 6: 79–81.

George, M.D. and T.W. Kureishy (1977). Trace metals in zooplankton from the west-coast of India, National Institute of Oceanography. *Tech. Rep.,* 02/77: 60–62.

Kress, N., B. Herut, E. Shefer and H. Hornung (1999). Trace element levels in fish clean and polluted coastal marine sites in Mediterranean sea, Read sea and North sea. *Helgol. Mar. Res.,* 53(3–4): 163–170.

Lau, S.S.S. and S.N. Lane (2002). Biological and chemical factors influencing shallow lake eutrophication: a long term study. *Science of Total Environment,* 288(3): 167–181.

Matkar, V.M., S. Ganapathy and K.C. Pillai (1981). Distribution of Zn, Cu, Zn and Fe in Bombay Harbour Bay. *Indian J. Mar. Sci.,* 10: 35–40.

Rowe, D.R. and E.F. Gloyna (1964). Radioactivity transport in water-transport of Zn in an aqueous environment. Tech. Report 4, USAEC, The University of Texas, p. 129–162.

Sarma, V.V., S.J.D. Varaprasad, G.V.M. Gupta and U. Sudakar (1996). Petroleum hydrocarbons and trace metals in Visakhapatnam harbour and Kakinada Bay, East Coast of India. *Indian J. Mar. Sci.,* 25: 148–150.

Satyanarayana, A. and I.M. Rao and B.R. Prasada Reddy (1985). Chemical oceanography of harbour and coastal environment of Visakhapatnam (Bay of Bengal): Part–I Trace metals in water and particulate matter. *Indian J. Mar. Sci.,* p. 139–146.

Srinivasa Rao, K.V. (1990). Biochemical and trace metal analysis of zooplankton and copepod fauna of Visakhapatnam coast. *Ph.D. Thesis,* Andhra University, pp. 178.

Valenta, P., H.W. Nurnberg, P. Klahre, H. Rutzel, A.G.A. Merks and S.J. Reddy (1983). A comparative study of toxic trace metals in the estuaries of the Oaster and Westersheldt and of the Sierra leone river. *Mahasagar-Bull. Nat. Inst. Oceanography,* 16(2): 109–126.

Chapter 11

Seasonal and Spatial Variations in Taxonomic Composition and Biomass of Zooplankton in Estuarine Waters and the Bay Environment of Coringa Mangroves, East Cost of India

N. V. Prasad

Division of Marine Biology, Department of Zoology, Andhra University, Visakhapatnam–530 003

ABSTRACT

Seasonal abundance and biomass of zooplankton were studied in the tropical estuarine and mangrove waters, from Coringa region. Taxonomic composition, abundance and biomass of zooplankton exhibited clear seasonal variations, being highest in pre-monsoon and lowest in monsoon. The overall mean abundance in pre-monsoon was 2.0 fold (mangrove region) and 1.5 fold (estuarine waters) higher than in monsoon. Altogether 25 diverse groups of zooplankton were observed. Over the study period, copepods overwhelmingly dominated the zooplankton community both in terms of species diversity and abundance (mean 67.04 per cent), followed by decapods larvae, bivalves, gastropods in the order of their abundance. Zooplankton biomass varied from 0.26 ml.m^{-3} to 4.91 ml.m^{-3}, whereas numerical density ranged from a minimum of 3,626 no.m^{-3} to a maximum of 1,29,717 no.m^{-3}. Species wise the zooplankton consisted of mainly backwater population, some to be freshwater forms and a few were endemic species. The seasonal variation of zooplankton biomass is positively correlated with environmental variables. The magnitude of changes in salinity was extremely impressive produce highly unstable conditions. This was largely responsible in the nature and heterogeneity of different zooplankton taxa.

Keywords: *Zooplankton, Composition, Biomass, Coringa mangroves.*

Introduction

Mangrove waterways are important coastal habitats in tropical and sub-tropical regions of the world, and represented a special case of estuarine environment characterized by a close association between macrophyte production and marine littoral communities (Por and Dor, 1984). Mangrove estuaries are very productive areas when compared to neighbouring coastal waters (Robertson and Blaber, 1992). Zooplankton by their sheer abundance and diversity constitute the most important community in the mangrove and estuarine waters. The high rate of zooplankton production influences enrichment of organic matter and plays a vital role in secondary and tertiary productions, represented by young instars of fishes (Franz *et al.,* 1984). In investigating food relationships in the waters and the rate of metabolism from the lower to the higher levels in the food web, basic may be gained from studying the density and biomass of main groups of zooplankton organisms, which build up this system of energy transfer in the ocean, estuaries, backwaters etc. (Uye *at al.,* 2000; Zhang and Wang, 2001). Though there are many studies on the zooplankton distribution, studies concerning zooplankton of mangrove waters are not adequately documented in the tropics, particularly in Indian coastal waters (Godhantaraman, 2001). The objective of the present investigation was to describe and compare the species composition, seasonal and spatial variation in biomass from estuarine and the bay environment of Coringa mangroves, southeast coast of India.

Materials and Methods

The bay-estuary complex of Coringa mangroves is situated near Kakinada, Andhra Pradesh (16 50′–17 00′N; 82 14′–82 22′E), along southeast coast of India. The bay-estuary complex of Coringa mangrove ecosystem including Kakinada Bay (north bay and south bay), Gaderu and Coringa rivers. Water and zooplankton samples were collected monthly from November 1997–October 1999. Altogether 15 GPS fixed locations, 4 in north bay (Stations 1–4), 3 in south bay (Stations 5–7), 5 in Gaderu river (Stations 8–12) and 3 in Coringa river (Stations 13–15) were selected. Hydrographical parameters such as temperature, salinity, dissolved oxygen and pH were measured by adopting standard methods (APHA, 1995). Zooplankton samples were collected by using a 120 μm mesh size net. Biomass of zooplankton was measured using the wet displacement volume method and aliquot method was employed for the enumeration of zooplankton (Wickstead, 1965). Comparisons of biomass by location and season were analysed by one-way analysis of variance (ANOVA). Seasons were defined as monsoon (June–September), post-monsoon (October–January) and pre-monsoon (February–May) based on the seasonal monsoons.

Results and Discussion

The bay-estuarine complex of Coringa mangrove ecosystem received the maximum rainfall from June to October coinciding with seasonal monsoon. The two monsoons namely the southwest and northeast monsoon brings the maximum rainfall for the area. Environmental parameters such as surface water temperature and salinity were highest in summer months (April–June) and lowest in monsoon months. The hydrographical variables exhibited greater fluctuations both in time and space with a marked variation from station to station. The Secchi depth was relatively high (mean 0.60–0.78 m) at (Stations 1 and 2) near the proximity of the sea than the other locations (Table 11.1). At the interior locations (Stations 9, 10, 11, 14 and 15) water transparency was least (mean < 0.40 m) obviously due to the shallow nature of the region and with turbidity caused by suspension of bottom sediment. The differences between surface and bottom temperature was generally insignificant, and it ranged between 0.4 to 1.2 C. This indicating lack of vertical stratification. The pH values revealed that these waters are more alkaline in nature. Relatively high dissolved oxygen values were observed at mangrove locations of Gaderu (mean 4.66 ml/l) than bay locations (mean 4.26 ml/l, north bay; 4.50 ml/l, south bay). In Gaderu the relatively high range of dissolved oxygen can be attributed to mangrove/phytoplankton activity and community respiration (Prasad, 2002; Prasad, 2003). Overall, salinity during the study period varied from total freshwater conditions (0.00 per cent 0, Station 15) to a normal sea water (34.18 per cent, Station 2). Seasonally, marked variation in salinity was observed. Spatially, the Kakinada Bay is characterized by relatively high salinity (mean > 27.00 per cent) during most part of the year. In Gaderu subject to freshwater influx salinity was mostly 18.00 per cent (mean). In Coringa, where the influx of freshwater even more, salinity was very low (mean < 5.00 per cent) mostly. The distribution of salinity is under the influence of physical factors like the river water discharge, tides and the prevailing circulation pattern existing in the area. In general, the north bay stations experience typical marine conditions during the most part of the year, whereas the south bay experience brackish water conditions due to the mixing of the low saline water from the river and high saline water of the adjacent bay. In Coringa mostly freshwater conditions were prevailed throughout the year.

Table 11.1: Hydrographical Characteristics (Range) of the Surface Water at Different Regions (Values in Parenthesis Denote Mean)

Parameter	*North Bay*	*South Bay*	*Gaderu*	*Coringa*
Secchi Depth (m)	0.30–1.90	0.24–0.70	0.14–0.60	0.26–0.40
	(0.75)	(0.45)	(0.36)	(0.31)
Temperature (°C)	24.8–31.4	25.4–34.8	25.4–31.8	29.2–32.4
	(28.26)	(27.92)	(28.94)	(30.67)
pH	6.2–7.9	6.7–7.9	6.7–7.8	6.9–7.1
	(7.09)	(7.13)	(7.15)	(7.04)
Salinity (%)	21.09–34.18	18.95–34.06	0–32.40	0–11.30
	(28.25)	(27.04)	(17.65)	(4.38)
D.O. (ml/l)	3.24–4.98	3.19–7.94	3.64–7.80	1.65–4.85
	(4.26)	(4.50)	(4.66)	{4.18)

Spatial and temporal distribution of zooplankton showed definite trends associated with environmental fluctuations. Range and mean values of biomass and numerical abundance of zooplankton at four different regions are listed in Table 11.2. Overall zooplankton biomass ranged from a minimum of 0.26 ml.m^{-3} to a maximum of 4.91 ml.m^{-3}. Whereas the numerical abundance fluctuated from 3,626 no.m^{-3} (Station 15) to 1,29,717 no.m^{-3} (Station 7). Altogether 25 diverse groups of zooplankton were encountered. Overall, copepods dominated the zooplankton during entire study period and constituted 67.04 per cent of the total zooplankton population. Abundance of varies zooplankton groups fluctuated in accordance with salinity regime. Zooplankton groups such as copepods, decapod larvae, chaetognaths, bivalves and gastropods were seen throughout the year in large numbers. Coelenterates, polychaete larvae, cladocerans were other major groups of zooplankton population, which showed a seasonal pattern, whereas ostracods, amphipods etc were obtained occasionally in sparse numbers.

The numerical abundance of copepods varied between 8,902 no.m^{-3} and 96,586 no.m^{-3} (mean 29,750 no.m^{-3}). Among copepods *Acartiidae* were the most successful copepods in Coringa mangrove waterways. The common species were *Paracalanus parvus, Eucalanus elongatus, Acrocalanus gibber, Centropages furcatus, Acartia erythraea, Corycaeus speciosus, Macrosetella gracilis, Eutusperpina acutifrons* dominated at the bay region, while the creek regions was predominated by *Acrocalanus* sp. The marine dominant species namely *Oncaea venusta, Oithona rigida, Sapphirina* sp., showed restricted distribution to north bay locations. The composition of copepods constituted by mixture of neritic and estuarine species with eryhaline estuarine forms dominated the counts. In general, copepods showed two peaks of abundance in a season. A major peak during July–October and minor peak during January–April. The major peak in Kakinada Bay coincided with the month of July–September. This could be due to the reason that the plankton population migrates towards the stable environment than the highly disturbed environment of estuarine area (Robertson and Blaber, 1992) *i.e.,* the mangrove region and the brackish water dominant southern part of the bay region. In the Gaderu and Coringa where the mangrove vegetation is dominant, peak abundance of copepods was observed during February–April coinciding pre-monsoon period.

Table 11.2: Range of Zooplankton Abundance and Biomass in Kakinada Bay, Gaderu and Coringa (Values in Parenthesis Denote Mean)

Region		Monsoon	Post–monsoon	Pre–monsoon
North bay	Abundance	7546–34575	8415–57718	10904–81830
	(no.m^{-3})	(20509)	(28479)	(31154)
	Biomass	0.70–2.55	0.76–1.54	0.66–2.82
	(ml.m^{-3})	(1.43)	(1.18)	(1.56)
South bay	Abundance	24626–77090	41276–94080	32623–29717
	(no.m^{-3})	(39070)	(50154)	(56984)
	Biomass	1.87–2.92	0.74–4.25	1.78–4.91
	(ml.m^{-3})	(2.50)	(2.77)	(3.18)
Gaderu	Abundance	10301–51935	11506–77490	32648–54672
	(no.m^{-3})	(24509)	(39441)	(47847)
	Biomass	0.84–1.86	0.55–4.30	0.91–2.87
	(ml.m^{-3})	(1.33)	(1.75)	(1.81)
Coringa	Abundance	3626–14979	7874–17907	4254–26428
	(no.m^{-3})	(8327)	(11385)	(13609)
	Biomass	0.46–0.67	0.26–0.50	0.28–1.15
	(ml.m^{-3})	(0.54)	(0.44)	(0.85)

Monsoon: June–September; Post-monsoon: October–January; Pre-monsoon: February–May.

Decapod formed the second largest group of zooplankton in order of abundance as well as percentage composition. This group was largely constituted by larval stages. Overall, brachyuran larvae formed major part of the decapod population, where the members of sergestidae outnumbered them. During the study, swarms of brachyuran zoeae were frequently appeared in the plankton samples at mangrove location of Gaderu, probably indicating mangrove habitat is an important breeding site for these brachyurans. Penaeidae was represented by protozoea and mysid stages of *Penaeus, metapenaeus,* and *Parapenaeiopsis.* The post larval stages of *Penaeus monodon, P. indicus, Metapenaeus monoceros* and *M. dobsoni* were noticed in Gaderu and Coringa, of which *M. monoceros* dominated by contributing 42 per cent of the total post larval population. Sergestidae was represented by *Lucifer hanseni, Acetes* sp., and *Sergestes* sp. Porcellanid and pagurid larvae were the representative of anomura, while zoea and megalopa represented brachyura. Scyllaridae, represented by phyllosoma larvae, were restricted only in Gaderu. In general, decapod larval abundance was high in mangrove locations (Gaderu and Coringa) relative to bay locations (north bay and south bay). Seasonally, post-monsoon period sustained higher decapod population than other seasons.

Among chaetognaths *Sagitta bedoti* was common at all stations, while *Sagitta enflata* was largely noticed towards the bay region. Other species like *S. pulchra, S. neglecta, S. robusta* were also observed in small number. In Gaderu and Coringa mangrove habitat, the chaetognaths abundance was low relative to bay stations. The high abundance of chaetognaths in the bay-estuary waterways of Coringa mangroves appears to correspond with higher salinities. The mature stages and adult forms of chaetognaths were present in large number during the peak abundance of copepods. Similar observations were made by Reeve *et al.* (1970) from Miami. Chaetognaths are generally known to be predaceous carnivores, which feed mainly on copepods (Chandra Mohan, 1963). In the present study, planktonic bivalves and gastropods were recorded in large numbers. Their average densities varied between 1,203 no.m^{-3} -87,300 no.m^{-3} and 6,371–12,233 no.m^{-3} respectively. It is noteworthy that the gastropod larvae were high in bay stations, near the proximity of the mangroves than the other locations. The detritus inputs from the surroundings mangrove mudflats make the waters nutrient rich and this particular character supports the high abundance of gastropod larvae in the bay region. It is also reported that molluscans like *Cerithidia* sp., *Murex hemifuses, Telescopium* sp., *Umbonium* sp., and *Nassarius* sp., were high in these areas (Dipti Raut, 1997). Bivalves and gastropods showed bimodal abundance with a primary peak during October and November (post-monsoon) and secondary peak during March and April (pre-monsoon).

Many other holo and mero plankton groups were also represented in zooplankton samples (Table 11.3), whose occurrence and abundance was sporadic, even though they collectively dominated zooplankton at times. The other taxa includes appendicularians, polychaete larvae, cladocerans, ostracods, mysids, medusae, amphipods, pteropods, isopods and fish larvae which occurred relatively small numbers. The hydromedusae and ctenophores appeared immediately after the monsoon sets in, and later their abundance gets decreased to minimum. The post-monsoon season was found to be favourable for the growth and activity of hydromedusae in the estuary. *Evadne* sp., and *Podon* sp., contributed to the cladocerans community. During monsoon months (August and September) swarms of cladocerans were observed at interior part of the creeks contributing 32.7 per cent to total zooplankton population at station 15 (Coringa). Among tunicates *Oikopleura* sp., was represented in good number. Usually, they were found to be associate phytoplankton abundance. Other tunicates like salps and doliolids were observed only at certain stations (Stations 1 and 2) in north bay. Mysids were common and were represented by *Mesopodopsis orientalis, Rhopalophthalmus kempi, R. indicus* and *Gastrosaccus muticus. Conchoecia* was the single ostracod species and more or less evenly distributed in the bay region. The amphipod *Hyperia* sp., being found only at north bay stations close to the neritic influence. Isopods and amphipods were found in fairly large number, while cumaceans and pteropods were rare in the area. Fish eggs and larvae were commonly observed at all stations and occasionally found in large number. Many invertebrate larval forms have been found in the Kakinada Bay, Gaderu and Coringa. They are mostly brachipod and echinoderm larvae. During the study, a gradual increasing trend in biomass was observed from monsoon period to pre-monsoon period. Silas and Pillai (1975) from the Cochin backwaters, reported high zooplankton density and biomass during pre-monsoon period and low during monsoon, with relatively more secondary peak during the post-monsoon period. In other Indian estuaries and backwater, Chandra Mohan (1963) from Godavari estuary; Nair and Tranter (1971) from Cochin backwater; Goswami and Singbal (1977), Padmavathi and Go swami (1996) from Mandovi-Zuari estuaries; Santha Kumari *et al.* (1999) from Dharantar estuary; Bhunia and Choudhury (1982) from Sundarban mangroves and Shanmugam *et al.* (1986) from Pichavaram mangroves have reported very poor zooplankton production during monsoon months.

Table 11.3: Percentage Composition (Overall) and Numerical Abundance (Range) of Major Zooplankton Groups

Group	*Abundance (no.m^{-3})*	*Percentage*
Copepods	8902–96586	67.04
Decapod larvae	3842–74418	7.56
Bivalves	1203–87300	6.40
Gastropods	6371–12233	4.38
Chaetognaths	487–7820	1.96
Lucifers	187–8580	1.12
Appendicularians	54–1627	0.67
Polychaete larvae	66–568	0.52
Cladocerans	15–8217	0.28
Ostracods	22–415	0.19
Mysids	17–302	0.12
Medusae	5–308	0.09
Amphipods	26–171	0.04
Fish larvae and eggs	12–129	0.03
Miscellaneous groups	47–634	9.60

A considerable variations in hydrographical conditions was noticed during the study period (Table 12.1). Salinity varied from almost freshwater to mere marine conditions (0.00–34.18 per cent). Such a large variation salinity plays a major role in controlling the zooplankton production. Rao (1977) reported the direct relationship between zooplankton production and salinity in Indian estuaries. Gopala Krishnan (1968) from the Hooghly-Matlah estuarine system also reported that the salinity play an important role in distribution of zooplankton in different regions of estuary. Since estuary being shallow, practically no longitudinal temperature gradient and vertical thermal stratification could be noticed. High abundance and biomass during pre-monsoon in Coringa mangrove waterways may be due to the comparatively stable hydrographical condition (particular salinity), whereas unstable hydrographical conditions prevailed (salinity 0.00–34.18 per cent) during monsoon (Table 11.1) which may account for poor zooplankton production. The high influx of freshwater during monsoon laden with silt result in a situation where the waters of mangrove dominated Gaderu and Coringa rivers experiences high turbidity conditions. This may also be a cause for the poor zooplankton production during this period.

Daniel and Prem Kumar (1965) from the eastern sector of Indian Ocean reported that the zooplankton production gradually decreases with increasing depth, where the offshore stations yield less abundance and biomass than the coastal waters. The results in the present study also revealed the fact that the Stations 1 and 2 in north bay (Kakinada Bay), which located in the proximity of the near-shore waters of Bay of Bengal had less abundance and biomass relative to other bay location. The depth profile seems to influence the zooplankton distribution as observed with regarding to stations 1 and 2. At Stations 1 and 2 where the depth was relatively more (> 10 m) and abundance, biomass was low as against the stations 3 and 4 in north bay. The recorded biomass was high at these two stations (3 and 4) where the depth never exceeded 6 meters.

Spatially, south bay support high zooplankton biomass (overall mean 2.81 $ml.m^{-3}$) followed by Gaderu (1.63 $ml.m^{-3}$), north bay (1.39 $ml.m^{-3}$) and Coringa (0.61 $ml.m^{-3}$) (Table 11.2). Similar trend was observed regarding numerical abundance. The high zooplankton production in south bay could be due to the contribution made by estuarine and freshwater species from the estuary and marine components through tidal incursion. Further to it, the south bay waters seem to act as an interface between the sea and mangrove environment. Another aspect of interest is that the bay receives nutrient rich waters from the surrounding small creeks traversing through the dense mangrove vegetation. The large scale ingress of nutrient rich waters from the surrounding mangroves makes the south bay into a highly suitable environment for the colonization of zooplankton community. In Coringa influenced by land drainage, with somewhat unstable conditions (salinity 0.00–11.30 per cent) (Table 11.1), the zooplankton production was low relatively. Horizontally, there is a decreasing tendency in the biomass values, when we move from the marine condition towards freshwater mixing region (Figure 11.1). Someto (1975) from the St. Margaret's Bay reported a periodic fluctuations in zooplankton production coincided with the tidal variations. Pillai and Pillai (1973) from the Cochin backwaters reported that the biomass fluctuation in according with the tides. However, effect of tides on the distribution of zooplankton did not indicate a definite pattern. During the monsoon period, zooplankton biomass in the inner locations of creek was more during ebb tide than flood. Whereas in pre-monsoon period showed a comparable trend like that of post-monsoon period with higher biomass of zooplankton during flood period than the ebb at most of the creek location.

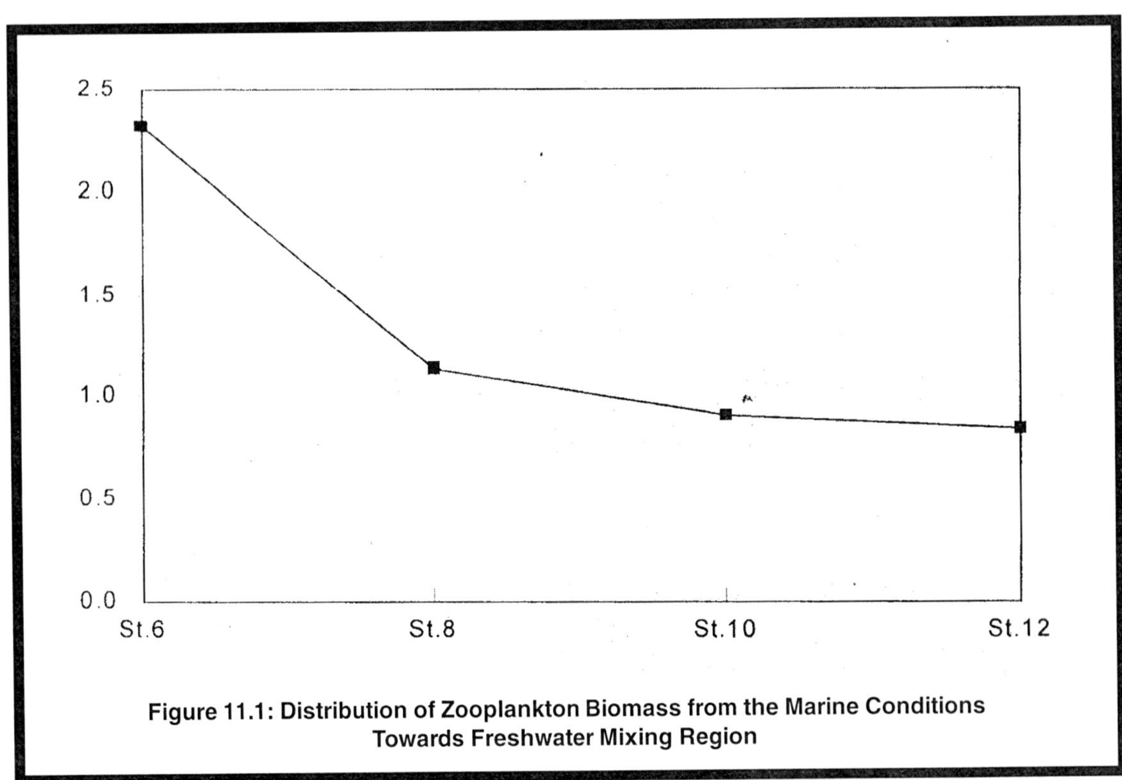

Figure 11.1: Distribution of Zooplankton Biomass from the Marine Conditions Towards Freshwater Mixing Region

The pre-monsoon season appear to be suitable for high production of zooplankton when salinity and temperature were fairly stable. In pre-monsoon period, the entire water column shows stable hydrographical conditions and the estuarine conditions become virtually an extension of adjoining sea, with high salinity and temperature values. Flood water influx during monsoon period showed significant effect on the zooplankton biomass distribution in Coringa mangrove waterways, since the dominant species constitute the bulk of the zooplankton were marine in origin. Consequent to the heavy rainfall during monsoon period and resultant large influx of freshwater into the mangrove region, the salinity in the Gaderu and Coringa drastically fall off and at most places reaches to that of freshwater conditions. Unless the population of these areas exhibit some sort of adaptative behaviour like migration or osmotic tolerance etc., their survival and retention may become very difficult. The movement of plankton population is controlled by the tidal currents and activities due to the freshwater down flow (Wooldridge and Erasmus, 1980). Probably on account of their mass movement and drift towards the bay region, where relatively higher salinities are encountered, higher zooplankton density and biomass was recorded near bay locations (south bay). This particular feature makes the bay to record higher biomass and abundance values even during the monsoon season. Following the monsoon decline the marine component gradually get established towards mangrove region in post-monsoon season with the invasion of seawater from the bottom zone towards the surface. Comparatively high salinity values were observed in Gaderu and Coringa during this period. This was reflected in the distribution of zooplankton.

During the study, a direct correlation between biomass and numerical abundance of zooplankton was observed. In general, high biomass values were coincided with high numerical abundance. However, at some stations, the zooplankton biomass has often shown lack of agreement with corresponding value of numerical abundance. This may be account for by the presence or absence or restricted distribution of certain large organisms such as medusae, ctenophores, siphonophores etc., reported to be responsible for the depletion of phytoplankton and copepods (Goswami, 1973; Desai *et al.*, 1983; Fransz *et al.*, 1984). Thus, the composition of zooplankton may also have a role in determining the biomass. In the present study, meroplankton component was high at mangrove location (Gaderu and Coringa), whereas relatively high holoplankton component at bay locations, was observed. In general, small sized forms such as copepods, crab zoeae, gastropod and bivalve larvae mostly accounted for zooplankton biomass. Whereas large sized forms such as medusae, ctenophores etc., contributed less in number.

Correlation for zooplankton biomass and population was positive and very significant at majority of the stations. The correlation was significant at 1 per cent level at Kakinada Bay ($r = 0.85$) and Gaderu ($r = 0.78$), while at Coringa ($r = 0.51$) it was only at 5 per cent level. Biomass also significantly correlated with copepods ($r = 0.94$), decapod larvae ($r = 0.75$), bivalves ($r = 0.71$) and adult crustacean ($r = 0.62$). High standing stock of zooplankton very often synchronized with phytoplankton abundance indicating a direct relationship though individual samples did not always show a proportionate correlation probably due to variation in composition of flora and fauna (Haridas *et al.*, 1973; Tiwari and Nair, 1993). In the study, it was also observed that the adults of different taxa of zooplankton constitute a major share of the pre-monsoon zooplankton biomass, whereas the larvae of benthos, nekton along with larvae of different zooplankton groups contributed major proportion of the post-monsoon biomass of zooplankton. During the present study, zooplankton biomass was significantly ($p < 0.05$) correlated with surface water salinity ($r = 0.72$). Other variables such as temperature, dissolved oxygen tested were showed insignificant ($p > 0.05$). This indicated that the salinity acts as a limiting factor in distribution of zooplankton biomass in the bay-estuarine waterways of Coringa mangrove ecosystem. It may be concluded that the ambient salinity and degree of stability achieved through neritic influence are prime influencing factors in limiting the zooplankton distribution in the bay-estuary complex of Coringa mangroves. Factors such as temperature, turbidity, dissolved oxygen, pH are of secondary in nature.

References

APHA (1995). *Standard Methods for the Examination of Water and Wastewater*, 19th edn. American Public Health Association Publication, Washington D.C., pp. 874.

Bhunia, A. and A. Choudhury (1982). Some ecological consideration for zooplankton in Chemagiri creek, Sagar islands (south), Sundarbans. *Mahasagar-Bull. Nat. Inst. Oceanogr.*, 15: 247–252.

Chandra Mohan, P. (1963). Studies on zooplankton of the Godavari estuary. *Ph.D. Thesis*, Andhra University, Visakhapatnam, pp. 167.

Daniel, A. and V.K. Prem Kumar (1965). The distribution of the standing crop of zooplankton in the eastern sector of the Indian Ocean during July–September 1962. *J. Mar. Biol. Assoc.*, India, 7: 440–452.

Desai, B.N., S.N. Gajbhiye, M. Jiyalal Ram and V.R. Nair (1983). Comparative account of zooplankton in polluted and unpolluted estuaries of Gujarat. *Mahasagar-Bull. Nat. Inst. Oceanogr.*, 16: 281–291.

Dipti Raut (1997). Benthic macro fauna of mangrove waters and the bay environment of Kakinada, east coast of India. *Ph.D. Thesis*, Andhra University, Visakhapatnam, pp. 118.

Fransz, H.G., J.C. Micquel and S.R. Gonzalez (1984). Mesozooplankton composition, biomass, and vertical distribution and copepod production in the stratified central north sea. *Netherlands J. Sea Res.*, 18: 82–96.

Gadhantaraman, N. (2001). Seasonal variations in taxonomic composition, abundance and food web relations of micro zooplankton in estuarine and mangrove waters, Parangipettai region, southeast coast of India. *Indian J. Mar. Sci.*, 30: 157–160.

Gopala Krishnan, V. (1968). Fishery resources of the Hooghly Matlah estuarine system and its relation to fisheries of Bay of Bengal. In: *Symposium on the Living Resources of the Seas around India*, CMFRI, Cochin, India.

Goswami, S.C. (1973). Observation on some planktonic groups of Kavaratti Atoll (Laccadives). *Proc. Nat. Sci. Acad.*, 39: 676–686.

Goswami, S.C. and S.Y.S. Singbal (1977). Ecology of Madovi and Zuari estuaries plankton community in relation to hydrographical conditions during monsoon months, 1972. *Indian J. Mar. Sci.*, 7: 33–38.

Haridas, P., M. Madhuparatap and T.S.S. Rao (1973). Salinity, temperature, oxygen and zooplankton biomass of the backwaters from Cochin to Alleppy. *Indian J. Mar. Sci.*, 2: 94–102.

Nair, K.K.C. and D.J. Tranter (1971). Zooplankton distribution along salinity gradient in the Cochin backwaters before and after monsoon. *J. Mar. Biol. Assoc.*, India 13: 210–230.

Padmavathi, G. and S.C. Goswami (1996). Zooplankton ecology in the Mandovi Zuari estuarine ecosystem of Goa, west coast of India. *Indian J. Mar. Sci.*, 25: 268–273.

Pillai, P.P. and A. Pillai (1973). Tidal influence on the diel variation of zooplankton with special reference to copepods in the Cochin backwaters. *J. Mar. Biol. Assoc.*, India, 15: 411–417.

Por, F.D. and I. Dor (1984). Foreword. In: *Hydrobiology of the Mangal,* (Eds.) F.D. Por and I. Dor. Dr. Junk Publications, The Hague, pp. 78.

Prasad, N.V. (2002). Characteristics of zooplankton and its production in the Bay-Estuary waterways of Coringa mangrove, east coast of India. *Ph.D. Thesis,* Andhra University, Visakhapatnam, pp. 196.

Prasad, N.V. (2003). Composition and abundance of meroplankton in Coringa mangrove ecosystem with special reference to aquaculture, Kakinada, Andhra Pradesh. *J. Aqua. Biol.,* 18: 29–34.

Rao, T.S.S. (1977). Salinity and distribution of brackish warm water zooplankton in Indian estuaries. In: *Proceedings of Symposium on Warm Water Zooplankton.* UNESCO/NIO, p. 196–204.

Reeve, M.R., J.E.B. Raymont and J.K.B. Raymont (1970). Seasonal biochemical composition and energy of Sagitta hipida. *Mar. Biol.,* 6: 357–364.

Robertson, A.I. and S.J.M. Blaber (1992). Plankton, epibenthos fish communities. In: *Tropical Mangrove Estuaries* (Eds.) A.I. Robertson and D.M. Alongi. American Geophysical Union, Washington, p. 173–224.

Santha Kumari, V., L.R. Tiwari and V.R. Nair (1999). Species composition, abundance and distribution of Hydromedusae from Dharamtar estuarine system, adjoining Bombay harbour. *Indian J. Mar. Sci.,* 28: 158–162.

Shanmugam, M., A.R. Kasinathan and S. Maramuthu (1986). Biomass and composition of zooplankton from Pichavaram mangroves, southeast coast of India. *Indian J. Mar. Sci.,* 15: 111–113.

Silas, E.G. and P.P. Pillai (1975). Dynamics of zooplankton in a tropical estuary (Cochin backwater), with a review on the zooplankton fauna of the environment. Bull. Dept. Mar. Sci., University of Cochin, 7: 329–355.

Someto, D.D. (1975). Tidal and diurnal effects on zooplankton samples variability in a nearshore environment. *J. Fisher. Res. Board of Canada,* 32: 347–366.

Tiwari, L.R. and V.R. Nair (1993). Zooplankton composition in Dharamtar creek and joining Bombay harbour. *Indian J. Mar. Sci.,* 22: 63 69.

Uye, S.I., N. Nagano and T. Shimazy (2000). Abundance, biomass, production and trophic roles of micro and net zooplankton in Ise Bay, Central Japan. *J. Oceanogr.,* 50: 389–398.

Wickstead, J.H. (1965). *An Introduction to the Study of Tropical Plankton.* Hutchinson tropical monograph, Hutchinson and Co., London, pp. 160.

Wooldridge, T.H. and T. Erasmus (1980). Utilization of tidal currents by estuarine zooplankton. *Estuar. Coast. Mar. Sci.,* 2: 107–114.

Zhang, W. and R. Wang (2001). Abundance and biomass of copepods nauplii and ciliates in Jlaozhou Bay. Oceanologia et limnol. sinica, 32: 280–287.

Chapter 12

Studies on Qualitative and Quantitative Plankton Production using Duck Droppings

S.K. Majhi, B.K. Mahapatra, K. Vinod and B.K. Mandal
Division of Fisheries, ICAR Research Complex for NEH Region, Barapani – 793 103, Meghalaya

ABSTRACT

Two different doses of duct droppings *i.e.* 0.25 g/lt and 0.5 g/lt were applied for plankton culture. On computation it has been found that a dose of 0.5 g/lt was ideal for both phytoplankton and zooplankton production. Among the phytoplankton, *Spirogyra* sp. and *Euglena* sp. and among the zooplankton, *Brachionus* sp. and *Lepadela* sp. were found dominating. The quantitative analysis revealed a phytoplankton population of 1229 nos./50 lit and a zooplankton population of 2225 nos./50 lit on the 30th day. On the 60th day, phytoplankton population was 2404 nos./50 lit and zooplankton population was 13810 nos./50 lit.

Keywords: *Duck excreta, Phytoplankton, Zooplankton.*

Introduction

The rationale behind integrated farming is to minimize wastes from the various sub-systems on the farm: wastes or by-products from each sub-system are used as inputs to other sub-systems to improve the productivity and lower the cost of production of the outputs of the various sub-systems (Edwards *et al.*, 1986). Integrated livestock fish farming system is a proven environmentally sustainable and economically viable technology that encompasses rational utilization of available resources and livestock such as pigs, ducks, chicken, etc (NACA, 1989). Different forms of integrated livestock fish farming *viz.* pig-fish, poultry-fish, duck-fish etc. have been evolved and popularized in India (Sarma and Pattanaik, 1985; Das, M.K., 1990; Ayyappan *et al.*, 1998; Borah *et al.*, 1998; Ghosh *et al.*, 2000 and Naskar *et al.*, 2002). Recently attention has been paid to develop the live feed culture utilizing the animal excreta from poultry, pig, duck, rabbit etc. The live feed containing balanced amino acid is an indispensable requisite for larval rearing of the captive bred fishes, prawn, ornamental fishes, etc. (Mahapatra and Datta, 1999; Mahapatra *et al.*, 1999 and Mahapatra *et al.*, 2000). In North Eastern Region of India, livestock farming is a prime activity of majority of the rural households. One or more of the animal components like piggery, poultry and duckery are invariably found in all houses (Bujarbaruah *et al.*, 1996). The social and economic conditions prevailing in NE region, integration of livestock may be the only source of fertilizer availability, at low cost, to make fish culture economically feasible and profitable.

Under this rural farming set up, fish farming would be more profitable if integrated with animal components and recycling animal excreta for the cause of plankton production and growth of the aquaculture industry. As the region has a very good scope for indoor aqua business especially for ornamental fish culture, the necessity of live feed culture technique is need of the hour. The present study was done to observe the qualitative as well as the quantitative plankton production using duck droppings.

Materials and Methods

Six cement cisterns (C1–C6) were used for the plankton culture. C1 and C2 were used as control, C3 and C4 for treatment I (E1), C5 and C6 for treatment II (E2). Pond water was used in all the experimental cisterns. Manuring was not done in control while in treatment I (E1), duck droppings @ 0.25 g/lit and in treatment II (E2), duck droppings @ 0.50 g/lit were applied. A water volume of 90,30,000 lt was maintained in all experimental cement cisterns. The study was carried out for a period of 60 days.

The plankton and water sampling were done on 30th and 60th day at 10:00 hrs. from each treatment and control. Air temperature, water temperature, pH, alkalinity, hardness, dissolved oxygen, free carbon-dioxide, chloride and total dissolved solids (TDS) were determined quantitatively using *Standard Methods* (APHA, 1989). The plankton samples were collected by filtering 50 litres of water from each cement cistern using a standard plankton net. The live samples were brought to the laboratory for preliminary observations and then were fixed in 5 per cent formalin and stored in glass bottles for detailed qualitative and quantitative estimation of plankton and the results have been presented based on the pooled data of the replication.

Results and Discussion

The qualitative analysis of plankton and physico-chemical properties of the water of the experimental cisterns are presented in Table 12.1 and 10.2, respectively. It has been observed that the dominance of zooplankton population increased with that of higher dose of duck droppings. In E2, the dosage of manure was high and the zooplankton reached its maximum both during 30th and 60th day sampling, which is also supported by the findings of earlier worker (Tsirtsis and Karydis, 1998). The quantitative analysis of phytoplankton is presented in Table 12.3 and zooplankton in Table 12.4. It has been found that in control the net plankton production were 1569 and 1478 numbers, out of which phytoplankton population were 1407 and 1355 and zooplankton population was 162 and 123 on 30th and 60th days, respectively. In E1, the net plankton population was 1762 and 4639 numbers, out of which the phytoplankton population was 773 and 1170 and zooplankton population was 989 and 3469 on 30th and 60th days. Similarly in E2, the net plankton population was 3454 and 16214, out of which phytoplankton population was 1229 and 2404 and zooplankton population was 2225 and 13810. It has been found that on 30th day among phytoplankton the population of *Scenedesmus* sp. was dominating in control and *Spirogyra* sp. in E1 and E2. Among zooplankton *Keratella* sp. was dominating in control and *Brachionus* sp. in E1 and E2. Similarly on 60th day sampling, among phytoplankton *Scenedesmus* sp. was found dominating in control and *Spirogyra* sp. in E1 and E2. Among zooplankton, *Testudinella* sp. in control and *Brachionus* sp. in E1 and E2 were dominating. The quantitative variation of phytoplankton and zooplankton is presented in Figures 12.1 and 12.2.

Table 12.1: Qualitative Plankton Production in Different Experimental Cisterns

Cement Cistern	*Phytoplankton*	*Zooplankton*
Control	*Scenedesmus* sp., *Pediastrum* sp., *Ankistrodesmus* sp., *Ulothrix* sp., *Spirogyra* sp., *Staurastrum* sp., *Tetrastrum* sp., *Placus* sp., *Fragilaria* sp.	*Keratella* sp., *Arulla* sp., *Nauplius larvae*, *Testudinella* sp., *Polyalthra* sp., *Calanoid copepod.*
E1	*Scenedesmus* sp., *Pediastrum* sp., *Ankistrodesmus* sp., *Ulothrix* sp., *Spirogyra* sp., *Staurastrum* sp., *Tetrastrum* sp., *Placus* sp., *Fragilaria* sp. *Cosmanium* sp., *Nitzschia* sp.	*Brachionus* sp., *Lepadella* sp., *Trichocera* sp., *Keratella* sp., *Arulla* sp., *Nauplius larvae*, *Testudinella* sp., *Polyarthra* sp., *Calanoid copepod.*
E2	*Scenedesmus* sp., *Pediastrum* sp., *Ankistrodesmus* sp., *Ulothrix* sp., *Spirogyra* sp., *Staurastrum* sp., *Tetrastrum* sp., *Placus* sp., *Fragilaria* sp. *Cosmanium* sp., *Nitzschia* sp., *Microsystis* sp., *Botryoccus* sp., *Ceratium* sp., *Navicula* sp., *Cryptomonas* sp.	*Brachionus* sp., *Lepadella* sp., *Trichocera* sp., *Keratella* sp., *Arulla* sp., *Nauplius larvae*, *Testudinella* sp., *Polyalrhra* sp., *Calanoid copepod*, *Cyclopoid copepod*, *Aureopsis* sp., *Philedema* sp., *Alora* sp., *Arulla* sp., *Discotracha* sp.

Table 12.2: Physico-chemical Parameters in Different Experimental Cisterns

Water Parameters	*Experimental Tanks*	*Air Temperature (°C)*	*Water Temperature (°C)*	*pH*	*Specific Conductivity (S/cm)*	*Dissolved Oxygen (ppm)*	*Free Carbon-dioxide (ppm)*	*Total Alkalinity (ppm)*	*Total Hardness (ppm)*	*Chloride (ppm)*
On 30th day	Control	21.5±1.5	27.0±1.5	6.52	19.8	6.0	14.1	5.1	12.019	0.01086
	E1	21.5±1.5	27.0±1.5	6.53	20.1	6.0	14.1	5.0	12.012	0.01086
	E2	21.5±1.5	27.0±1.5	6.53	20.1	6.0	14.0	5.0	12.012	0.01065
On 60th day	Control	24.0±1.5	27.5±1.5	7.2	11.50	8.3	12.0	5.0	8.009	5.85
	E1	24.0±1.5	27.5±1.5	7.3	11.61	9.5	12.0	5.0	8.008	5.85
	E2	24.0±1.5	27.5±1.5	7.3	11.60	9.8	12.0	5.0	8.008	5.85

Table 12.3: Quantitative Record of Zooplankton (no./50 lit) During the Study Period

	Quantitative Record of Zooplankton on 30th Day			*Quantitative Record of Zooplankton on 60th day*		
Species	*Control*	*E1*	*E2*	*Control*	*E1*	*E2*
Dicranophorus sp.	07	12	20	05	10	28
Euchlamus sp.	10	00	00	00	08	10
Conochiloides sp.	13	18	27	10	15	35
Colurella sp.	12	10	16	09	05	20
Calanoid copepod	00	03	02	00	00	01
Cyclopoid copepod	01	00	01	00	00	02
Nauplius	02	00	00	05	00	00
Macrothrix sp.	01	01	01	00	03	04
Difflugia sp.	00	00	01	00	00	01
Polyarthra sp.	02	10	20	00	15	25
Testudinella sp.	19	158	186	25	78	180
Lecane sp.	00	03	00	00	00	01
Arulla sp.	00	02	01	00	00	00
Ciliates	00	00	00	00	00	01
Brachionus sp.	12	527	1019	10	1992	7855
Alora sp.	00	08	06	01	00	07
Philedema sp.	00	02	00	02	00	00
Lepadella sp.	05	220	830	07	1320	5525
Trichocera sp.	00	00	01	00	00	05
Aureopsis sp.	00	02	03	01	00	09
Keratella sp	78	13	90	48	18	100
Discotracha sp.	00	00	01	00	05	01
Total	**162**	**989**	**2225**	**123**	**3469**	**13810**

Table 11.4: Quantitative Record of Phytoplankton (no./50 lit) During the Study Period

	Quantitative Record of Phytoplankton on 30th Day			*Quantitative Record of Phytoplankton on 60th day*		
Species	*Control*	*E1*	*E2*	*Control*	*E1*	*E2*
Scenedesmus sp.	1000	68	25	680	63	102
Ankistrodesmus sp.	130	08	15	380	07	00
Pediastrum sp.	200	90	43	173	45	212
Selanastrum sp.	09	00	01	05	00	00
Ulothrix sp.	10	00	01	04	00	00
Staurastrum sp.	00	00	00	00	00	03
Coelastrum sp.	00	01	08	01	05	20
Tetrastrum sp.	01	02	00	05	06	00

Contd...

Table 11.4–Contd...

	Quantitative Record of Phytoplankton on 30th Day			Quantitative Record of Phytoplankton on 60th day		
Species	*Control*	*E1*	*E2*	*Control*	*E1*	*E2*
Placus sp.	00	00	01	00	00	00
Fragillaria sp.	00	00	05	00	01	15
Navicula sp.	04	00	08	00	05	25
Cryptomonas sp.	00	04	00	00	00	00
Dictyosphaerium sp.	00	00	01	00	00	00
Euglena sp.	01	53	140	05	41	280
Spirogyra sp.	20	280	589	50	700	980
Microsystis sp.	08	83	124	10	100	290
Botryoccus sp.	04	100	82	01	57	225
Cosmanium sp.	01	55	154	00	95	200
Ceratium sp.	19	29	32	41	45	52
Total	**1407**	**773**	**1229**	**1355**	**1170**	**2404**

This base line information on plankton production using duck droppings definitely helps in developing package of practice on live feed culture of plankton under controlled condition.

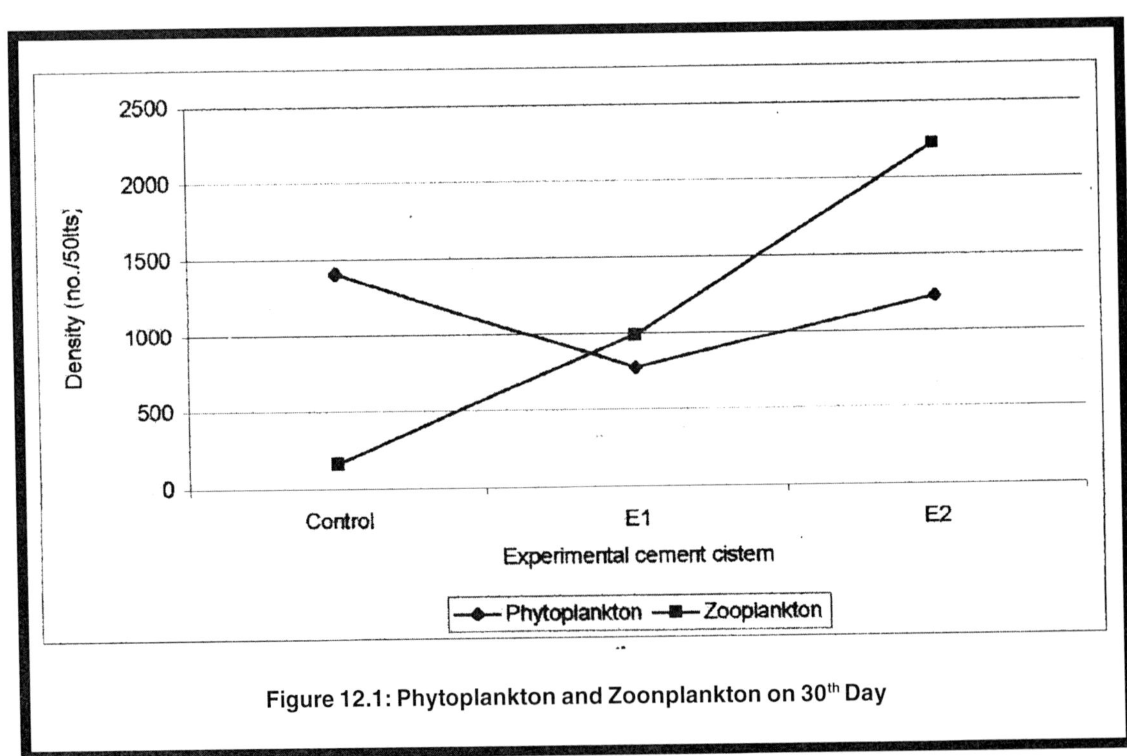

Figure 12.1: Phytoplankton and Zoonplankton on 30th Day

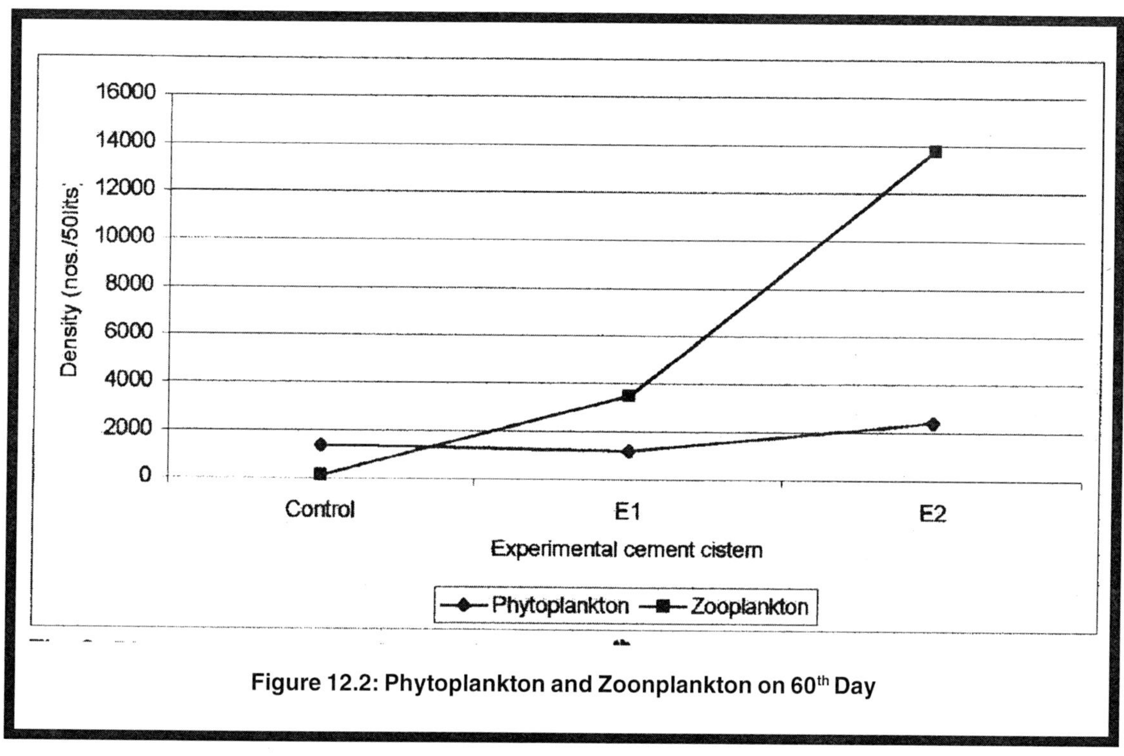

Figure 12.2: Phytoplankton and Zoonplankton on 60th Day

Acknowledgement

The authors are thankful to the Director, ICAR Research Complex for NEH Region for providing necessary facilities to carry out the work and also to the authorities of NATP-CGP for financial assistance.

References

APHA (1989). *Standard Methods for the Examination of Water and Wastewater,* 17th edn. American Public Health Association, Washington, DC.

Ayyappan, S., K. Kumar and J.K. Jena (1998). Integrated fish farming practices and potentialities. *Fishing Chimes,* 18(1): 15–19.

Borah, B.C., A.K. Bhagowati, U.K. Baruah, H.K. Gogoi and S. Nath (1998). Integrated poultry-pig-fish farming. *J. Inland Fish. Soc.,* India, 30(1): 39–46.

Bujarbaruah, K.M., A. Das and S.K. Nanda (1996). *Status of Animal Husbandry in Meghalaya.* Published by ICAR Research Complex for NEH Region. Barapani, Meghalaya.

Das, M.K. (1990). Integrated fish farming with poultry, duckery and livestock. In: *Technologies for Inland Fisheries Development,* (Eds.) V.V. Sugunan and U. Bhaumik. Published by CICFRI, Barakpore, West Bengal, pp. 59–68.

Edwards, P., K. Kaewpaitoon, E.W. McCoy and C. Chantachaeng (1986). Pilot small-scale crop/ livestock/fish integrated farm. AIT Research Report, 184, Air, Bangkok, Thailand, pp. 131.

Ghosh, A.K., B.K. Mahapatra and T.K. Baguli (2000). Integrated approach for enhancing, fish production. Souvenir: State level Seminar-cum-Workshop on integrated approach for enhanced fish production. Feb. 24–25, Dept. of Fisheries, Govt. of Tripura.

Mahapatra, B.K. and N.C. Datta (1999). Food preference and growth performance of the exotic cat fish *Clarias gariepinus* (Burchell). *Environment and Ecology*, 17(1): 207–210.

Mahapatra, B.K., A. Datta, A. Basu, U.K. Dey and K.K. Sengupta (1999). Observations on the spawning and rearing of angelfish, *Pterophyllum scalare* (Lichenstein). In: *Eco-friendly Management of Resources for Doubling Fish Production: Strategies for 21st Century*, (Eds.) M. Sinha, Dhirendra Kumar and P.K. Katiha. Proceedings of the National Seminar, December 22–23, 1999. Inland Fisheries Society of India, Barrackpore, p. 102–106.

Mahapatra, B.K., K.K. Sengupta, U.K. Dey, G.C. Rana, A. Dutta, A. Basu and S. Saha (2000). Controlled Breeding and Larval rearing of *Clarias batrachus* (Linn.) for mass-scale propagation. *Fishing Chimes*, 19(10&11): 97–102.

NACA (1989). *Integrated Fish Farming in China*. Network of Aquaculture Centers in Asia and the Pacific (NACA). Technical Manual No. 7.

Naskar, S., B.K. Mahapatra and A. Gas (2002). Eco-friendly management of pig waste through integrated Pig-Fish farming: A case study. National Seminar on "*Environmental Biology and Fish Biology*", February 1–3, 2002, Visva-Bharati, Santiniketan, p. 64–65.

Sarma, A.L.N. and P.K. Pattanaik (1985). Ecological studies on zooplankton of freshwater ponds in and around Bhubaneswar. *J. Environ. Biol.*, 6(4): 245–256.

Tsirtsis, G. and M. Karydis (1998). Evaluation of phytoplankton community indices for detecting eutrophic trends in the marine environment. *Environmental Monitoring and Assessment*, 50: 255–269.

Chapter 13

Spacio-temporal Trends in Phytoplankton Primary Production in Marikanave Reservoir

Syed Fasihuddin & E.T. Puttaiah***

**Department of Botany, Government Arts and Science College, Karwar – 581 301*

***Department of Environmental Science, Jnanasahyadri, Kuvempu University, Shankarghatta – 577 451*

ABSTRACT

Spacio-temporal studies on primary productivity by phytoplankton in three sectors of Marikanave reservoir revealed low productivity. The GPP varied between 0.18 to 43.42 $mgCm^3\ d^{-1}$. NPP ranged between 0.10 to 22.69 $mgCm^3\ d^{-1}$. CR was between 0.20 to 21.2 $mgCm^3d^{-1}$, and P : R ratio ranged between 0.28 to 5.01. The sector III exhibited a stage of collapse. The reduced production rate is attributed to low level of water fluctuation that resulted in macrophytic dominance. Hence, primary production is deviated from phytoplankton to aquatic macrophytes. The results obtained in the present study have been compared with the productivity values of some reservoirs of India.

Keywords: *Reservoir, GPP, NPP.*

Introduction

The man-made aquatic ecosystems play a vital role in inland fishery development in India due to their vast area and huge production potential. These ecosystems have to be studied individually for their proper exploitation and conservation. Odum (1956) developed the idea of determining P : R ratio in order to define a community type and also the concept of autotrophy and heterotrophy. Ganf and Horne (1975) used P : R ratio for determining the trophic status of waterbodies. The P : R ratio more than one signifies autotrophic nature of production (Cole, 1979). Whereas Vollenweider (1974) used GPP as a criteria for classifying water bodies on tropic nature as, oligotrophic (0.065–0.3 g Cm^{-2} d^{-1}), mesotrophic (0.25–1.0 gCm^{-2} d^{-1}) and eutrophic (1.0–8.0 g Cm^{-2} d^{-1}).

Marikanave, also called Vanivilas Sagar dam in Karnataka, has been built across the river Vedavathi in 1907 in Chitradurga district known for water scarcity and frequent famines of the past. It is located in the geographical ordinates of 13 20′ latitude and 75 16′ longitude.

Materials and Methods

The reservoir was arbitrarily divided in to three transverse sectors *viz.* Sector–I, Sector–II and Sector–III (Figure 13.1). Light and dark bottle method (Vollenweider, 1974 and Trivedy *et al.,* 1987) was adopted for the measurement of primary productivity in terms of gross primary production (GPP), net primary production (NPP) along with community respiration (CR). The bottles were incubated for a period of 12 hours from sunrise to sunset. The results were expressed as mgCm^3 d^{-1}. The study has been carried out in the reservoir from February 2001 to January 2002.

Results

Data on seasonal and sectoral range in gross primary production net primary production, community respiration and P : R ratio is presented in Table 13.1. The monthly fluctuations of all these parameters are depicted in Figure 13.2.

Table 13.1: Seasonal and Sectoral Range in Primary Productivity

Sector	*GPP(mgCm³d⁻¹)*	*NPP(mgCm³d⁻¹)*	*CR(mgCm³d⁻¹)*	*P : R Ratio*
		Pre-monsoon		
I	29.08–43.32	16.61–22.69	9.02–21.2	2.01–4.21
II	6.21–9.21	4.12–5.08	2.07–4.17	2.25–2.20
III	0.72–0.81	0.21–0.51	0.29–0.49	1.41–2.75
		Monsoon		
I	3.16–21.00	0.68–13.22	2.48–7.78	1.27–2.69
II	0.38–6.11	0.16–4.02	0.22–3.26	1.43–2.92
III	0.18–0.60	0.10–0.32	0.09–6.31	6.28–2.00
		Post-monsoon		
I	23.38–27.81	16.14–18.26	7.22–9.55	2.91–3.23
II	5.42–7.52	3.21–4.34	2.21–3.23	2.34–2.94
III	0.38–0.71	0.18–0.45	0.20–0.39	1.82–2.95

Gross Primary Production

It is the total rate of photosynthesis including the organic matter used up in respiration during the period of measurement. The GPP values are found to be in the range of 0.18 to 43.32 mgCm^3 d^{-1} in the reservoir. The sectoral production values varied between 3.16 and 43.32 mgCm^3 d^{-1} (Sector–I), 0.38

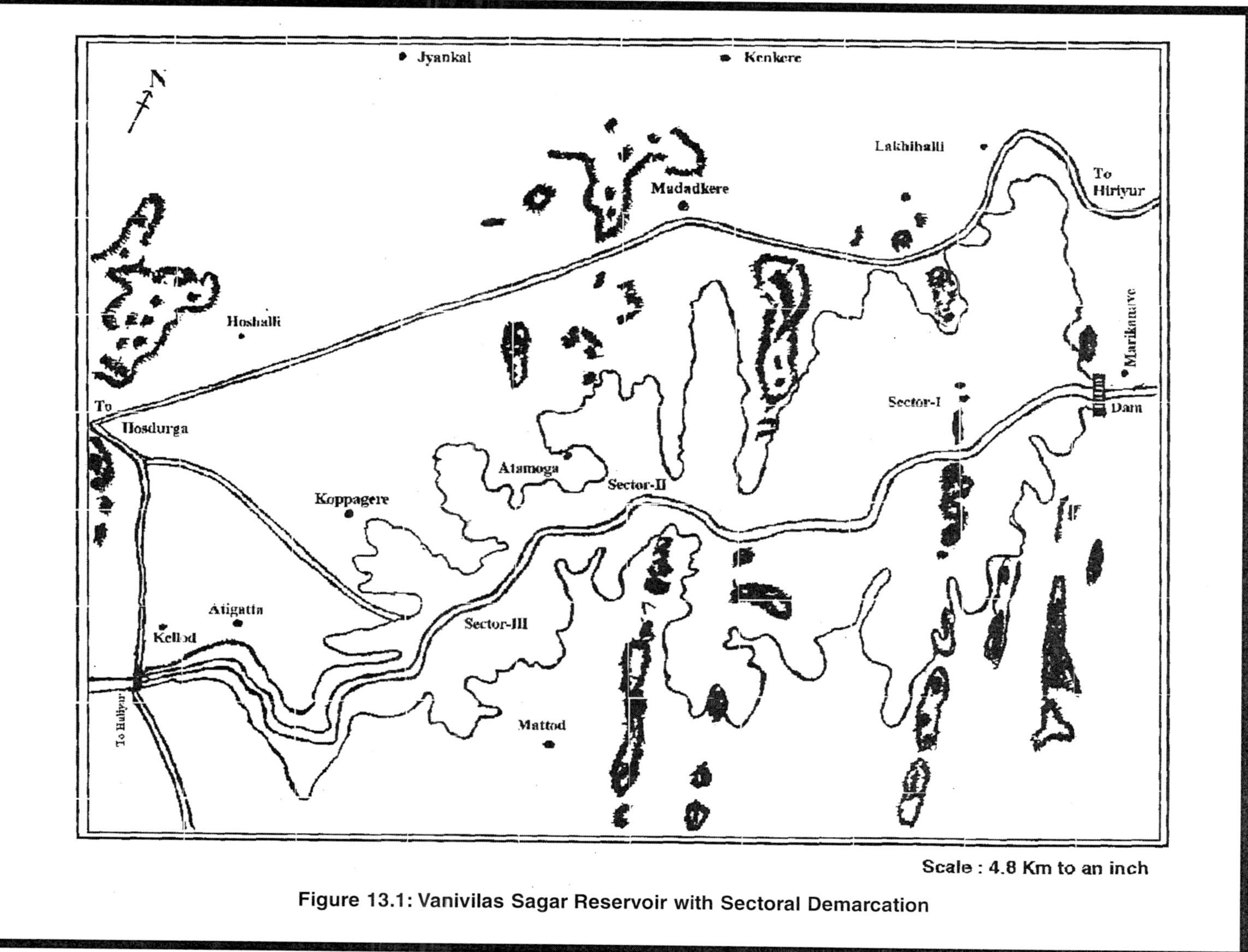

Scale : 4.8 Km to an inch

Figure 13.1: Vanivilas Sagar Reservoir with Sectoral Demarcation

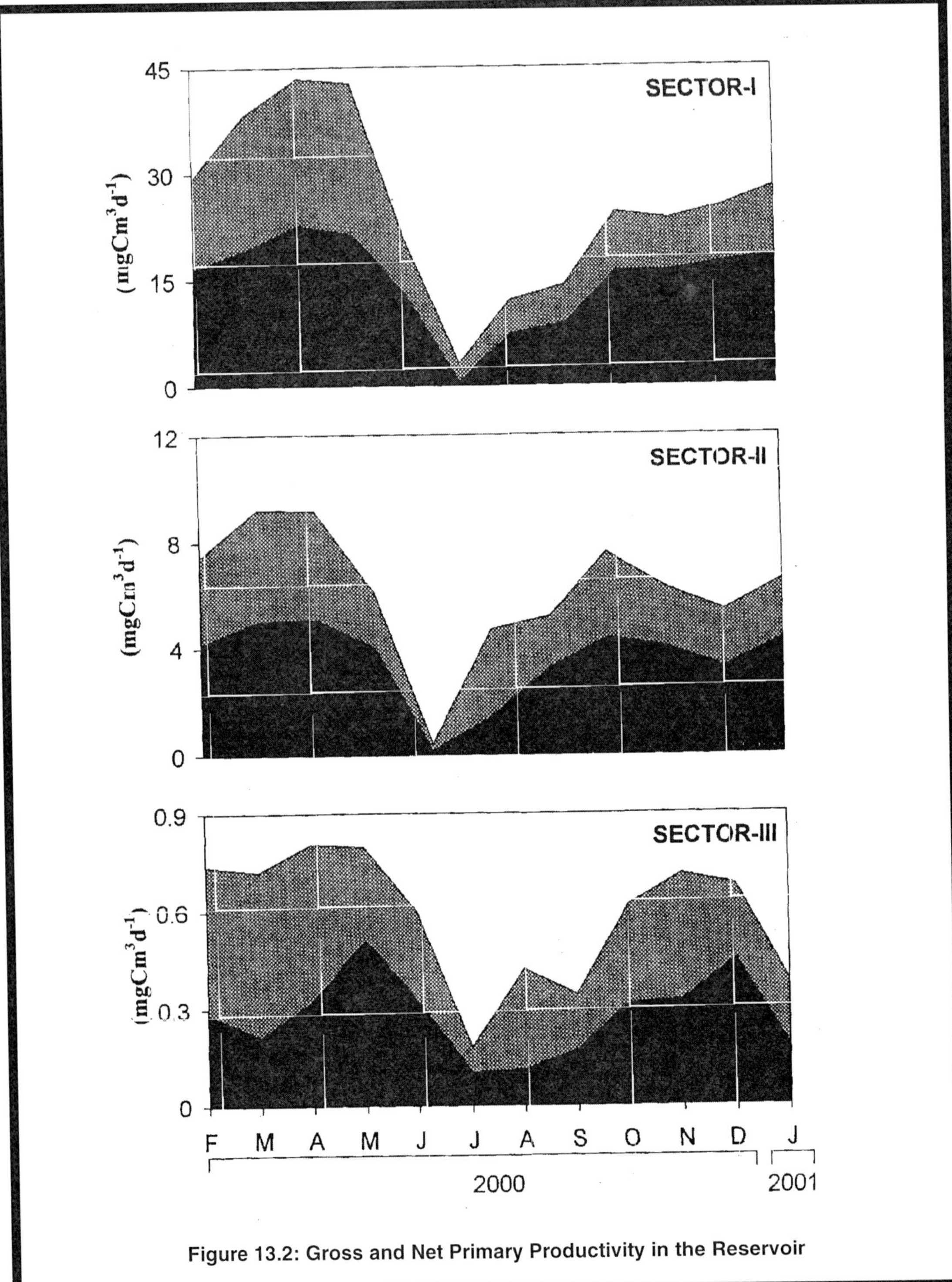

Figure 13.2: Gross and Net Primary Productivity in the Reservoir

and 9.21 mgCm^3 d^{-1} (Sector–II) and 0.18 and 0.81 mgCm^3 d^{-1} (Sector–III). The seasonal values varied in the range of 0.72 to 43.32 mgCm^3 d^{-1}, 0.18 to 21.00 mgCm^3 d^{-1} and 0.38 to 27.81 mgCm^3 d^{-1} during pre-monsoon, monsoon and post-monsoon months respectively. The rate of gross production was high in pre-monsoon at Sector-I, where as third sector exhibited a stage of collapse (0.18 to 0.60 mgCm$_3$ d^{-1}).

Net Primary Production

The net production is the rate of storage of organic matter in cells in excess of respiratory utilization by phytoplankton during the period of measurement. The values in the reservoir were found to be in the range of 0.10 to 22.69 mgCm^3 d^{-1}. The sectoral values ranged between 0.68 to 22.69 mgCm^3 d^{-1}, 0.16 to 5.08 mgCm^3 d^{-1} and 0.10 to 0.51 mgCm^3 d^{-1} in Sectors I, II and III respectively. Maximum was recorded in sector I and the other two sectors showed low values.

The seasonal values varied in the range of 0.21 to 22.69 mgCm^3 d^{-1}, 0.10 to 13.22 mgCm^3 d^{-1} and 0.18 to 18.26 mgCm^3 d^{-1} during pre-monsoon, monsoon and post-monsoon months respectively. Net productivity was more during pre-monsoon season. Where as the monsoon season was found to be imparting negative effect at all the sectors, which showed poor net production.

Community Respiration

The community respiration rate fluctuated between 0.09 to 21.20 mgCm^3 d^{-1} in the reservoir. The sectoral and seasonal rates followed the same trends as that of GPP and NPP. The sectoral respiration rate was in the range of 2.48 to 21.2 mgCm^3 d^{-1} (Sector–I), 0.22 to 4.17 mgCm^3 d^{-1} (Sector–II), and 0.09 to 6.31 mgCm^3 d^{-1} (Sector–III) in the reservoir. Peak value was recorded at Sector–I (9.02 mgCm^3 d^{-1}), but least CR was at Sector–III (0.39 mgCm^3 d^{-1}).

The seasonal CR values varied in the range of 0.29 to 21.2 mgCm^3 d^{-1}, 0.09 to 7.78 mgCm^3 d^{-1} and 0.20 to 21.2 mgCm^3 d^{-1} during pre-monsoon, monsoon and post-monsoon months respectively.

P : R Ratio

The gross primary production to community respiration ratio varied from 0.28 to 4.21 in the reservoir. The sectoral ratio ranged between 1.27 to 4.21 (Sector–I), 1.43 to 2.20 (Sector–II) and 0.28 to 2.95 (Sector–III). The seasonal range was found to be in between 1.41 to 5.04 during pre-monsoon, 0.28 to 2.69 in monsoon and 1.82 to 3.23 in post-monsoon seasons. The P : R ratio showed seasonality with low values during monsoon season of the study period.

Discussion

The capacity of synthesizing organic matter by any aquatic ecosystem depends largely on physico-chemical parameters like temperature, nutrients and solar radiation besides producer populations. Water bodies having N_2 fixing blue green algae represent very rarely as limiting factor, likewise free CO_2 will be limiting where there is a permanent bloom of blue green algae, especially *Microcystis* sp., (Pathak, 1979). The wide fluctuations in limno-chemical features also impart considerable seasonal variations in the rate of primary productivity (Ramakrishnia and Das, 1998) in the reservoirs.

Likens (1975) observed a close relationship between temperature and productivity. In the present study high production rate during pre-monsoon months at high temperature in Sector–I indicates the existence of close relationship between these two. However, Sectors–II and III did not show such a relationship and which may probably due to shallowness with aquatic macrophytes. They utilize solar radiation and nutrients leaving very little of the two inputs for photosynthetic activities by phytoplankton. Thus, the energy is diverted and low production was noticed.

The size and shape of the reservoirs also influence their productivity (Rawson, 1952 and 1953; Symon *et al.,* 1965). Some deeper tropical reservoirs of Tamil Nadu developed blooms and showed high production (Sreenivasan, 1969). Contrary to this, the shallow reservoirs like Mettur and Bhavaninagar showed poor productions. Therefore, in the present study the Sector–I being little deep with moderate aquatic weeds showed comparatively higher production. Correlation between transparency and primary productivity in some reservoirs of Andra Pradesh is phenomenal (Das, 2002). No such relationship was observed in Tawa reservoir (Srivatsava *et al.,* 1985). In the present investigation, a strong negative relationship was established (r = –0.363). Niranjan (1994) and Patil (1998) have reported similar relationship.

Carbonate and bicarbonate alkalinity is another factor is influencing the rate of production particularly in tropical reservoirs to a great extent. Alkalinity below 500 mg^{-1} indicate low photosynthetic rate (Das, 2002). In the reservoirs of Andra Pradesh seasonal variations in alkalinity showed significant relationship with gross and net production. However, Sreenivasan (1965) opined that alkalinity is not always the limiting factor and found high rate of production in Amaravathy with low alkalinity. However, in the present study a strong positive relationship was observed (r =0 585).

Franklin (1969) opined that the nutrients acts as limiting factors for productivity. Sabu and Azis (1995) also emphasised on strong positive relationship between phosphorus and productivity. Das (2002) reported a direct correlation as a limiting factor for primary production in Idukki reservoir. In the present investigation nitrates and phosphates showed a strong positive relationship (r = 0.475 and 0.512) with gross primary production. Banerjee and Ray (1979) observed sharp level of fluctuation and abrupt drawdown of water have a destabilising effect on phytoplankton community and primary production. However, in the present study it was inferred that the low level of water fluctuation *i.e.,* inflow 586.44 MM^3 and outflow 440.32 MM^3 (average for past five years) together with physico-chemical profile made the environment ideal for the development of aquatic macrophytes. Hence, it is inferred that the production process is deviated from phytoplankton to aquatic weeds thereby reducing the primary productivity. A comparative account on primary production status in some Indian reservoirs is presented in Table 13.2

Table 13.2: Primary Production Status in Some Indian Reservoirs

Sl.No.	*Reservoir*	*Source*	*Unit*	*Primary Production*
1.	Amaravathy	Sreenivasan (1965 a)	$gO_2m^2d^{-1}$	0.38–18.3 (GPP)
2.	Aliyar	Sreenivasan (1965 a)	$gO_2m^2d^{-1}$	0.278–0.699 (GPP)
3.	Sandynulla	Sreenivasan (1965 a)	$gO_2m^2d^{-1}$	6.1 (GPP)
4.	Tungabhadra	Ray and Banerjee (1979)	$mgCm^3hr^{-1}$	16.2–189.3 (GPP)
5.	Nagarjuna Sagar	Sugunan and Pathak (1986)	$mgCm^3d^{-1}$	90.1–205 (GPP)
6.	Yerrakalava	Das (2002)	$mgCm^3hr^{-1}$	41.66–104.16 (NPP)
7.	Wyra	Das (2002)	$mgCm^3d^{-1}$	875–2500 (GPP)
8.	Bhavani Sagar	Anon., (1981)	$mgCm^3d^{-1}$	830.7 (GPP)
9.	Dhom	Trivedy (1993)	$mgCm^3d^{-1}$	20.83–145.80 (GPP)
10.	Manasa Sarovar	Adolia (1991)	$mgCm^3d^{-1}$	542–6000 (GPP)
11.	Kolar	Sugunan (1995)	$mgCm^3d^{-1}$	441–490 (GPP)
12.	Gandhi Sagar	Rao and Choube (1990)	$mgCm^3hr^{-1}$	26.75–40.83 (GPP)
13.	Hirakud	Das *et al.* (1993)	$mgCm^3hr^{-1}$	900–2250 (GPP) 765–2025 (NPP)
14.	Rihand	Singh and Desai (1980)	$mgCm^3d^{-1}$	387.04 (GPP) 161.44 (NPP)
15.	Vanivilas Sagar	Present study	$mgCm^3d^{-1}$	0.18–43.42(GPP) 0.10–22.69 (NPP)

Acknowledgements

One of the authors (SF) is thankful to UGC for providing teacher fellowship under 9th plan period. Thanks are due to Kuvempu University for providing facilities.

References

Banerjee, R.K. and P. Ray (1979). Soil and water quality of Tungabhadra reservoir as indices of biological productivity. Lecturer delivered at the Summer Institute on Culture and Capture Fisheries of Man-made Lakes in India. Central Inland Fisheries Research Institute, Barrackpore, India, July–August 1979, pp. 46–53.

Cole, G.A. (1979). *Text Book of Limnology*, 2nd ed. The C.V. Mosby Co., USA.

Das, A.K. (2002). Phytoplankton Primary Production in some Selected Reservoirs of AP. *Geobios*, 29: 52–57.

Dash, M.C., P.C. Mishra, G.K. Kar and R.C. Das (1983). Hydrobiology of Hirakuddam reservoir. In: *Ecology and Pollution of Inland Lakes and Reservoirs*, (Eds.) P.C. Mishra and R.K. Trivedy. Ashish Publishing House, New Delhi, p. 317–338.

David, A., P. Ray, S.V. Govind and Rajagopal (1968). Limnology and fisheries of the Tungabbadra reservoir. Central Inland Fisheries Research Institute, Barrackpore, November 1969, 13: 188.

Devaraj, K.V., H.S. Mahadeva and A.A. Fazal (1988). Hydrobiology of Haemavathy reservoir. *Proceedings Asian Fisheries Society*, Mangalore, p. 323–327.

Dwivedi, R.K., S.J. Karmachandini and H.C. Joshi (1986). Limnology and productivity of Kulgarhi Reservoir, M.P. *J. Inland Fish Soc.*, India, 18(2): 65–70.

Gnaf, T. and A.J. Horne (1975). Diurnal stratification, photosynthesis and nitrogen fixation in a shallow equatorial lake. *Freshwater Biology*, 4: 13–39.

Franklin, T. (1969). Phytoplankton of Bhavanisagar. *Madras J. Fish*, 5: 112–118.

Gopinath, (1995). Ecological studies on Hemavathi River Basin. *Ph.D. Thesis*, University of Mysore.

Khatri, T.C. (1985). A note on the limnological characters of the Idukki reservoir. *Indian J. Fish*, 32(2): 267–269.

Likens, G.E. (1975). Primary production of inland aquatic ecosystems. In: *The Primary Productivity of the Biospheres*, (Eds.) H. Lieth and R.H. Whittakar. Springer-Verlag, New York.

Niranjan (1994). Ecology of Suvarnamukhi and Chikkahole reservoirs. *Ph.D. Thesis*, University of Mysore.

Odum, H.T. (1956). Primary production in flowing waters. *Limnol. Oceanogr.*, 1: 102–117.

Pathak, V. (1979). Evaluation of productivity in Nagarjunasagar reservoir (Andhra Pradesh) as a function of hydrological and limnochemical parameters. *J. Inland Fish Soc.*, India, 11(2): 49–68.

Patil, S.K. (1998). Studies on Physico-chemical aspects of Tunga River. *Ph.D. Thesis*, Kuvempu University.

Ramakrishniah, M. and S.K. Sarkar (1982). Plankton productivity in relation to certain hydrological factors in Konar reservoir (Bihar). *J. Inland Fish Soc.*, India, 14(1): 58–68.

Rawson, D.S. (1952). Mean depth and fish production in large lakes. *Ecology*, 33: 513.

Rawson, D.S. (1953). Morphometry as a dominant factor in the productivity of large lakes. *Verh. Int. Verenin. Theor. Angew. Limnol.*, 12: 164–175.

Sabu Thomas and P.K. Abdul Azis (1995). A study on the primary production in peppara reservoir. *Proc. of Seventh Kerala Science Congress*, Palakkad, p. 76–77.

Sreenivasan, A. (1965a). Limnology of tropical impoundments III Limnology and productivity of Amaravathy reservoir. *Hydrobiologia*, 26(3): 501–516.

Sreenivasan, A. (1969). Limnology of tropical impoundments: A comparative study of the major reservoirs of Tamil Nadu. In: *Seminar on Ecology and Fisheries of Freshwater Reservoirs*. ICAR, November 27 to 29, 1969, CIFRI, Barrackpore.

Srivastava, N.P., M. Ramakrishniah and A.K. Das (2000). Ecology and Fisheries of Taws Reservoir. Bull. No. 100. CIFRI. Barrackpore.

Syed Fasihuddin (2002). Ecological studies on Vanivilas Sager reservoir, Chitradurga district. *Ph.D. Thesis*, Kuvempu University.

Trivedy, R.K., P.K. Goel and C.L. Trisal (1987). *Practical Methods in Ecology and Environmental Science*. Environmental Publications, Karad.

Vollenweider, R.A. (1969). *A Manual on Methods for Measuring Primary Productivity in Aquatic Environment*. IBP Hand Book No. 12 Blackwell Scientific Pub., UK.

Chapter 14

Study of Algal Communities of Sonvad Dam of Dhule as Indicators of Organic Water Pollution

S.N. Nandan & D.S. Jain***

**S.S. V.P.S's L.K. Dr. P.R. Ghogrey Science College, Dhule – 424 005 (M.S.)*

***Gangamai Education Trusts Arts, Commerce and Science College, Nagaon, Dhule – 424 004 (M.S.)*

E-mail: jaindevendras@rediffmail.com

ABSTRACT

Algae are involved in water pollution in number of significant ways. There are few studies on algal exploitation as to biological indicators of water quality and pollution. As a consequence of rapidly expanding industrialization and excessive pollution growth most drivers lakes, streams, dams and rivers and other water bodies are being increasingly polluted. It is now well known that the most important use of algae can be correlated with water pollution studies. In Maharashtra State very few workers had paid attention on ecology of algae with relation to pollution.

It was thought worth to study algal communities which are being used as indicators of organic pollution in Sonvad dam of Dhule district (M.S.). In present study, twenty genera like *Navicula, Nitzschia, Synedra, Gomphonema, Cyclotella, Oscillatoria, Phormidium, Microcystis, Lyngbya, Chlorella, Scenedesmus, Ankistrodesmus, Closterium, Pandorina, Melosira, Spirogyra, Pediastrum, Euglena, Phacus* and *Trachelomonas* were observed. The total score of each station of dam and river was greater than 20 indicating the confirmed high organic pollution. The genera like *Euglena, Oscillatoria, Scenedesmus, Navicula, Nitzschia,* and *Microcystis,* are found in organically polluted were in present study as supported by earlier workers. Thus algae are used as indicators and alleviators of organic water pollution.

Introduction

Rai and Kumar (1977) made observations on the seasonal variation in the algal communities of pond polluted with fertilizer factory effluents. Gunale and Balkrishnan (1981) have used algae as biomonitors of eutrophication in the study of Pavana, Mula and Mutha rivers flowing through Poona City. Jagdale *et al.* (1987) studied pollution of Godavari river at Nanded.

Looking to the literature cited more information is needed with reference to ecology of algae in Maharashtra. Ecological study of algae particularly of lake, ponds and reservoir like dams with relation to pollution in Maharashtra are scanty. It was thought worth to study algal communities which are being used as indicator of pollution in Sonvad dam of Dhule district of Maharashtra (India).

Materials and Methods

The Sonvad dam is situated at Dongargaon of taluka Shindkheda of Dhule district. Two stations of Sonvad dam and one station of Bhat river away from Sonvad dam were selected for the purpose of collecting the algal and water samples at monthly interval.

The chemical analysis of water samples were made for physico-chemical parameters according to methods of APHA (1975). The qualitative study of 4 groups of algae was made according to Whiton (1969). The pollution tolerant genera were recorded for 2 stations of Sonvad dam and one station of Bhat river. The assessment of water quality of 3 stations was made by algal pollution tolerant genera according to Palmer (1969). The identification of algal taxa was made with the help of standard monographs (Hustedt, 1930; Pochmann, 1942; Hubner-Pestalozzi, 1955; Pringsheim, 1956; Desikachary, 1959; Randhawa, 1959; Ramanathan, 1964; Philipose, 1967; Iyengar and Desikachary, 1981; Gonzalves, 1981; and Sarode and Kamat, 1984).

Results and Discussion

The pollution tolerant genera of algae from 3 stations are listed in descending order according to Palmer (1969). In present study thirty one pollution tolerant genera of algae were observed (Table 14.1). In present study twenty genera like *Navicula, Nitzschia, Synedra, Gomphonema, Cyclotella, Oscillatoria, Phormidium, Microcystis, Lyngbya, Chlorella, Scenedesmus, Ankistrodesmus, Closterium, Pandorina, Melosira, Spirogyra, Pediastrum, Euglena, Phacus* and *Trachelomonas* were considered for assessing water quality of study area (Plates 14.1 and 14.2). By using Palmer's index of pollution for rating of water samples as high or low organically polluted at 3 stations was used for assessing water quality.

The results are as shown in Table 14.1. Out of 20 genera, 12, 14 and 17 genera were observed at Stations 1, 2 and 3 respectively (Table 14.2). The total score of each station was greater than 20 indicating the confirmed high organic pollution (Plate 14.3). The degree of organic pollution was greater at station SB-III as compared to other two stations.

Palmer (1969) has shown that genera like *Scenedesmus, Oscillatoria, Microcystis, Navicula* and *Euglena* are found in organically polluted waters as supported by Goel *et al,* (1986). Similar genera were recorded in present investigation.

Table 14.1: Pollution Tolerant Genera of Algae from 3 Stations of Sonvad Dam in Order of Decreasing Emphasis (Palmer, 1969)

Sl.No.	*Genus*	*Group**	*Total Points*	*Stations*		
				SD–I	*SD–II*	*SD–III*
1.	*Euglena*	F	172	–	+	+
2.	*Oscillatoria*	B	161	+	+	+
3.	*Scenedesmus*	G	112	+	+	+
4.	*Chlorella*	G	103	–	+	–
5.	*Nitzschia*	D	98	+	+	+
6.	*Navicula*	D	92	+	+	+
7.	*Synedra*	D	58	–	+	+
8.	*Ankistrodesmus*	G	57	+	+	–
9.	*Phacus*	F	57	–	+	+
10.	*Phormidium*	B	52	+	+	+
11.	*Gomphonema*	D	48	+	+	+
12.	*Cyclotella*	D	47	–	+	–
13.	*Closterium*	G	45	+	+	+
14.	*Pandorina*	F	42	+	–	–
15.	*Microcystis*	B	39	+	+	+
16.	*Spirogyra*	G	37	+	+	+
17.	*Anabena*	B	36	–	+	–
18	*Pediastriun*	G	35	+	+	+
19.	*Trachelomonas*	F	34	–	–	+
20.	*Fragilaria*	D	33	+	+	+
21.	*Ulothrix*	G	33	–	–	+
22.	*Surirella*	D	33	+	–	–
23.	*Lyngbya*	B	28	–	–	+
24.	*Spirulina*	B	25	+	+	+
25.	*Cymbella*	D	24	–	+	+
26.	*Coelastrium*	G	24	–	+	+
27.	*Cladophora*	G	24	–	–	+
28.	*Melosira*	G	22	+	+	+
29.	*Pinnularia*	D	18	+	+	+
30.	*Cosmarium*	G	17	+	–	–
31.	*Gonium*	G	16	+	–	–

*: F: Flagellates, D: Diatoms, B: Blue greens, G: Greens.

SD–I and SD–II: Stations of Sonvad dam; SB–III: Station of Bhat River.

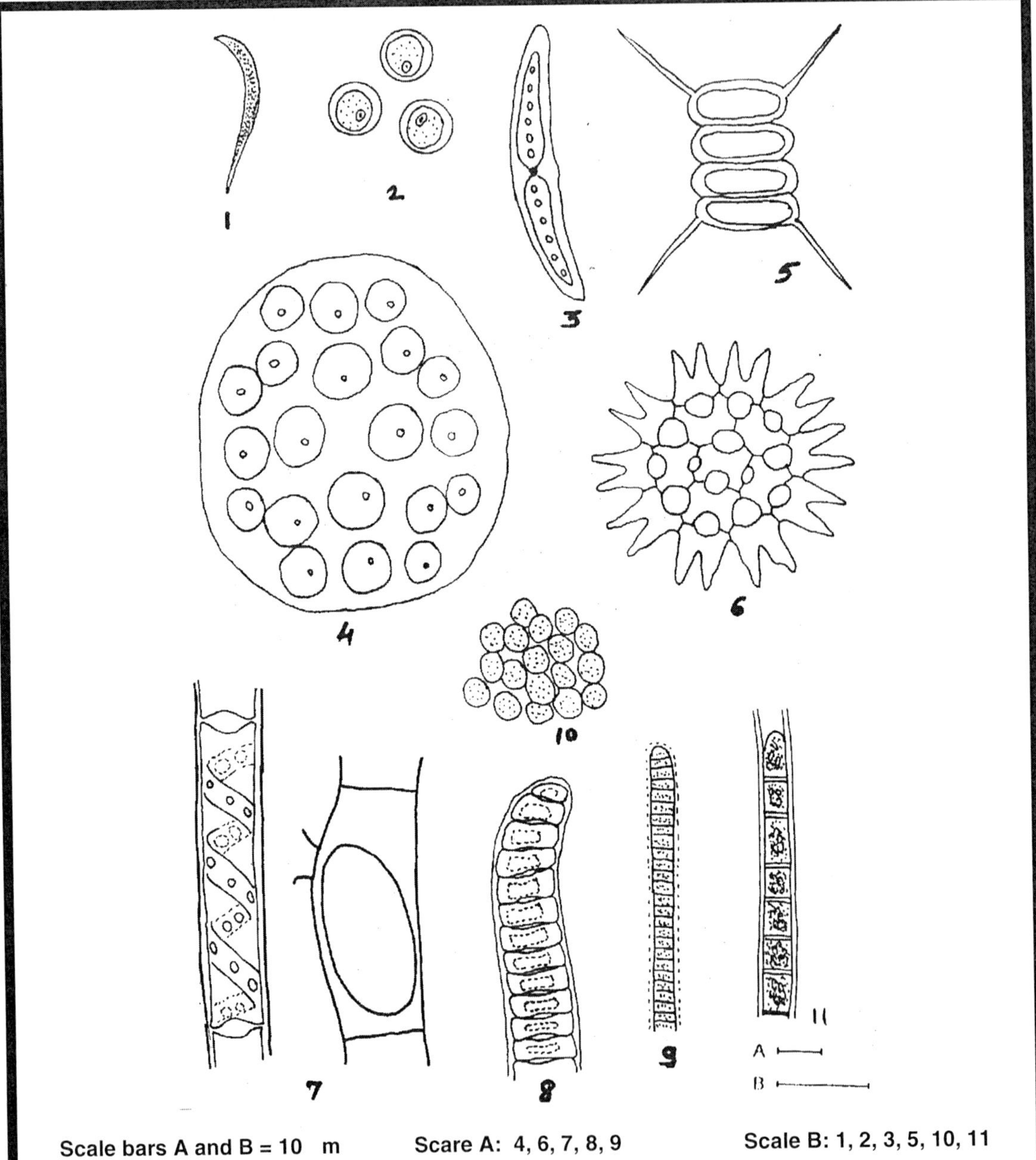

Plate 14.1: Algal Taxa Showing Pollution index at 3 Stations of Sonvad Dam

(1) ***Ankistrodesmus convolutus cord.***; (2) ***Chlorella vulgaris Beijerink***; (3) ***Closterium moniliferum (Bory) Ehr.***; (4) ***Pandorina morum (mull.) Bory.***; (5) ***Scenedesmus quadricauda (Turpin) Brebisson***; (6) ***Pediastrum duplex Mayen V. reticulatum Lagerheim***; (7) ***Spirogyra gracilis (Hassall) Kuetz.***; (8) ***Oscillatoria perornata Skuja***; (9) ***Phormidium ambiguum Gom.***; (10) ***Microcystis flo-aquae (Wittr) Kirchn.***; (11) ***Lyngbya borgerii Lemm.***

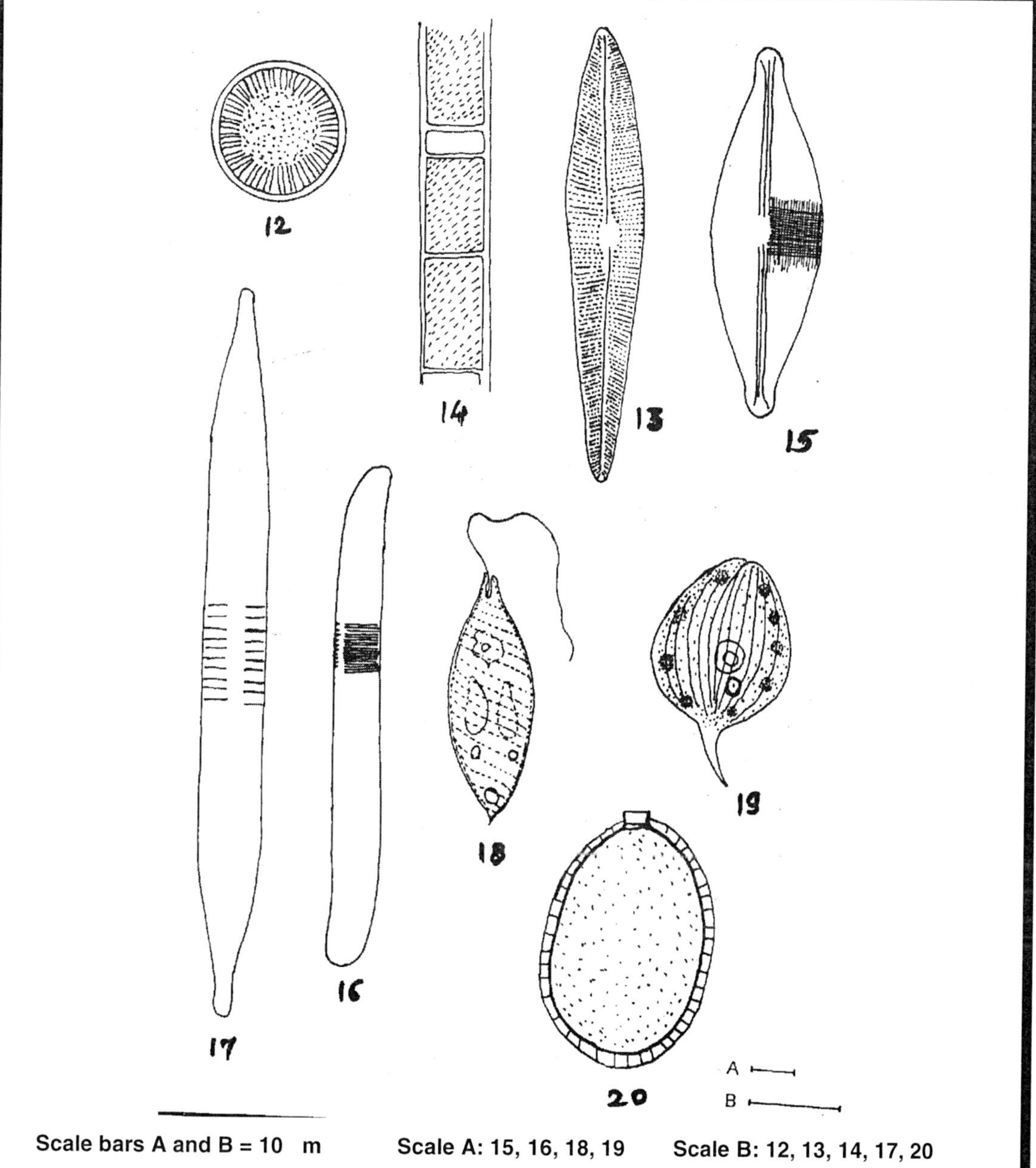

Scale bars A and B = 10 m Scale A: 15, 16, 18, 19 Scale B: 12, 13, 14, 17, 20

Plate 14.2 Algal Taxa Showing Pollution index at 3 Stations of Sonvad Dam

(12) ***Cyclotella catenata Brun.***; (13) ***Gomphonema lanceolatum Ehr.***; (14) ***Melosira granulata (Ehr) Ralfs.***; (15) ***Navicula cuspidata Kutz. Vv ambiqua (Ehr) Cleve.***; (16) ***Nitzschia obtusa W. Smith v. Scalpelliformis Grun.***; (17) ***Synedra ulna (Nitz.) Ehr. F. Staurodestitula Pant.***; (18) ***Euglena clavata Skuja***; (19) ***Phacus orbicularis Hubner.***; (20) ***Trachelomonas lefevrei Defl.***

Table 14.2: Pollution Index of Algal Genera Used for Assessing Water Quality at 3 Stations of Sonvad Dam (Palmer, 1969)

Sl. No.	Group/Genera	Palmers Pollution Index		
		I	II	III
1.	Cyanophyceae			
	1. *Oscillatoria*	4	4	4
	2. *Phormidium*	1	1	1
	3. *Microcystis*	1	1	1
	4. *Lyngbya*	–	–	1
2.	Chlorophyceae			
	5. *Chlorella*	–	–	3
	6. *Scenedesmus*	4	4	4
	7. *Ankistrodesmus*	–	–	2
	8. *Closterium*	1	1	1
	9. *Pandorina*	1	–	–
	10. *Spirogyra*	1	1	1
	11. *Pediastrium*	1	1	1
3.	Bacillariophyceae			
	12. *Navicula*	3	3	3
	13. *Nitzschia*	3	3	3
	14. *Synedra*	–	2	2
	15. *Gomphonema*	1	1	1
	16. *Cyclotella*	–	1	–
	17. *Melosira*	1	–	–
4.	Euglenineae			
	18. *Euglena*	–	5	5
	19. *Phacus*	–	2	2
	20. *Trachelomonas*	–	–	1
	Total Score	**22**	**30**	**36**

The epilethic and epiphytic algae may form excellent indicators of pollution (Round, 1965). In present study, presence of *Oscillatoria, Phormidium* and *Lyngbya* as epilethic algae and certain diatoms like *Gomphonema, Cymbella* and *Navicula* as epiphytic algae were recorded. They are indicators of pollution following the view of Palmer (1969). Similar observations were encountered by Hosmani and Bharati (1980) and More and Nandan (2000).

Acknowledgement

Authors are grateful to Principal, B.M. Patil, for providing the facility for investigation.

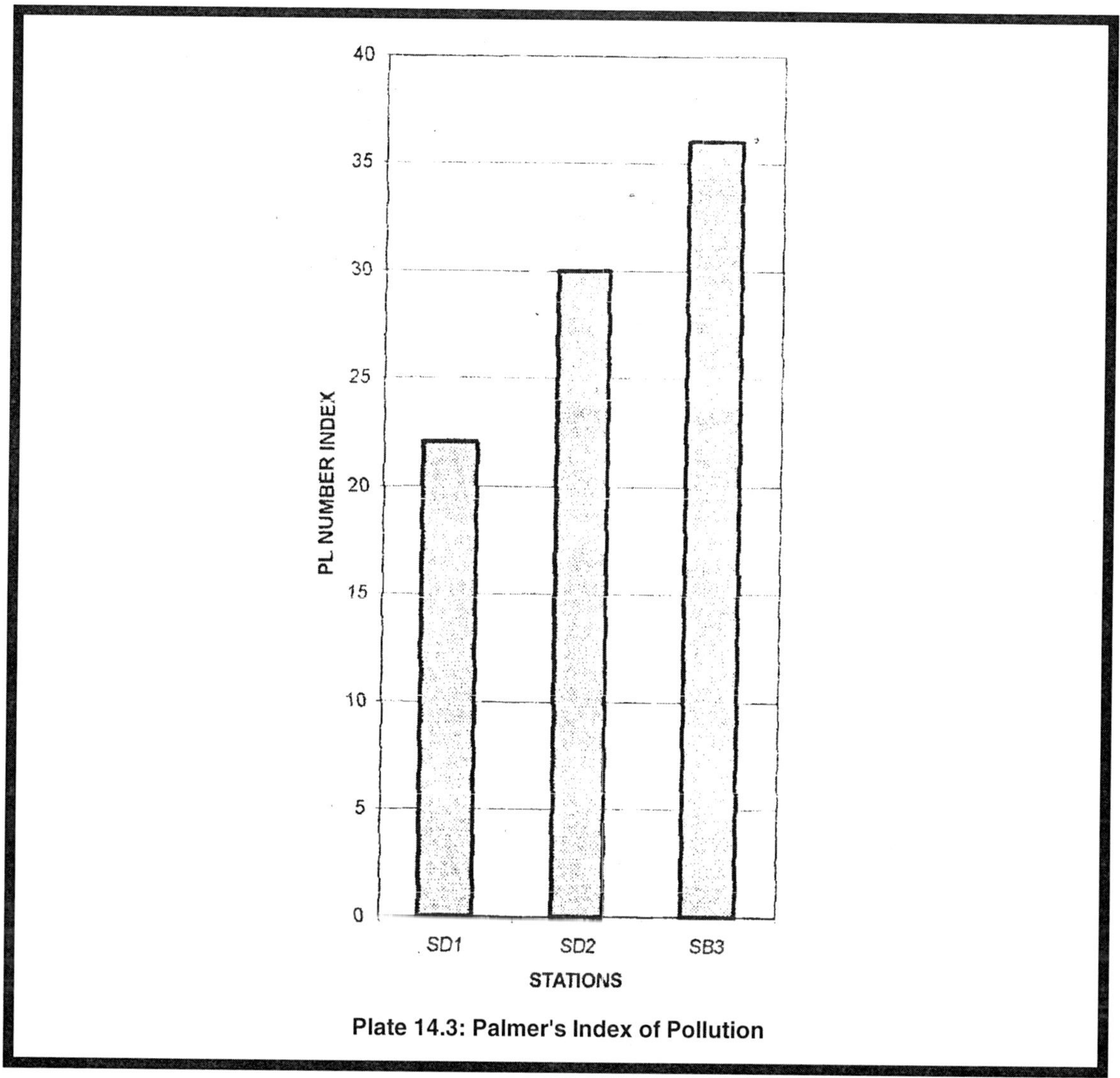

Plate 14.3: Palmer's Index of Pollution

References

APHA (1975). *Standard Methods for the Examination of Water and Wastewater*, 14th Ed. American Public Health Association, New York.

Desikachary, T.V. (1959). *Cyanophyta*. ICAR Monographs on Algae, New Delhi, pp. 686.

Gael, P.K., S.D. Khavakar, A.Y. Kulkarni and R.K. Trivedy (1986). Limnological studies of few fresh water bodies in south western Maharashtra with special reference to their chemistry and phytoplankton. *Pollution Research*, 5(2): 79–84.

Gonzalaves, E.A. (1981). *Oedogoniales*. ICAR Monographs, New Delhi, p. 1–757.

Gunale, V.R., M.S. Balakrishnan (1981). Biomonitoring of eutrophication in the Pavana Mula and Mutha rivers flowing through Poona. *Indian J. Env. Hlth.*, 23(4): 316–322.

Hosmani, S.P., S.G. Bharati (1980). Limnological studies in pond and lakes of Dharwar: Comparative phytoplankton ecology of four water bodies. *Phykos,* 19(1): 27–43.

Hubner-Pestlozzi (1955). Die Binnengewasser Dr. August Thienemann Band. XVI 4Teli.

Hustedt, F. (1930). Bacillariophyta (diatomaceae) In: *A. Pascher's Die Susswasser Flora Mitteleuropas,* 40: 1–466.

Jagdale, M., M.M. Salunke, Y.B. Yabute, S.N. Kodaskar and K.N. Ghuge (1984). Pollution of Godavari river water at Nanded. *Pollution Research,* 3: 83–84.

More, Y.S. and S.N. Nandan (2000). Hydrobiological study of algae of Panzara river (Maharashtra). *Ecology Env. & Cons.,* 1: 99–103.

Palmer, C.M. (1969). A composite rating of algae tolerating organic pollution. *Journal Phycology,* 5: 78–82.

Philipose, M.T. (1967). *Chlorococcales.* ICAR, New Delhi, p. 1–356.

Poachman, A. (1942). Symposis der Gattung. Phacus Arch. Protistenk, 95(2): 81–252.

Pringsheim, E.G. (1956). Contribution towards monograph of the genus Euglena. *Nova. Acta. Leopodina,* 18: 1–168.

Ramanathan, K.R. (1964). *Ulotrichales.* ICAR, New Delhi, p. 1–188.

Randhawa, M.S. (1959). *Zygnemaceae.* ICAR, New Delhi, pp. 478.

Rai, L.C. and H.D. Kumar (1977). Studies on the seasonal variation in the algal communities of pond polluted with fertilizer factory effluent. *Indian J. Ecol.,* 4(2): 124–131.

Round, F.E. (1965). *The Biology of the Algae.* Edward Arnold Publication, London, pp. 269.

Sarode, P.T. and N.D. Kamat (1984). *Freshwater Diatoms of Maharashtra.* Saikripa Prakashan, Aurangabad, pp. 1–338.

Chapter 15

Biological Evaluation of Water Quality Using Phytoplankton on Phutala Lake and Well in Nagpur

Arun K. Pandey & Rakesh K. Pandey***

**R.K.D.F. Homoeopathic Medical College, Bhopal (M.P.)*

***School of Environmental Biology, A.P.S. University, Rewa (M.P.)*

ABSTRACT

Increase in biological oxygen demand, eutrophication and extinction of several species of water animals due to toxic residues are the major problems that have emerged because of non-stop and careless release of industrial effluents into water bodies.

Increased Environmental Awareness and concern regarding severe effects of non-biodegradable toxic chemicals like pesticides, detergents, PAHs, has made concerns to look for 'bugs' which can degrade them.

Biological studies are essential part of water pollution control programmes and are designed to provide information for a variety of water management functions.

The following communities of aquatic organisms are common in aquatic ecosystem: plankton, periphyton, macrophyton, macro invertebrates, fish amphibians, aquatic reptiles, water birds, and mammals.

Keywords: *Increase in biological oxygen demand (IBOD), Increase environmental awareness (IEA), Biological studies (BS).*

Introduction

Phytoplankton is present in all types of standing water bodies as well as in the middle and lower reaches of rivers. It is responsible for the primary production of the biomass in the freshwater zones and for the biogenic oxygenation of the water during the day time.

Phytoplankton consists by chlorophyll-bearing organisms. These are blue green algae, green algae, other algae and coloured flagellates. Algae float in the water, coloured flagellates swim and show migrations. Occurrence of phytoplankton depends on light and shows a stratification in the vertical zone of water column. A horizontal zoning is apparent especially in long artificial reservoirs.

Analysis of Phytoplankton

Materials/Equipments

1. Compound microscope with calibrated eyepiece micrometer
2. Micro slides and cover slips
3. Micropipet
4. Beakers
5. Clinical centrifuge with graduated centrifuge tube
6. Measuring cylinder.

Methodology

1. 15 ml of the sample is centrifuged for 20 minutes at 1,000 g. The settled material is used for the examination of phytoplankton.
2. Sample is analysed qualitatively to determine the species composition as well as quantitatively to determine the number of organisms per unit volume.
3. Qualitative analysis should be done by observing and identifying the organisms in number of fields under the microscope and listing the species.
4. Quantitative estimation can be done by different methods of counting outlined below.

Calculation

Calculate the number of organisms per ml as follows:

$$No/ml = \frac{C \quad A_1}{A_2 \quad S \quad V}$$

where,

C = No. of organisms counted

A_1 = Area of cover slip, mm^2

A_2 = Area of one strip, mm^2

S = No. of strips counted, and

V = Volume of sample under the cover slip, ml.

Algae Important in Water Supplies

1. *Hydrodictyon*
2. *Peridinum*
3. *Dinobryon*
4. *Volvox*
5. *Aphanizomenon*
6. *Gomphosphaeria*
7. *Anacystis*
8. *Anabaena*
9. *Synedra*
10. *Ceratium*
11. *Nitella*
12. *Tabellaria*
13. *Pandorin*

Polluted Water Algae

1. *Anacystis*
2. *Chlorella*
3. *Oscillatoria*
4. *Phacus*
5. *Euglera*
6. *Gomphonema*
7. *Arhorospira*
8. *Anabaena*
9. *Lyngbya*
10. *Chlamydomonas*
11. *Cartenia*
12. *Stigeoclonium*
13. *Phormidum*
14. *Chlorognium*
15. *Nitzschia*
16. *Tetraedron*
17. *Spirogyra*

Table 15.1: Occurrence of Phytoplankton at Different Station on Phutala Lake and a Well in Nagpur

Sl. No.	*Family/Species*	*Sampling Station Phutala–I*	*Sampling Station Phutala–II*	*Sampling Station In Well–III*
1.	**Cynophyceae**			
	Chroococcus	P	P	P
	Microcystis	P	P	P
	Oscillatoria	–	–	–
	Anabaena	P	–	–
	Phormidium	–	P	–
2.	**Chlorophyceae**			
	Chlorococcum	P	P	P
	Chlorella	P	P	P
	Chroomonus	P	P	P
	Closterium	P	P	P
	Cosmarium	–	–	–
	Scenedesmus	P	P	–
	Selenastrum	P	–	–
	Golenkimia	P	P	–
3.	**Pyrrhophyceae**			
	Perdinum	P	P	P
4.	**Euglenophyceae**			
	Euglena	P	P	–
5.	**Bacillariophyceae**			
	Navicula	P	–	–
	Total No. of Species	**13**	**11**	**7**

Table 15.2: Community Structure of Phytoplankton Population in Water Samples of Phutala Lake and Well in NEERI Campus at Nagpur

Station No.	*Total No. of Phyto-plankton in per ml. (Lackey Method)*	*Percentage Composition of Different Groups*					*Palmer's Pollution Index*
		Cynophy-ceae	*Chlorophy-ceae*	*Pyrrhpy-ceae*	*Euglenophy-ceae*	*Bacilla-riophyceae*	
I	2078	34.1	58.1	3	3.4	1.4	12
II	1956	40.9	51.2	4.7	3.2	–	14
Well III	1109	54.2%	43%	2.8%	–	–	4

Results and Discussion

Samples of phytoplankton were collected from Phutala lake and a dugwell. Phutala lake situated near Bharat Nagar area adjacent to P.K.V. agricultural fields in the heart of Western Nagpur for water quality assessment. A dug well situated in side NEERI campus and pollution with respect to water quality assessment and pollution status of dug well water.

From sampling station No. 1, total phytoplankton count was observed to be 2078 algae per ml of water sample. A total of 13 algal species were recorded, out of which 3 species belonged to cyanophceae, 7 species to chlorophyceae, and one species each to pyrrhophyceae, Euglenophyceae and Bacillariophyceae. The Palmer's index value was observed to be 12 (Tables 13.1 and 13.2). In case of sampling station No. 2, total phytoplankton count was observed to be 1956 algae per ml of water sample, a total of 11 algal species were recorded, out of which 3 species belong to Cynophyceae, 6 species to Chlorophyceae, and one species each to pyrrhophyceae and Euglenophyceae, the Palmer's pollution index value was observed to be 14.

It has been observed that in the dug well water three Phytoplankton groups were present. Total phytoplankton count was observed to be 1109 algae per ml or water sample. A total of 9 algal species were observed out of which 3 species belonged to Cynophyceae, 5 species of Cholorophyceae and 1 species of Pyrrhophyceae. The Palmer's pollution index value was observed to be 4.

References

Bala-Trepat, J.M., P.A. Williams and W.C. Evans (1971). *J. Bacteriol.*, 117: 1153–1157.

Barker, J.E. and R.J. Thompson (1973). *Environmental Protection Technology Series* No. EPA–R2–73–167 April, pp. 29.

Cooper, R.L. (1977). *Effluent and Water Treatment*, 17: 230–237.

Curds, C.R. (1971). *Water Research*, 5: 1049–1066.

Chapter 16

Diversity of Zooplankton in Rakasakoppa Reservoir of Belgaum, North Karnataka

B.N. Sunkad

Department of Zoology, R.L. Sc. Institute, Belgaum – 590 001

ABSTRACT

The present investigation was undertaken to study the biodiversity of zooplankton in Rakasakoppa reservoir of Belgaum (North Karnatak). A total 27 species belonging to 4 groups have been identified. Rotifera (15 species), cladocera (7 species), copepoda (4 species) and ostracoda (1 species) contributed to zooplankton richness in the reservoir. Statistical analysis of the data revealed that zooplankton groups were positively/negetively correlated with some physical, chemical and biological parameters.

Keywords: *Zooplankton diversity, Seasonal fluctuation, Reservoir.*

Introduction

Zooplankton constitutes a vital link in the food chain of an ecosystem and fish yield is to a great extent depend on their abundance. The abundance of zooplankton is more or less governed by the interaction of number of physical, chemical and biological processes. Fresh water bodies have gained much importance in recent years mainly because of their multiple uses. There are numerous natural and artificial ponds, reservoirs and lakes in the Indian subcontinent (Rao, 1975). Several workers have attempted to study the hydrobiological profile of varied water bodies (Singh, 2000; Kaushik and Sharmla, 1994; Patil and Karikal, 2001; Sunkad and Patil, 2003) and diversity of the organisms. The need of water is increasing day by day invariably due to increase in population, urbanization etc. Simultaneously, the water quality of standing water becoming more and more unfit to mankind due to interference. Thus, water quality assessment can be done either monitoring the physico-chemical properties of water or by analysing inhabiting biota.

In the present study much emphasis was given to investigate the zooplankton diversity of the peremial water reservoir of Belgaum in Karnataka state. The results were obtained are discussed in light of available literature with comments on recorded ecological correlations.

Materials and Methods

The observations, apart of general limnological survey, study of biodiversity was undertaken in the Rakasakoppa reservoir located 74 22′ East longitude and 15 49′ North latitude in Belgaum district. It is a major source for supply of drinking water to the Belgaum city. Reservoir has been constructed across the Markandeya river and the catchment area is about 20.8 sq. kms with a maximum depth 75 feet.

The water samples were collected in a clean polyethelene containers fortnightly from February 1998 to January 1999. Water analysis was done as per the procedures of standard methods (APHA, 1991). The samples were collected by filtering known volume of sample through the bolting nylon cloth and were preserved in 4 per cent formalin. The zooplankton species were identified following Battish (1992) and their numbers were enumerated by using "Sedgewick rafter cell" expressed in number of organisms per litre.

Results and Discussion

Abiotic factors of the sampled reservoir (Table 16.1) exhibited range of water temperature was from 20.9–34.4 C. pH was slightly alkaline throughout the study period and the values are ranged between 7.02–8.12. DO (6.68–8.9 mg/l). BOD ranged between 0.43–0.96 mg/l. Chloride (11.87–25.59 mg/l), Total hardness (2.2–9.4 mg/l) and total alkalinity was 24.8–34.2 mg/l. It is also indicated that concentrations are well below the permissible limits.

Zooplankton species survive under a wide range of environmental conditions, their growth and intensity depend on a number of physical chemical and biological factors. Monthly fluctuations of zooplankton population recorded for 12 months (February 98 to January 99) are presented in Table 16.2. Zooplankton community in the reservoir represented by four groups *viz.* Rotifera, cladocera, copepoda and ostracoda.

Total 27 species were documented, out of which 15 species belongs to rotifera, 7 species to cladocera,4 species to copepoda and 1 species to ostracoda (Table 16.3).

Rotifera forms the largest group accounted 35.46 per cent of the total zooplankton population (Table 16.2). This group depicted higher qualitative diversity (15 species) than other groups. Seasonal abundance showed maximum mean number in summer (335.25 per cent org/l) Brachionus formed the dominant genus among the rotifera group. Genus Brachionus represented by 6 species such as *Brachionus calyciflorus, B. forficula, B. angularis, B. caudatus, B. quadridentatus* and *B. falcatus. Brachionus caudatus* fluctuated between 23 org/l (July) to 96 org/l (April). *B. forficula* fairly abundant. *B. angularis* found throughout the study period except June and August. *B.quadridentatus* fluctuated between 20 org/l (January) to 4 and org/l (April). *B. falcatus* ranged from 6 to 26 org/l. *B. calyciflorus* was fairly abundant throughout the year and varied between 7 to 53 org/l.

Table 16.1: Monthly Variations in Physico-chemical Factors in Rakasakoppa Reservoir

Parameters	*Feb.–98*	*March*	*April*	*May*	*June*	*July*	*August*	*Sept.*	*Oct.*	*Nov.*	*Dec.*	*Jan.*
W. Temp	29.1	31.2	34.3	34.4	28.6	23	23.6	23.9	23.8	23.1	20.9	22
A. Temp	33	35.3	39.5	36.8	31.9	26.8	27.9	28.6	28.8	28.6	28.1	29
pH	8.12	7.56	7.64	7.8	7.26	7.14	7.22	7.02	7.18	7.96	7.96	8.1
	± 0.09	0.37	0.32	0.18	0.11	0.18	0.14	0.1	0.28	0.23	0.26	0.24
DO	8.9	7.8	7.4	8.2	7.3	7.59	6.8	6.68	7.74	8.4	8.9	8.8
	±0.21	0.35	0.29	0.14	0.12	0.17	0.12	0.33	0.15	0.1	0.15	0.29
BOD	0.58	0.44	0.64	0.72	0.76	0.96	0.46	0.48	0.43	0.67	0.44	0.45
	±0.24	0.05	0.26	0.13	0.05	0.26	0.05	0.1	0.08	0.18	0.08	0.13
Na	3	4.2	6.6	10.5	11.8	10.5	6.8	4.8	7	3.9	3.5	3.2
	±0.5	0.44	1.94	1.45	0.75	0.5	0.83	0.44	1	0.74	0.35	0.44
K	0	0	0	2.5	1.7	1	0.4	0	0.6	0.4	0	0
	±0	0	0	0.39	0.29	0.7	0.54	0	0.54	0.54	0	0
CO_2	4.79	3.79	5.19	4.99	3.99	5.39	3.19	2.99	2.39	4.19	6.39	2.19
	±0.44	0.83	0.83	0.2	0.2	0.54	0.44	0.11	0.89	0.44	0.89	0.4
Cl	22.68	18.88	20.99	22.49	21.99	25.59	20.98	21.79	20.99	20.38	20.47	11.87
	±1.60	1.24	0.7	0.61	1.45	1.55	0.78	1.48	1.41	0.82	0.35	0.81
TH	6.4	8.4	8	6.8	6.2	9.4	7.6	7	6	4.8	2.2	5.2
	±0.84	0.54	0.37	1.78	1.64	1.51	0.54	0.7	0	0.7	0.44	1.3
TA	29.6	32	31.2	28.4	24.8	30	28.4	29	25.2	31.2	34.2	33.8
	±1.67	3.16	2.79	0.5	1.02	2.89	2.25	1.1	0.96	2.33	1.54	0.46
PO_4	2.63	3.7	4.3	5.35	4.85	2.9	3.5	3	2.96	2.7	2.68	2.6
	±0.48	0.57	1.15	1.54	0.94	0.74	0.5	0.15	0.15	0.73	0.35	0.15

All parameters expressed in mg/l except W. Temp., A. Temp. and pH.

±: Standard deviation.

DO: Dissolved oxygen, BOD: Biochemical oxygen demand, Na: Sodium, K: Potassium, Cl: Chloride, TH: Total hardness, TA: Total alkalinity, PO_4: Phosphate.

Table 16.2: Month-wise Distribution of Zooplankton Groups (Nos./l)

Months	*Rotifer*	*Cladocera*	*Copepod*	*Ostracoda*
February 98	396 ± 31.68	86 ± 31.94	175 ± 20.61	8 ± 31.62
March	284 ± 21.64	78 ± 18.43	168 ± 21.72	0 ± 0
April	382 ± 36.22	0 ± 0	137 ± 22.29	0 ± 0
May	280 ± 28.48	0 ± 0	168 ± 21.95	0 ± 0
June	216 ± 21.58	312 ± 30.06	158 ± 13.92	0 ± 0
July	174 ± 27.12	442 ± 48.22	162 ± 26.19	0 ± 0
August	164 ± 22.98	392 ± 22.22	156 ± 16.06	70 ± 20.26
September	191 ± 29.96	428 ± 34.23	191 ± 14.59	92 ± 31.01
October	210 ± 41.23	320 ± 29.08	238 ± 16.79	84 ± 15.68
November	205 ± 30.66	280 ± 16.95	296 ± 31.11	56 ± 25.35
December	288 ± 21.54	243 ± 21.0	298 ± 41.37	70 ± 33.97
January	281 ± 28.90	213 ± 24.22	268 ± 29.69	0 ± 0

Table 16.3: Distribution of Various Species of Zooplankton (Numbers/l)

Zooplankton	*Feb.–98*	*March*	*April*	*May*	*June*	*July*	*August*	*Sept.*	*Oct.*	*Nov.*	*Dec.*	*Jan.*
Rotifera												
Brachionus forficula	42	38	28	34	24	23	26	0	0	28	15	19
B. angularis	34	21	36	27	0	0	0	17	30	22	26	22
B. caudatus	82	69	96	48	60	23	50	90	0	0	0	36
B. calyciflorus	20	10	7	13	9	10	20	32	53	23	38	32
B. falcatus	16	15	22	24	08	0	0	26	22	0	06	08
B. quadra-dentatus	36	38	48	26	23	0	0	0	0	0	34	20
Keratella tropica	28	23	44	26	18	23	0	0	0	0	0	0
K. cochlaris	32	26	28	31	26	12	0	0	0	0	0	0
Cephalodella gibba	12	04	0	0	0	00	0	0	24	16	19	21
Trichocera cylinderica	20	09	07	0	0	0	0	0	0	34	36	20
Asplanchna priodenta	38	19	0	0	0	0	0	0	43	32	58	32
Filinia longiseta	36	12	34	23	22	42	0	0	0	0	0	48
Polyarthura vulgaris	0	0	0	0	0	16	27	26	38	50	56	23
Ascomorpha sps.	0	0	0	14	18	11	16	0	0	0	0	0
Lacane monostyla	0	0	32	14	08	14	25	0	0	0	0	0
Cladocera												
Daphnia carinata	44	30	0	0	186	209	144	192	162	98	71	47
Alona rectangula	10	17	0	0	0	31	42	38	28	27	0	0
Moina macro-copa	0	0	0	0	40	48	68	23	26	42	56	44
Bosminopsis dietersi	20	18	0	0	0	80	56	104	84	69	58	54
Ceriodaphnia cornuta	12	13	0	0	60	46	54	54	20	44	0	0
Macrothrix laticornis	0	0	0	0	26	28	10	17	0	0	0	0
Streblocercus sps.	0	0	0	0	0	0	18	0	0	0	0	0
Copepoda												
Mesocyclops leuckarti	15	20	25	0	0	0	0	0	0	45	44	20
Tropocyclops prasinus	25	13	20	42	28	20	0	0	0	45	38	48
Rhinodiaptomus indicus	60	45	35	26	54	62	46	67	96	96	94	88

Keratella formed the subdominant genus consists two species *K. trophica* and *K. cochlaris*. They formed 5.27 and 5.04 per cent of total rotifer group. They found abundantly in summer and in rest of the season absent or some time appeared negligible numbers. *Asplanchna* sps. formed 7.2 per cent of total rotifer group. These were totally absent in winter season. *Polyarthura* sps. were absent in monsoon and appeared in June, formed peak in December. Polyartura abundant during winter indicating its preference for clear water (Pandey *et al.*, 1994). *Trichocera cylinderica, Cephalodella* sps., *Filinia* sps., *Ascomorpha* sps. and *Lacane lunaris* (monostyla) were found less or negligible numbers. *Lacane* and *Keratella* sps. appear to be associated with semi-polluted and polluted waters (Singh and Pandey, 1990).

Rotifers have versatile capacity to survive in different environments, generally they are abundant in summer months indicating direct relationship with high temperature (Singh, 2000). Similar observations were found in the present study also. Data on the zooplankton group was subjected to the statistical analysis and depicted in Table 16.4. Rotifers were correlated with A Temp. Maximum number of rotifers found in summer indicating the influence of temperature. This observation is an concurrence with the work of Singh (2000).

Table 16.4: Simple Correlation Coefficient Tests Between Different Groups of Zooplankton and Other Parameters

Parameters	*'r' Value*
Rotifera v/s pH	0.6774*
Rotifera v/s W. Temp	0.5722*
Cladocera v/s W. Temp	– 0.7927**
Cladocera v/s pH	– 0.6343*
Copepoda v/s Ca	– 0.7714**
Copepoda v/s DO	0.6059*
Copepoda v/s PO_4	– 0.6072*
Copepoda v/s T. Hardness	– 0.8370
Copepoda v/s Mg	– 0.7687**

Cladocers are important components of food web and an integral link in aquatic food chain in fresh waters (Battish and Kumari, 1986). Cladocerans constituted the second largest group of zooplankton documented 7 species namely *Daphnia carinata, Alona rectangular, Moina macrocopa, Bosminopsis dietersi, Ceriodaphnia cornuta, Macrithrix* sps. and *Streblocerus* sps.

Daphnia carinata present throughout the study period except April and May. It fluctuated between 10 to 209 org/l. *Alona rectangula* found in February and March with small density, disappeared from April to June and again reappeared in July reached maximum numbers in September.

Bosminopsis dietersi was fluctuated between 18 to 104 org/l. It was observed throughout the study period except April to June. *Moina macrocopa* ranged between 23 to 68 org/l. Other species of cladocera are *Macrothrix* and *Streblocerus* found in negligible numbers. Seasonally cladocerans were found in rainy season followed by winter. During summer the low density could be due to more dense growth of rotifers thus avoiding competition and negatively correlated with rotifer population. Cladocerans were correlated with ostracoda (r = 0.5704) significant at 5 per cent level.

Copepoda, the third subdominant group of zooplankton represented by *Mesocyclops leuckarti, Tropocyclops prasinus, Rhinodiaptomous indicus* and Naupilous larva. This group was recorded throughout the study period and found abundantly in winter season.

Mesocyclops leuckarti appeared from February to April and disappeared from May to October again made its appearance in November. Similar observation was made by Kaushik and Sharma (1994). *T. prasinus* ranged from 13 org/l to 48 org/l. *R. indicus* was present throughout the study period. It fluctuated between 26 org/l to 96 org/l. Nauplius larva found throughout the study period and exhibited bimodal pattern.

Ostracoda the smallest group of zooplankton consist only *cypris* sps. It appeared in February and disappeared from March to July, again appeared in August forms a peak in September (92 org/l). Thereafter decline in the population density. Statistically ostracoda positively correlated with cladocera ($r => 0.5$) and with water temperature ($r => 0.5$) significant at 5 per cent level.

Overall results emerged from the analysis of Rakasakoppa reservoir reveals that the reservoir appears to undergo the state of eutrophication, if proper steps not taken. In this context, there is an immediate think about keeping the aquatic habitat not only clean but also hygienic. Therefore interference of human beings should be limited that to only for drinking purpose and regular monitoring is essential

Acknowledgement

The author B.N. Sunkad is grateful to UGC, New Delhi for awarding Teacher Fellowship under faculty Improvement Programme.

References

APHA (1991). *Standard Methods for the Examination of Water and Wastewater*, 18th Edition, New York, U.S.A.

Battish, S.K. (1992). *Freshwater Zooplankton of India.* Oxford and IBH Publ. Co. Pvt. Ltd., London, U.K.

Battish, S.K. and Paravinder Kumari (1986). Effect of physico-chemical factors on the seasonal abundance of cladocera in typical pond at village of Raqba, Ludhiana. *Indian. Ecol.,* 13(1): 146–151.

Kaushik, K.S. and D.N. Sharma (1994). Trophic status and rotifer fauna of certain water bodies in central India. *J. Env. Biol.,* 16(4): 283–291.

Pandey, B.N., A.K. Jha, P.K.L. Das and K. Pandey (1994). Zooplanktonic community in relation to certain physico-chemical factors of Koshi Swamp, Parina, Bihar. *Environ. and Ecol.,* 12(3): 563–567.

Patil, H.S. and S.M. Karikal (2001). Zooplankton diversity of Bhutnal reservoir at Bijapur-Karnataka state. In: *Water Quality Assessment, Biomonitoring and Zooplankton Diversity,* (Ed.) B.K. Sharma, pp. 236–249.

Rao, V.S. (1975). An ecological study of three fresh water ponds of Hyderabad, India III: The phytoplankton. *Hydrobiologia,* 47: 319–337.

Singh, D.N. (2000). Seasonal variation of zooplankton in a tropical lake. *Geobios,* 27(2–3): 97–100.

Singh, U.N. and Shakuntala Pandey (1990). Water quality of polluted water bodies of North Bihar. *Environ. and Ecol.,* 8(1): 300–310.

Sunkad, B.N. and H.S. Patil (2003). Water quality assessment of Rakasakoppa reservoir of Belgaum, Karnataka. *Indian J. Ecol.,* 30(1): 106–109.

Chapter 17

Impact of Sewage Disposal and Agricultural Waste on the Freshwater Quality and Phytoplankton Density of a Tropical Stream in the Western Ghats in India

M. Ganesan,# N. Jayabalan* & K. Jegatheesan***

*Department of Plant Science, School of Life Sciences, Bharathidasan University, Tiruchirappalli – 620 024, Tamil Nadu

** Department of Botany, ANJA College, Sivakasi – 620 024, Tamil Nadu

#E-mail: markganesh@hotmail.com

ABSTRACT

Water quality of small stream flowing through the Western Ghats has been assessed. The water samples were collected from upstream and down stream of a small town of Watrap. The physico-chemical parameters like pH, temperature, salinity, alkalinity, chloride, sulfate, nitrate, phosphate, silicate, dissolved oxygen, biological oxygen demand of both station were compared. Water quality at down stream of the town was highly disturbed by sewage disposal. BOD showed a four-time increase and other nutrients like sulfate, phosphate, and silicate showed two-fold increase. The primary production was also studied. And very low primary productivity values were obtained in the down stream. The phytoplankton density was also observed and the density of phytoplankton was mostly affected by the sewage disposal.

Keywords: *Freshwater quality, Phytoplanktons, Stream, Western ghats.*

Introduction

Water is the most important single substance for man's survival. Like other forms of biological life, human beings are also extremely dependent upon water for drinking, cooking, bathing etc. Anyone can survive much longer without food than he can without water. So, the water quality studies play a vital role in our life system. If the water is bad plants and animals will not grow or reproduce. Animals and plants stressed because of poor water quality and are also prime targets for pathogens and parasites (Trivedy and Goel, 1984).

Now a day's sewage problem is one of the dangerous problems. The sewage water usually contains large amount of inorganic and organic substances and enormous amount of sledge. So, this sewage becomes growth media for the certain harmful disease-producing microbes (Wetzel, 2000). If there is a chance for the mixing of this sewage water into fresh water system this will become a polluted one. In our study we examined the both upstream (Station A) and downstream (Station B) water of the small stream, which was entered in to the city Watrap. The upstream water is directed from the foothills of Western Ghats and any polluting agent did not affect it. Only after the entry into the city the water was severely affected. So in our study the water was examined before and after the mixing of sewage water and agricultural wastes.

Materials and Methods

The water samples were collected in the upstream and downstream of the town. The procedure for the water sample collection and preservation of water sample techniques were followed as described in the APHA (1989). The pH of the water was measured on the spot by using gun pH meter. The Electrical Conductivity (EC) of water was measured by the conductivity bridge (Allen, 1974). The Dissolved Oxygen, Biological Oxygen Demand, Total Solids, Total Suspended Solids, Total Dissolved Solids, Total Nitrates, Sulphate, Inorganic Phosphate, Chloride and Silicates were measured by using the standard methods established for the examination of water in APHA and AWWA (1989). Total hardness and concentration of Calcium, Magnesium Alkalinity and Salinity were determined by standard volumetric method of Trivedy and Goel (1984). The primary productivity studies were analysed by using the standard method called Light and Dark bottle method (Garren and Garat, 1969). For the estimation and identification of phytoplankton density the Chemocytometer was used. Both surface water and the bottom water were collected for the estimation of phytoplanktons. All estimation was carried out using five replicates. The data presented are mean of five independent determinations.

Results and Discussion

Physico-chemical Parameters

In the present work there was no greater deviation in the temperature of the water in both Stations A and B. Only 26 C in Station B and 25 C in station A were obtained. But there was great deviation in the pH and EC values in Station B compared to Station A. The obtained pH value of Station A was 6.9 and the pH in the Station B was 7.2 and this indicated the presence of high H^+ ions in the water. This H^+ ion concentration was increased due to the disposal of other inorganic materials in to the water. The Electrical Conductivity was also increased in Station B from 0.098 to 0.162 μmoh/cm. Mitra (1993) obtained similar increasing results. This was also due to the presence of more amounts of chemicals and solid materials. The Total Solids including the suspended solids and dissolved solids were highly increased in the station B due to the accumulation of sewage sludge in the water. The obtained results were listed in the Table 17.1.

Table 17.1: The Effect of Sewage on the Physico-chemical Parameters of River Periyar in Both the Station A and Station B

Sl.No.	*Physico-chemical Parameters*	*Station A Upstream*		*Station B Downstream*	
		Mean ± S.E.	*Range*	*Mean ± S.E.*	*Range*
1.	Atmospheric temperature (°C)	26 ± 0.233	25.5–26.4	25 ± 0.247	24.5–25.7
2.	Water temperature (°C)	22 ± 0.554	21.3–23.0	21 ± 0.462	21.5–23.0
3.	pH	6.9 ± 0.187	6.75–7.1	7.2 ± 0.168	7.0–7.4
4.	ECC (moh/cm)	0.098 ± 0.008	0.090–0.10	0.162 ± 0.004	0.14–0.18
5.	Total solids (mg/l)	2.2 ± 0.102	2.1–2.35	4.7 ± 0.24	4.3–4.8
6.	Total Dissolved Solids (mg/l)	0.4 ± 0.05	0.35–0.5	0.7 ± 0,08	0.5–0.85
7.	Total Suspended Solids (mg/l)	1.8 ± 0.098	1.7–2.0	4.0 ± 0.19	3.85–4.35
8.	Salinity (mg/l)	3.472. ± 0.232	3.10–3.50	6.240 ± 0.462	6.0–6.35
9.	Total hardness (mg/l)	26 ± 0.563	24–27	32 ± 0.453	30–34.5
10.	Magnesium hardness (mg/l)	12 ± 0.345	11–13	14 ± 0.36	13–15.5
11.	Calcium hardness (mg/l)	14 ± 0.220	12–15	18 ± 0.254	16.5–18.5
12.	Dissolved Oxygen (DO) (mg/l)	7.6 ± 0.321	7.2–7.8	4.1 ± 0.124	3.8–4.35
13.	Biological Oxygen Demand (BOD) (mg/l)	3.942 ± 0.097	3.10–4.0	0.992 ± 0.085	0.92–1.05
14.	Chloride (mg/l)	1.905 ± 0.123	1.80–2.10	0.826 ± 0,431	0.79–0.89
15.	Sulphate (mg/l)	1.36 ± 0,175	1.20–1.48	2.95 ± 0.146	2.81–2.99
16.	Phosphate (mg/l)	1.24 ± 0.104	1.15–1.39	2.14 ± 0.097	2.11–2.16
17.	Silicate (mg/l)	1.54 ± 0.112	1.35–1.76	1.94 ± 0.120	1.90–1.99
18.	Nitrate (mg/l)	0.15 ± 0.002	0.12–0.16	0.43 ± 0.009	0.39–0.45

SE (±): Standard Error.

Regarding the Salinity, about 3.47 mg/l in Station A and 6.24 mg/l in Station B were obtained. The two-fold increase was obtained in the salinity value of the affected water. This type of increase in salinity values were also obtained by Goel, *et al.* (1980). This shows the excess of salinity producing substances present in the water. The hardness of water is due the presence of magnesium and calcium ions present in the water. The increased value of hardness in Station B was due to the presence of more amounts of calcium and magnesium ions in the water. Calcium and magnesium are the substances that are entered into the aquatic ecosystem due to the degradation of animal wastes by bacteria (Jegatheesan *et al.,* 1998). About 32 mg/l of total hardness was observed in the station B and only 26 mg/l of total hardness was observed in the station A. In both the station the Calcium Hardness was higher than Magnesium Hardness. The Dissolved Oxygen content of the water is maintained by the phytoplanktons present in the water. If the phytoplankton density was affected the Dissolved Oxygen content of the water was also affected and it would affect the quality of the water (Trivedy *et al.,* 1984). The phytoplankton density of both the stations A and B was listed in the Table 17.2. In our present work in the Station B about 4.1 mg/l of DO was observed. This indicated the polluted nature of the water. It has been widely accepted that the any water containing below the level of 7 mg/l of Dissolved Oxygen has to be considered as polluted (Mortimer, 1971). Biological Oxygen Demand is the another parameter, which can determine the nature of the water body. In our present work about 3.942 mg/l of

BOD was obtained but in the Station A. But in the station B only 0.992 mg/l was obtained. If the BOD value of the water decreases it reflects the polluted nature of water (BIS, 1991 and WHO, 1993).

Table 17.2: The Effect of Sewage on the Phytoplankton Density of River Periyar in Both Station A and Station B

Sl.No.	*Name of the Species*	*Station A*		*Station B*	
		Mean	*Percentage*	*Mean*	*Percentage*
(I)	**Cyanophyceae**				
1.	*Anabaena* sp.	1300	5.41	–	–
2.	*Microcystis* sp.	500	2.08	500	2.38
3.	*Nostoc* sp.	500	2.08	–	–
4.	*Synchosystics* sp.	3200	13.3	1500	7.14
(II)	**Chlorophyceae**				
5.	*Chlorella* sp.	200	0.83	100	0.47
6.	*Closterium* sp.	1300	5.41	700	3.33
7.	*Gonium* sp.	600	2.5	400	1.90
8.	*Oedogonium* sp.	300	1.25	–	–
9.	*Pediastrum duplex*	1200	5.00	400	1.90
10.	*Pediastrum simplex*	1300	5.41	400	1.90
11.	*Pithophora* sp.	100	0.41	–	–
12.	*Scenedesmus* sp.	1100	4.58	700	3.33
13.	*Schroederla* sp.	1300	5.41	300	1.42
14.	*Spirogyra* sp.	300	1.25	100	0.47
15.	*Stigeoclonium* sp.	100	0.41	–	–
16.	*Treubaria* sp.	400	1.66	100	0.47
17.	*Tetraedron* sp.	1600	6.66	1000	4.76
18.	*Tetrastrum* sp.	200	0.83	100	0.47
19.	*Ulothrix* sp.	100	0.41	–	–
20.	*Volvox* sp.	100	0.41	100	0.47
21.	*Zygnema* sp.	300	1.25	100	0.47
(III)	**Bacillariophyceae**				
22.	*Asterionella* sp.	708	2.91	1000	4.76
23.	*Cocconeis* sp.	1100	4.58	1300	6.19
24.	*Diatoma* sp.	1300	5.41	1500	7.14
25.	*Epithemia* sp.	100	0.41	100	0.47
26.	*Fragilaria* sp.	500	2.08	400	1.90
27.	*Frustulia* sp.	100	0.41	100	0.47
28.	*Gomphonema* sp.	1500	6.25	1800	8.57
29.	*Melosira* sp.	300	1.25	400	1.90
30.	*Navicula* sp.	1000	4.16	900	4.20
31.	*Pinnularia* sp.	300	1.25	400	1.90
32.	*Synedra* sp.	500	2.08	400	1.90
33.	*Tabellaria* sp.	100	0.41	200	0.95
(IV)	**Euglenophyceae**				
34.	*Euglena* sp.	500	2.08	3500	16.6
	Total	**24000**	**100**	**21000**	**100.0**

Table 17.3: The Effect of Sewage on the Biological Parameters of River Periyar in Both the Station A and Station B

Sl.No.	*Biological Parameters (Primary Productivity Values)*	*Station A Upstream*	*Station B Downstream*
1.	Gross Primary Productivity (GPP) (mgc/m^3/hr)	1.26 ± 0.09	0.19 ± 0.004
2.	Net Primary Productivity (NPP) (mgc/m^3/hr)	0.882 ± 0.002	0.130 ± 0.004
3.	Community Respiration (CR) (mgc/m^3/hr)	0.812 ± 0.006	0.060 ± 0.001
4.	GPP/NPP ratio	1.429 ± 0.067	2.16 ± 0.423

The inorganic content of the water such as nitrate, sulfate, phosphate, silicate and chloride were analysed. The substances like nitrate, sulfate and phosphate shows two fold increase from the normal level. The increasing amount of nitrate, sulfate and phosphate was due to the presence of vegetable and agricultural wastes. The silicate amount was increased from 1.24 mg/l to 1.94 mg/l and this was due to the presence of more amount of sludge in the affected water. Nitrate content of the water also increased from 0.15 mg/l to 0.43 mg/l. This may be due to the presence of agricultural wastes present in the water. The two-fold increase was observed in the level of sulfate. In the station B about 2.95 mg/l of sulfate was observed. Only the chloride content of water was decreased in the station B because, the chloride content is dependent on the amount of chemicals present in the water. The more amount of chemicals leads to the decrease in the chloride content (APHA, 1989). In our present work about 1.905 mg/l of chloride in station A and 0.826 mg/l of chloride in B were obtained. Some previous investigators (Mitra and Gupta, 1996) reported that the level of chloride in the water body was completely affected if the sewage water entered in to it.

Biological Parameters

Among the biological parameters the study of Primary Productivity played a vital role. The amount of Gross Primary Productivity (GPP), Net Primary Productivity (NPP) and Community Respiration (CR) can determine the nature of any water. The very low level of GPP, NPP and CR values were obtained in the downstream *i.e.* Station B. One of the previous investigator (Sayceswarei and Singaracharya, M.A., 1995) reported that the amount of GPP, NPP and CR values were decreased when it was polluted by sewage or other wastes. Only about 0.19 mg/c/hr/m^3 of GPP, 0.13 mg/c/hr/m^3 of NPP and 0.06 mg/c/m^3/hr of CR were obtained in the Station B. This was due to the presence of large amount of chemicals and less quantity of phytoplankton in the water (Gupta *et al.,* 1990). The chlotphycean members were severely affected by the pollution and the well growth of *Euglenophycean* members were obtained in the Station. *Euglenophycean* members can grow if the more amounts of chemicals are present in the water. The same results were obtained by Jegatheesan *et al.* (1998). One of the Limnologist Polishchuk (1989) reported that the human disturbance can affect the diversity of phytoplankton and the quality of the water. The effect of sewage on the phytoplankton density was listed in the Table 17.2. From the obtained results we conclude that the members of Chlorophyceae and Bacillariophyceae were mainly affected. The primary productivity studies were also done in both the stations. The obtain results were presented in the table and it clearly shows that reduction in the primary production values in the Station B compared to Station A. Our present work proves that the downstream (Station B) of the River Periyar is highly affected by the sewage and agricultural wastes. So, the water become highly contaminated and polluted one. In the banks of this river there is no any other factories or industries. So, the wastes from the city above pollute the stream. If the municipality of the city take the proper action the stream will escape from the action of polluting agent or from the sewage problem.

References

Allen, S.E. (1974). *Chemical Analysis of Ecological Materials.* Blackwell Scientific Publications, Oxford.

APHA, AWWA, WPCF (1989). *Standard Methods for the Examination of Water and Wastewater,* 16th edn. American Public Health Association, Washington, D.C.

BIS (1991). *Indian Standard Drinking Water Specification,* 1st Revision. Bureau of Indian Standards, New Delhi.

Gaarder, T.H. and H. Garat (1927). Investigation of the Plankton in the Oslo Fjord. *Rapp. P.V. Renu. Commn. Int. Expor. Scient. Mer. Medeterr,* 42: 1–48.

Goel, P.K., B. Gopal and R.K. Trivedy (1980). Impact of sewage on freshwater ecosystem: Physico-chemical characters of water and their seasonal changes. *Int. J. Ecol. Environ. Sci.,* 6: 91–116.

Gupta, S.K., A. Mitra and S. Adhikari (1990). Post irrigation affect of Calcutta. Sewage effluent on soil and vegetation. Natural seminar on development in soil science 1996, Anand proceedings, pp. 203.

Jegatheesan, K. (1998). Hydrological profile of a sector of river Cauvary. *Ph.D. Thesis.* Mangalore University, p. 22–61 (Unpublished).

Mitra, A. (1993). Characteristics of Calcutta sewage and sledges and their effects on soil properties and plant nutrients with special reference to heavy metals. *Ph.D. Thesis,* University of Calcutta.

Mitra, A. and S.K. Gupta (1996). Distribution of nutrients and heavy metals in different soil horizons, under sewage irrgated condition. Natural seminar on development in soil science 1996, Anand proceedings, pp 203.

Mortimer, C.H. (1971). Chemical changes between sediments and water in the great lakes speculation on probable regulatory mechanism. *Limnol. Oceanography,* 16: 357–404.

Polishchuk, L.K. (1999). Contribution analysis of disturbance caused changes in phytoplankton diversity. *Ecology,* 80(2): 721–725.

Rosen, G. (1981). Phytoplanktons indicators and their relation to certain chemical and physical factors. *Limnologia,* 13(a): 263–290.

Sayeeswari, A., M.A. Singaracharya (1995). Productivity studies in Balal Lake, Bodhan, A.P. India. *Environ. Ecol.,* 13(1): 208–212.

Trivedy, R.K. and P.K. Goel (1984). *Chemical and Biological Methods for Water Pollution Studies.* Environmental Publication, Karad.

Wetzel, R.G. (2000). Freshwater ecology. *Limnology,* 1(6): 3–9.

WHO (1993). *Guidelines for Drinking Water Quality,* Vol. 1: Recommendations, 2nd edn. World Health Organization, Geneva.

Chapter 18

Studies on Zooplankton Distribution in the Coastal Waters of Dakshina Kannada, West Coast of India

T.V. Ramana and M.P.M. Reddy

College of Fishery Science, Muthukur – 524 344, Nellore District, Andhra Pradesh

ABSTRACT

This article describes the monthly distribution of zooplankton in the Arabian Sea off Mangalore, off Malpe and off Gangolli covering the entire Dakshina Kannada coast. The quantitative distribution of zooplankton revealed maximum values in September and minimum values in January/April for both wet weight and dry weight for the entire region during 1979–80 and 1987–88. The qualitative distribution of different zooplanktonic taxa found in the present investigation has been discussed.

Keywords: *Zooplankton, Distribution, Dakshina kannada.*

Introduction

It is a well-known fact that plankton forms the major food item of fishes. Particularly, the role of zooplankton becomes very important both as consumer of phytoplankton and as contributors to fish. Therefore, detailed knowledge on zooplankton distribution and their intensities is necessary for a correct understanding of the fishery resources of any coast. The Dakshina Kannada coast along the west coast of India supports a rich fishery consisting of sardines, mackerels and anchovies of varied intensities in different periods of the year. Some information is available on the zooplankton distribution along some parts of the Dakshina Kannada coast, but these are restricted to 20–30m depth (Suresh and Reddy, 1978 and Benkappa *et al.,* 1979). However, detailed information is not available up to 50 m depths along the important regions of the Dakshina Kannada coast. Hence an intensive study on the zooplankton distribution in the Arabian Sea covering the entire Dakshina Kannada coast was carried out.

Materials and Methods

The present study was carried out along three sections perpendicular to the coast one off Mangalore, one off Maple and one off Gangolli in the Arabian Sea along the Dakshina Kannada Coast (Figure 18.1). These sections were selected in such a way that they cover the southern, central and northern regions of the entire hundred-kilometer length of the Dakshina Kannada coast. These

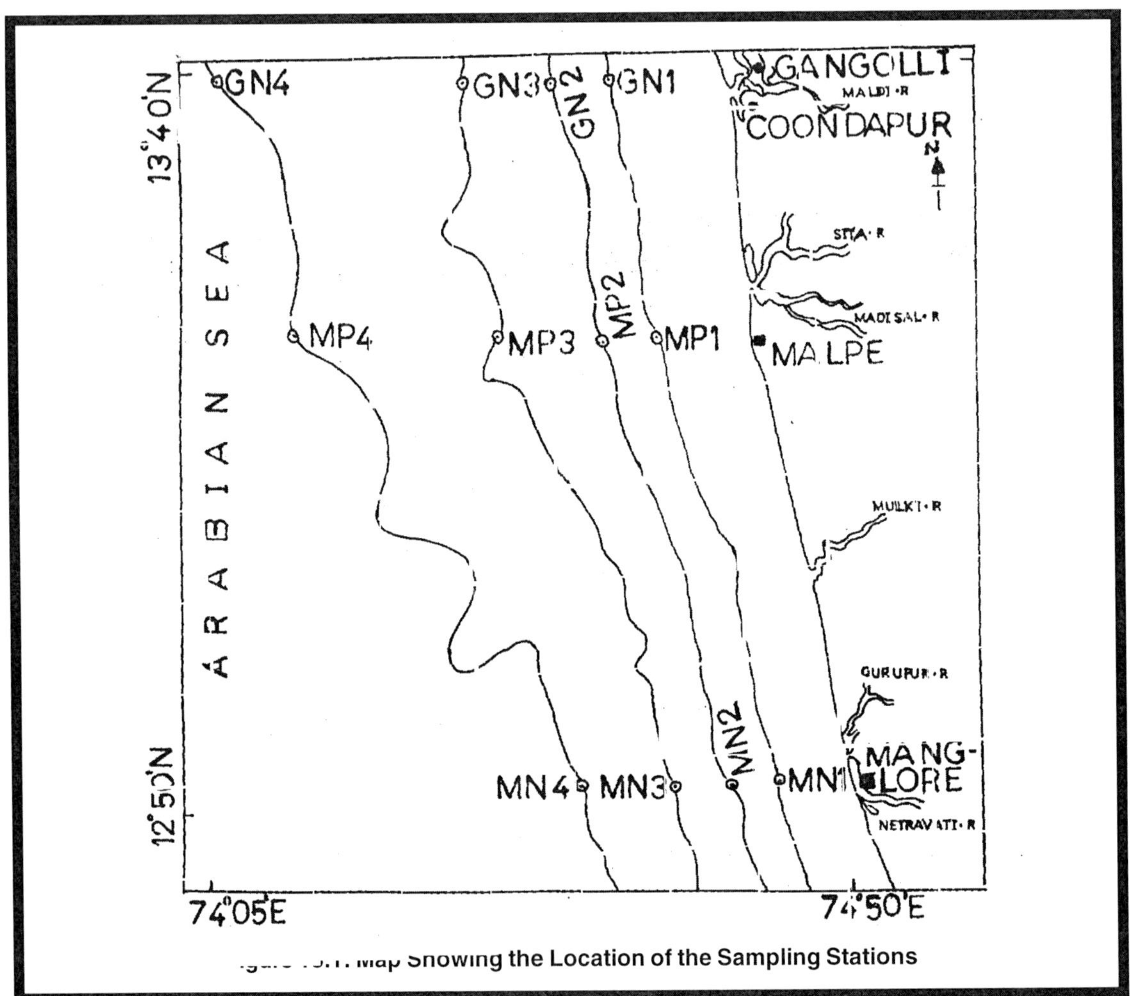

...Map Showing the Location of the Sampling Stations

particular sections were selected also since Mangalore, Malpe and Gangolli are the major fish landing centers along the Dakshina Kannada coast. Four stations one at each depth *viz.* 20, 30, 40 and 50 m were selected along each section. Stations at 20 m depth were MN1, MP1 and GN1; while those at 30 m depth were MN2, MP2 and GN2. Similarly, the stations at 40 m and 50 m depths were MN3, MP3 and GN3 and MN4, MP4 and GN4 respectively.

Zooplankton samples were collected at monthly intervals from those twelve stations during the cruises of the fishing vessel, M.F.V Dolphin of the College of Fisheries. The collections were spread over the period, April 1987 to May 1988 except in the months of June, July and August when collections could not be made due to unfavourable weather conditions. At all the stations vertical hauls of zooplankton were made using a Heron-Tranter net. The zooplankton samples were collected with a net having a bolting silk of mesh size of 200 μ. In the laboratory, the zooplankton samples were divided in to two sub-samples with the help of a Folsom Plankton Splitter for quantitative and qualitative analysis. From one sub-sample, the wet weight and dry weight of zooplankton were determined following the standard methods. Estimations of qualitative composition of the zooplankton were done with the other sub-sample. First the macro plankton were separated and counted and then the samples were made up to a known volume. From this, 1 ml was taken and the number of different groups of zooplankton was counted. The values were represented in number/100 m^3. In addition to the 1987–88 data, zooplankton biomass, both, wet weight and dry weight data (unpublished) collected by the Department of Fishery Oceanography, College of Fisheries during the period from May 1979 to May 1980 off Mangalore, off Maple and off Gangolli were used in the present study to investigate the quantitative distribution of zooplankton along Dakshina Kannada coast in that period.

Results and Discussion

Quantitative Distribution

The quantitative distribution of zooplankton (in terms of wet weight and dry weight) of the columnar waters is given Tables 18.1 and 18.2 for 1979–80. During this period, the biomass of zooplankton in terms of wet weight fluctuated from 19.04 to 1,224.1 mg/m^3 in the section off Mangalore, from 32.94 to 640.80 mg/m^3 in the section off Maple and from 11.25 mg/m^3 to 723.25 mg/m^3 in the section off Gangolli. The highest values occurred during September at all the stations and the lowest values occurred generally in January or in April. A secondary peak in wet weight was found generally in February/November. The general trend observed in the values of dry weight of zooplankton was almost similar to that of wet weight. During the period 1979–80, the zooplankton biomass dry weight values varied from 5.91 to 186.28 mg/m^3 in the section off Mangalore and from 1.08 to 148.50 mg/m^3 in the section off Gangolli.

During the period 1987–88, the zooplankton wet weight biomass values ranged between 11.00 and 723.88 mg/m^3 in the section off Mangalore; between 10.33 and 538.24 mg/m^3 in the section off Malpe and between 7.59 and 684.25 mg/m^3 in the section off Gangolli (Table 18.3). In general, the monthly distribution showed two peaks, the major peak in September and the minor peak generally in February. In this period also the trend in the dry weight variation was very similar to that of wet weight (Table 18.4). The zooplankton biomass dry weight values fluctuated between 1.63 and 96.29 mg/m^3 in the section off Mangalore, between 1.97 and 84.52 mg/m^3 in the section off Malpe and between 1.07 and 92. 18 mg/m^3 in the section off Gangolli.

Table 18.1: Quantitative Distribution of Zooplankton (Wet weight mg/m³) in the Columnar Waters at Different Stations Located Along Dakshina Kannada Coast

Section	*Period*	*1979*							*1980*				
	Station	*April*	*May*	*Aug.*	*Sept.*	*Oct.*	*Nov.*	*Dec.*	*Jan.*	*Feb.*	*March*	*April*	*May*
I	MN1	70.35	136.14	175.16	1017.9	168.62	132.09	145.61	27.63	190.16	89.78	140.44	341.52
	MN2	149.62	142.10	166.48	1224.1	137.05	87.67	121.37	61.52	118.24	114.26	116.08	164.74
	MN3	71.12	216.7	176.82	546.1	270.52	73.54	78.64	40.46	114.18	89.75	137.56	158.09
	MN4	95.34	136.1	155.86	468.5	140.41	61.55	58.19	19.04	233.17	38.23	96.38	89.64
II	MP1	99.02	160.78	297.60	640.80	53.72	226.35	182.47	184.96	220.63	130.88	109.18	555.68
	MP2	94.44	132.35	429.40	551.40	99.69	120.62	286.04	92.35	295.98	188.22	94.64	224.87
	MP3	32.94	73.64	287.24	387.62	64.16	309.84	151.16	167.52	162.85	57.65	74.38	169.38
	MP4	36.88	48.20	209.33	480.63	138.04	174.09	113.88	72.88	40.93	36.14	62.64	272.06
III	GN1	88.69	95.66	201.63	723.25	76.59	108.80	36.92	16.19	453.62	47.54	15.12	58.19
	GN2	20.47	39.12	106.34	422.12	124.38	71.63	97.66	37.22	64.57	40.93	11.25	29.73
	GN3	17.55	19.67	83.74	140.92	75.72	96.52	53.12	19.54	37.16	30.75	13.36	14.66
	GN4	12.88	11.49	41.62	96.09	131.92	40.86	37.61	14.09	30.62	35.82	11.78	11.63

Table 18.2: Quantitative Distribution of Zooplankton (Dry Weight mg/m³) in the Columnar Waters at Different Stations Located Along Dakshina Kannada Coast

Section	*Period*	*1979*							*1980*				
	Station	*April*	*May*	*Aug.*	*Sept.*	*Oct.*	*Nov.*	*Dec.*	*Jan.*	*Feb.*	*March*	*April*	*May*
I	MN1	19.72	29.63	34.68	151.72	28.55	34.52	46.28	10.75	57.07	15.89	29.52	35.92
	MN2	32.16	27.74	25.69	186.28	31.96	16.71	44.03	22.62	34.46	29.06	23.08	14.46
	MN3	14.78	36.52	39.58	108.32	65.82	20.19	24.25	17.44	30.82	12.64	19.96	20.24
	MN4	22.15	27.71	28.91	73.16	30.76	11.92	39.07	5.91	43.71	7.38	12.38	11.09
II	MP1	19.08	22.40	57.61	73.64	11.66	43.75	36.70	43.20	33.87	17.64	20.72	61.19
	MP2	19.47	29.93	66.54	68.67	25.73	19.82	48.09	31.08	46.17	36.19	19.17	29.85
	MP3	8.48	11.18	48.18	58.48	18.65	56.29	25.61	26.48	23.19	8.53	12.56	14.39
	MP4	3.56	9.78	32.17	67.18	29.03	29.01	24.53	12.97	6.88	4.67	10.29	29.84
III	GN1	28.95	24.77	48.12	148.56	28.86	37.21	13.72	1.79	100.62	15.32	2.54	32.20
	GN2	7.49	12.55	29.61	35.42	34.33	28.47	33.91	4.25	14.59	14.51	4.92	14.38
	GN3	7.36	7.32	20.19	35.15	25.12	31.07	18.62	2.26	6.54	9.93	4.31	7.62
	GN4	1.08	5.36	17.82	30.56	39.42	12.85	12.68	2.74	5.17	11.69	2.37	5.08

Qualitative Distribution

The abundance and occurrence of various groups of zooplankton along the three sections during the period form April 1987 to May 1988 are presented in Tables 18.5 to 18.16.

Table 18.3: Quantitative Distribution of Zooplankton (Wet Weight mg/m³) in the Columnar Waters at Different Stations Located Along Dakshina Kannada Coast

Section	*Period*	*1987*							*1988*			
	Station	*April*	*May*	*Sept.*	*Oct.*	*Nov.*	*Dec.*	*Jan.*	*Feb.*	*March*	*April*	*May*
I	MN1	35.63	119.20	723.68	73.87	58.20	77.40	26.70	163.80	49.95	77.27	126.76
	MN2	29.18	62.17	315.36	40.44	63.10	47.58	37.20	76.52	56.28	24.48	59.26
	MN3	19.65	34.53	94.51	78.40	31.82	23.22	27.00	84.28	37.68	21.88	29.94
	MN4	12.47	22.94	72.08	42.15	65.26	19.50	21.60	75.63	36.84	11.00	19.00
II	MP1	47.63	108.63	538.24	106.14	92.65	83.67	24.02	253.68	62.45	25.10	68.16
	MP2	21.59	74.12	276.18	111.35	72.08	45.20	29.28	75.68	51.60	23.32	39.38
	MP3	14.68	32.65	92.62	63.75	34.75	37.62	21.00	55.54	42.60	11.74	28.62
	MP4	10.33	20.11	66.17	84.86	27.62	29.00	15.52	38.42	35.66	11.00	14.56
III	GN1	29.65	87.62	684.25	71.96	58.08	56.80	23.40	98.10	69.12	21.00	46.34
	GN2	15.17	59.18	326.34	55.23	91.54	48.35	20.61	83.92	40.96	13.48	18.59
	GN3	11.68	30.55	110.65	62.16	55.66	37.90	22.41	42.63	42.00	15.66	11.57
	GN4	7.59	24.67	70.33	53.43	47.35	42.46	18.68	38.42	41.35	15.25	12.00

Table 18.4: Quantitative Distribution of Zooplankton (Dry Weight mg/m³) in the Columnar Waters at Different Stations Located Along Dakshina Kannada Coast

Section	*Period*	*1987*							*1988*			
	Station	*April*	*May*	*Sept.*	*Oct.*	*Nov.*	*Dec.*	*Jan.*	*Feb.*	*March*	*April*	*May*
I	MN1	17.65	30.18	96.29	20.14	20.72	21.75	3.54	63.55	16.34	24.29	44.35
	MN2	6.22	12.66	49.42	12.97	24.16	18.56	8.45	10.65	15.97	8.56	10.28
	MN3	3.17	7.15	27.35	22.72	8.65	6.54	2.15	12.29	11.00	2.14	5.29
	MN4	1.63	4.29	19.10	13.00	17.61	3.87,	2.04	0.75	8.95	1.94	6.93
II	MP1	10.88	27.62	84.52	30.33	31.08	32.87	1.97	50.62	18.82	5.54	21.96
	MP2	5.72	10.38	37.19	34.46	28.49	17.26	2.05	5.58	16.31	4.32	12.70
	MP3	3.19	6.54	21.64	23.14	12.85	7.40	5.63	7.54	15.31	2.36	5.97
	MP4	2.25	3.11	15.55	25.13	8.74	6.54	7.74	5.55	11.68	2.00	3.05
III	GN1	11.08	23.47	92.18	21.00	14.16	18.50	4.69	22.20	15.00	4.94	18.56
	GN2	5.24	9.45	48.34	15.79	29.37	14.67	5.05	11.20	14.50	6.37	7.37
	GN3	3.02	4.83	25.64	23.00	16.17	9.78	3.00	11.00	21.67	5.05	2.36
	GN4	2.07	3.05	17.42	18.01	12.86	12.75	1.96	6.17	14.32	3.10	1.07

Medusae

During the period of investigation, the abundance of medusae fluctuated from 0–2432/100 m³ along Mangalore section, from 0–2154/100 m³ along Malpe section and between 0 and 3540/100 m³ along Gangolli section. Generally, higher number of medusae occurred during November/December.

Table 18.5: Monthly Variations in Different Groups of Zooplankton (no/100m^3) in Columnar Waters at Station MN_1 (Section I)

Period	*1987*						*1988*				
Groups	*April*	*May*	*Sept.*	*Oct.*	*Nov.*	*Dec.*	*Jan.*	*Feb.*	*March*	*April*	*May*
Medusae	–	–	–	–	–	–	–	–	–	–	–
Ctenophores	–	–	–	–	–	–	–	–	–	–	–
Siphonophores	–	–	–	–	–	–	–	–	–	–	–
Chaetognaths	4200	2600	12524	2440	–	16624	27	60925	316	2840	1652
Polychaete larvae	200	600	–	–	13600	72570	800	14560	–	679	259
Evadne	–	7925	–	–	150000	–	–	–	58724	–	79450
Penilia	–	22500	235324	334725	24700	12600	–	–	–	12000	12800
Copepods	100000	345000	665560	629000	764650	212500	233235	1504675	508970	124000	580000
Copepodites	400000	225000	185735	104350	485700	–	79450	1200000	640625	156000	214260
Lucifer	1150	200	198	80	–	1200	–	3250	179	500	150
Decapod larvae	300	28000	200000	100400	105200	–	35042	111860	40575	3000	17300
Ostracods	–	–	–	–	–	–	–	–	–	–	–
Gastropod larvae	–	–	16200	27000	–	–	–	–	–	–	–
Bivalve larvae	–	–	–	–	–	–	–	–	–	–	–
Oikopleura	–	25000	–	–	70960	50600	65960	58000	67000	–	17200
Salpids	3670	–	–	–	–	–	–	–	–	–	–
Doliolids	–	–	–	–	–	–	–	–	–	–	–
Fish eggs	–	–	–	–	–	–	–	–	–	–	–
Fish larvae	120	–	312	–	–	175	–	64	–	120	–

Table 18.6: Monthly Variations in Different Groups of Zooplankton (no/100m^3) in Columnar Waters at Station MN_2 (Section I)

Period	*1987*						*1988*				
Groups	*April*	*May*	*Sept.*	*Oct.*	*Nov.*	*Dec.*	*Jan.*	*Feb.*	*March*	*April*	*May*
Medusae	80	–	367	–	420	2432	–	850	980	104	–
Ctenophores	–	–	–	965	1140	–	18	900	–	–	–
Siphonophores	–	–	–	690	–	–	112	–	–	–	–
Chaetognaths	8020	2640	7500	350	320	18250	120	28820	2545	5440	2210
Polychaete larvae	40000	–	–	–	–	60725	–	7000	120871	20170	–
Evadne	–	56000	–	10400	64770	–	–	9064	–	6275	57000
Penilia	–	80000	5200	145270	22515	50613	–	240000	120000	4620	32000
Copepods	640000	500000	407000	730610	345830	527400	530000	760000	616390	398820	645000
Copepodites	80000	140000	6744	–	20950	46900	130000	640000	50830	104830	193400
Lucifer	416	–	1500	–	1700	–	420	2000	615	255	–
Decapod larvae	14000	–	–	–	62900	22000	36000	98370	–	56400	–
Ostracods	–	–	–	–	–	–	–	–	–	–	–
Gastropod larvae	–	–	–	–	–	2000	–	–	–	–	–
Bivalve larvae	–	–	–	–	–	–	–	–	–	–	–
Oikopleura	18600	–	32000	12600	50480	–	41440	48619	245700	49000	42530
Salpids	–	2780	–	–	–	–	2690	–	210	–	–
Doliolids	–	–	–	202	–	–	119	–	–	–	–
Fish eggs	–	–	–	–	–	–	–	–	–	–	–
Fish larvae	80	–	512	216	–	–	–	116	–	62	–

Table 18.7: Monthly Variations in Different Groups of Zooplankton (no/100m³) in Columnar Waters at Station MN_3 (Section I)

Period	*1987*						*1988*				
Groups	*April*	*May*	*Sept.*	*Oct.*	*Nov.*	*Dec.*	*Jan.*	*Feb.*	*March*	*April*	*May*
Medusae	156	56	116	–	210	1359	–	214	163	24	45
Ctenophores	–	–	–	26	116	–	–	415	–	–	–
Siphonophores	2356	–	–	416	–	–	–	–	22	–	–
Chaetognaths	24313	1658	6584	1648	12295	32798	4260	43116	4510	525	3074
Polychaete larvae	11871	–	12260	–	–	24999	3260	–	–	17300	–
Evadne	24751	28571	–	–	32600	–	–	–	26412	26735	27750
Penilia	28640	18714	12372	4325	48270	26420	–	62200	17540	34700	42936
Copepods	248175	164285	178684	155800	290741	130900	122000	440000	121439	283050	263192
Copepodites	74570	14680	53672	45350	143630	35245	16270	253174	142474	84887	21556
Lucifer	–	57	125	260	–	–	140	320	–	–	71
Decapod larvae	126	–	–	–	–	12750	156	13840	17000	1475	–
Ostracods	–	–	–	–	–	–	–	–	–	–	–
Gastropod larvae	–	–	–	–	–	2456	–	–	–	–	–
Bivalve larvae	–	–	–	–	–	7186	–	–	14700	–	–
Oikopleura	22587	2057	28615	25493	4365	77531	11216	12000	88840	14850	–
Salpids	–	4650	–	–	–	–	–	–	–	–	–
Doliolids	–	–	–	–	–	–	–	–	–	–	–
Fish eggs	–	–	–	–	–	–	–	–	–	–	–
Fish larvae	28	–	217	85	–	–	–	12	28	14	–

Table 18.8: Monthly Variations in Different Groups of Zooplankton (no/100m³) in Columnar Waters at Station MN_4 (Section I)

Period	*1987*						*1988*				
Groups	*April*	*May*	*Sept.*	*Oct.*	*Nov.*	*Dec.*	*Jan.*	*Feb.*	*March*	*April*	*May*
Medusae	168	21	110	76	612	242	–	179	146	–	–
Ctenophores	–	–	–	14	32	–	–	115	10	–	–
Siphonophores	–	–	112	228	12	–	–	–	–	–	–
Chaetognaths	–	5072	11250	2544	14240	2714	4717	15590	6120	–	3910
Polychaete larvae	3625	–	–	–	–	–	4500	–	21795	7105	–
Evadne	–	20625	–	4300	19435	–	–	–	–	–	15256
Penilia	12055	5626	2860	–	109815	–	–	105000	–	12510	10500
Copepods	100200	125897	246245	162460	342340	335324	37000	432000	75280	201360	183405
Copepodites	2650	35600	24362	12305	20565	21324	12350	226884	68750	20169	–
Lucifer	–	200	–	–	–	–	–	250	–	–	–
Decapod larvae	–	–	–	–	–	7500	4300	8625	–	–	–
Ostracods	–	–	–	–	–	–	–	–	–	–	–
Gastropod larvae	–	–	–	–	–	1260	2500	–	–	–	–
Bivalve larvae	–	1025	–	–	–	–	–	–	–	–	–
Oikopleura	5261	–	24116	48297	–	37310	42395	48000	53918	6740	5120
Salpids	–	3511	–	–	–	–	–	–	–	–	–
Doliolids	–	–	–	–	–	–	–	–	–	–	–
Fish eggs	–	–	–	–	–	–	–	–	–	–	–
Fish larvae	–	–	65	–	–	–	126	–	16	–	–

Table 18.9: Monthly Variations in Different Groups of Zooplankton (no/100m^3) in Columnar Waters at Station MP$_1$ (Section II)

Period	*1987*						*1988*				
Groups	*April*	*May*	*Sept.*	*Oct.*	*Nov.*	*Dec.*	*Jan.*	*Feb.*	*March*	*April*	*May*
Medusae	–	–	814	300	2154	204	122	199	–	467	–
Ctenophores	–	–	–	–	–	–	–	–	–	–	–
Siphonophores	–	–	726	1206	–	–	146	–	–	–	–
Chaetognaths	720	4800	37400	2060	69720	27321	3890	75440	12058	1132	2448
Polychaete larvae	20000	–	–	–	–	32120	26573	–	–	21961	–
Evadne	–	72000	12840	–	225296	–	–	34625	–	–	93710
Penilia	26000	14000	25160	–	55000	–	–	–	65700	14260	27270
Copepods	526000	1000000	1242200	700000	1645565	451810	14852	645845	1316900	743170	186300
Copepodites	28300	40000	29209	–	89700	36000	34120	1829065	93615	465900	–
Lucifer	123	176	142	–	345	–	–	466	–	155	–
Decapod larvae	75000	–	7600	–	14700	21122	12240	–	–	–	33500
Ostracods	–	–	–	–	–	–	–	–	–	–	–
Gastropod larvae	–	–	–	–	–	–	–	–	–	–	–
Bivalve larvae	–	–	–	–	–	–	–	–	–	–	22000
Oikopleura	26000	31000	1745	125	42572	10860	19630	294727	35510	37000	14700
Salpids	–	8760	–	–	–	–	–	–	–	–	4168
Doliolids	–	–	–	–	–	–	–	–	–	–	–
Fish eggs	–	–	–	–	–	–	–	–	–	–	–
Fish larvae	100	–	148	–	–	–	–	–	–	–	–

Table 18.10: Monthly Variations in Different Groups of Zooplankton (no/100m^3) in Columnar Waters at Station MP$_2$ (Section II)

Period	*1987*						*1988*				
Groups	*April*	*May*	*Sept.*	*Oct.*	*Nov.*	*Dec.*	*Jan.*	*Feb.*	*March*	*April*	*May*
Medusae	–	–	160	–	275	200	–	145	–	200	–
Ctenophores	–	–	–	–	146	–	–	–	200	140	–
Siphonophores	–	–	365	246	–	–	–	–	–	–	–
Chaetognaths	2112	1266	70621	8770	60680	92621	2930	126657	4680	346	5500
Polychaete larvae	13650	–	15000	–	–	24000	–	–	–	31600	–
Evadine	–	87860	167000	137260	224620	–	20000	–	37900	–	176000
Penillia	15600	10000	60000	–	160000	–	–	77500	212900	37000	56000
Copepods	412250	1412220	442000	565420	645900	353800	342800	529000	78130	231640	412600
Copepodites	44000	12000	84000	67000	364000	260400	25000	1041326	87810	430400	320000
Lucifer	45	88	–	–	–	–	–	58	150	–	135
Decapod larvae	24650	44332	45350	55000	51000	–	830	16960	4000	74255	–
Ostracods	–	–	–	–	–	–	–	–	–	–	–
Gastropod larvae	–	–	–	–	–	–	–	–	–	–	–
Bivalve larvae	–	12000	–	–	–	–	–	–	–	–	–
Oikopleura	53000	1000	29000	–	39000	64500	12200	74000	–	54730	63720
Salpids	–	11445	–	–	–	–	340	–	–	–	–
Doliolids	–	–	–	–	–	–	–	–	–	–	–
Fish eggs	–	–	–	–	–	–	–	–	–	–	–
Fish larvae	35	–	116	–	–	–	–	16	–	–	–

Table 18.11: Monthly Variations in Different Groups of Zooplankton (no/100,^{3}) in Columnar Waters at Station MP$_3$ (Section II)

Period	1987						1988				
Groups	*April*	*May*	*Sept.*	*Oct.*	*Nov.*	*Dec.*	*Jan.*	*Feb.*	*March*	*April*	*May*
Medusae	–	–	126	–	1120	276	–	114	362	–	–
Ctenophores	–	–	–	–	–	–	–	–	–	–	–
Siphonophores	–	–	214	148	–	–	–	–	–	–	–
Chaetognaths	100	–	2518	2008	15000	8950	7865	31194	12461	825	–
Polychaete larvae	4172	–	8274	21472	7055	–	–	14382	–	28867	–
Evadne	–	–	14000	–	106122	–	–	17421	21436	–	34000
Penilia	42800	24700	2095	–	119500	10095	–	42900	7820	–	–
Copepods	342570	85714	665420	584855	143495	184657	71773	742840	21420	112850	142260
Copepodites	15000	14582	36857	21758	23615	56902	6592	245950	7400	214130	150400
Lucifer	100	–	120	184	179	37	–	–	–	120	–
Decapod larvae	52957	–	210	64	–	3287	3650	–	–	67209	–
Ostracods	–	–	–	–	–	–	–	–	–	–	–
Gastropod larvae	–	–	–	–	–	–	8775	–	–	–	–
Bivalve larvae	–	–	–	–	–	–	–	–	–	–	–
Oikopleura	35714	–	14610	12879	–	96000	19430	114570	7521	17950	1770
Salpids	–	–	–	–	–	–	–	–	–	–	–
Doliolids	–	–	–	–	–	–	–	–	–	–	–
Fish eggs	–	–	–	–	–	–	–	–	–	–	–
Fish larvae	–	–	128	–	–	–	–	–	–	–	–

Table 18.12: Monthly Variations in Different Groups of Zooplankton (no/100m^3) in Columnar Waters at Station MN$_4$ (Section II)

Period	1987						1988				
Groups	*April*	*May*	*Sept.*	*Oct.*	*Nov.*	*Dec.*	*Jan.*	*Feb.*	*March*	*April*	*May*
Medusae	442	35	216	–	592	438	–	259	279	465	27
Ctenophores	–	–	–	–	–	25	–	–	–	–	–
Siphonophores	–	–	550	446	–	–	–	–	67	–	–
Chaetognaths	495	1895	9695	–	21214	18596	8450	60570	17404	256	2466
Polychaete larvae	–	9526	2000	–	4804	7814	6820	–	8205	7213	–
Evadne	9526	–	–	–	–	–	–	–	–	–	18432
Penilia	36214	17840	9600	–	–	5216	–	8290	98421	7041	–
Copepods	136845	168421	180000	90159	190059	71505	172000	163472	62752	85646	197580
Copepodites	42361	16780	14000	6445	12325	30276	24745	239412	25856	35541	–
Lucifer	–	51	–	–	–	–	639	–	–	–	490
Decapod larvae	–	–	–	–	6745	10435	–	–	–	–	–
Ostracods	–	–	–	–	–	–	–	–	–	–	–
Gastropod larvae	–	9500	–	–	–	9500	–	–	–	–	7257
Bivalve larvae	–	–	–	–	–	–	–	–	–	–	5220
Oikopleura	17212	11526	7200	6400	4325	27900	–	37294	8421	24740	–
Salpids	–	3500	–	–	–	–	–	–	–	–	7340
Doliolids	–	69	–	–	–	–	–	–	–	–	590
Fish eggs	–	–	–	–	–	–	–	–	–	–	–
Fish larvae	–	–	145	–	–	–	–	21	–	–	–

Table 18.13: Monthly Variations in Different Groups of Zooplankton (no/100m³) in Columnar Waters at Station GN_1 (Section III)

Period	*1987*						*1988*				
Groups	*April*	*May*	*Sept.*	*Oct.*	*Nov.*	*Dec.*	*Jan.*	*Feb.*	*March*	*April*	*May*
Medusae	–	–	–	–	–	–	–	–	–	–	–
Ctenophores	–	–	–	–	–	–	–	4900	–	–	–
Siphonophores	–	–	–	–	–	–	–	–	–	–	–
Chaetognaths	8020	2440	3845	1669	6779	11920	8618	49400	135	4835	1831
Polychaete larvae	2000	–	11000	–	–	24000	–	17200	–	200	369
Evadne	–	160000	129600	120000	145000	–	17000	219670	20042	17540	180400
Penilia	–	180000	168310	667271	13310	14000	–	–	174600	–	12000
Copepods	580000	500000	416560	192820	729560	486312	634321	1500000	402200	145906	540289
Copepodites	212000	260000	126850	386235	129760	17217	185345	638690	345600	132000	325626
Lucifer	120	–	200	–	271	134	–	1165	204	214	120
Decapod larvae	16000	–	12600	20000	83264	–	69121	86300	39000	69700	34086
Ostracods	–	–	–	–	–	–	–	–	–	–	–
Gastropod larvae	–	–	–	60000	–	–	–	–	–	–	–
Bivalve larvae	–	–	–	–	–	–	–	–	–	–	–
Oikopleura	60000	75000	6400	51600	14906	134550	140600	20000	47000	–	51200
Salpids	–	1180	–	–	–	–	–	–	–	–	–
Doliolids	–	–	–	–	–	–	–	–	–	–	–
Fish eggs	–	–	–	–	–	–	–	–	–	–	–
Fish larvae	80	–	215	47	–	14	–	–	200	138	–

Table 18.14: Monthly Variations in Different Groups of Zooplankton (no/100m³) in Columnar Waters at Station GN_2 (Section III)

Period	*1987*						*1988*				
Groups	*April*	*May*	*Sept.*	*Oct.*	*Nov.*	*Dec.*	*Jan.*	*Feb.*	*March*	*April*	*May*
Medusae	–	1186	284	34	380	2176	–	298	350	–	–
Ctenophores	–	–	–	–	16	–	–	–	–	–	–
Siphonophores	–	–	3200	2110	–	–	–	–	–	–	–
Chaetognaths	120	–	2640	1371	–	8950	4965	30185	1159	925	1579
Polychaete larvae	4712	–	8980	2185	12190	16755	8145	14491	–	18120	2142
Evadne	–	4000	7260	–	24262	–	–	17600	21869	–	8635
Penilia	–	–	–	–	139675	3110	–	94180	7490	–	30889
Copepods	324758	321000	614660	125053	721550	469345	48773	972269	5969	429163	212678
Copepodites	150000	14380	28825	21756	556285	85902	18592	270927	2046	167130	–
Lucifer	130	210	180	275	454	37	–	411	–	144	–
Decapod larvae	–	–	265	128	89	7280	7193	–	–	17299	–
Ostracods	–	–	–	85	3200	–	–	–	–	–	–
Gastropod larvae	–	–	–	57	–	–	5875	–	–	–	–
Bivalve larvae	–	–	–	–	420	–	–	–	–	–	–
Oikopleura	9287	–	6498	12789	–	95291	14430	14750	7490	154245	12000
Salpids	–	9670	196	180	–	–	160	16	–	–	–
Doliolids	–	–	–	–	–	–	17	–	–	–	–
Fish eggs	–	–	–	103	–	–	–	–	–	–	–
Fish larvae	–	–	160	–	–	18	42	–	–	–	–

Table 18.15: Monthly Variations in Different Groups of Zooplankton (no/100m^3) in Columnar Waters at Station GN_3 (Section III)

Period	*1987*						*1988*				
Groups	*April*	*May*	*Sept.*	*Oct.*	*Nov.*	*Dec.*	*Jan.*	*Feb.*	*March*	*April*	*May*
Medusae	18	15	239	23	190	1167	22	566	40	45	55
Ctenophores	–	–	–	–	–	–	94	–	–	–	–
Siphonophores	–	1560	360	765	–	–	–	–	–	–	–
Chaetognaths	156	2761	12978	5278	35929	8324	6214	58400	156	153	5209
Polychaete larvae	–	–	1025	1000	–	2160	–	–	794	1791	–
Evadne	–	35600	14000	18920	56911	–	–	–	–	–	2166
Penilia	–	38330	6270	5201	45504	–	–	104000	–	–	14668
Copepods	212953	360777	270800	170400	499652	58721	33565	470702	141886	140886	96870
Copepodites	21655	4284	14650	7292	82560	15217	25675	408846	104905	80796	2166
Lucifer	–	16	275	340	195	–	–	625	–	–	–
Decapod larvae	–	–	–	–	8700	–	6795	–	6250	6250	8231
Ostracods	–	–	–	–	–	–	–	–	–	–	–
Gastropod larvae	–	27090	–	–	–	–	8019	–	2116	2166	–
Bivalve larvae	–	–	–	–	–	–	–	–	–	–	–
Oikopleura	–	–	13860	7635	4200	13600	4500	166000	10403	–	4260
Salpids	1172	4000	–	–	–	45	250	–	–	–	1910
Doliolids	–	–	–	–	–	–	–	–	–	–	–
Fish eggs	–	–	–	–	–	–	–	–	–	–	–
Fish larvae	–	–	190	–	–	–	–	–	–	–	–

Table 18.16: Monthly Variations in Different Groups of Zooplankton (no/100m^3) in Columnar Waters at Station GN_4 (Section III)

Period	*1987*						*1988*				
Groups	*April*	*May*	*Sept.*	*Oct.*	*Nov.*	*Dec.*	*Jan.*	*Feb.*	*March*	*April*	*May*
Medusae	35	115	386	444	475	3540	23	–	472	790	–
Ctenophores	–	–	–	–	–	–	–	–	–	–	–
Siphonophores	–	–	–	–	–	–	40	400	–	123	30290
Chaetognaths	50	2625	10840	25006	50594	6400	2965	20560	41813	16693	–
Polychaete larvae	6144	–	–	–	15504	4475	–	4205	4220	56000	8968
Evadne	–	4611	3200	8300	–	–	21300	–	4000	–	–
Penilia	–	20830	38338	42425	20000	–	14200	8305	42838	4500	–
Copepods	204611	766760	162000	50600	80896	90557	440109	50420	383363	162502	11338
Copepodites	150000	26000	28615	18000	42838	45520	–	4560	8350	87912	–
Lucifer	–	–	–	–	–	–	–	–	–	–	–
Decapod larvae	11500	1116	22530	1445	–	50700	–	8450	4260	17100	12060
Ostracods	–	–	–	–	–	–	–	–	–	–	–
Gastropod larvae	400	–	–	–	–	2155	14200	–	200	–	–
Bivalve larvae	–	–	–	–	–	–	–	–	–	–	–
Oikopleura	20800	8345	22400	33500	15504	10422	–	–	–	23295	–
Salpids	–	216	120	75	–	–	3056	160	–	–	–
Doliolids	–	–	–	–	–	–	–	–	–	–	–
Fish eggs	–	–	–	–	–	–	–	–	–	–	–
Fish larvae	–	–	135	–	–	–	–	–	–	–	–

Chaetognaths

In the columnar waters, the number of Chetognaths fluctuated between 0 and 60,925/100 m^3 along Mangalore section, between 0 and 126,657/100 m^3 along Malpe section and from 0 to 58,400/100 m^3 along Gangolli section. In general, two peaks were observed for this group, one in February and the other during November/December.

Penilia

Peak occurrence of Penilia was observed during February/October/March in the Mangalore section, during March/October in the Malpe section and during October/February/March in the Gangolli section. The numbers of Penilia fluctuate from 0 to 3,34,725/100 m^3 along Mangalore section, from 0 to 2,12,900/100 m^3 along Malpe section and between 0 and 6,67,271/100 m^3 along Gangolli section.

Copepods

This group formed the major constituent of the zooplankton throughout the study period. Generally, the copepod population more during May 1987, February 1988 and November 1987. Least numbers were found during January at majority of the stations. The numerical abundance of this group varied from 37,000 to 1504675/100 m^3 along Mangalore section; from 14852 to 1646565/100 m^3 along Malpe section and between 11338 and 1500000/100 m^3 along Gangolli section.

Copepodites

This group formed the major constituent only next to their adults *i.e.* copepods. The general trend of variation showed maximum values during February at all the stations except at 30 m and 50 m depth stations in the Gangolli section, which recorded peak values during October. Minimum values were recorded during January 1988/May 1987 at majority of the stations. Numerical abundance of this group varied from 0 to 1200000/100 m^3; from 0 to 1829065/100 m^3 and from 0 to 638690/100 m^3 along Mangalore section, Malpe section and Gangolli section respectively.

Lucifers

The peak abundance of lucifers was registered during February at all the stations. Numerical abundance of lucifers varied form 0 to 3250/100 m^3 along Mangalore section; from 0 to 639/100 m^3 along Malpe section and between 0 and 1165/100 m^3 along Gangolli section, in the columnar waters along Dakshina Kannada coast.

Decapod Larvae

Deep-water stations registered less numbers of these larvae than the shallower stations. The number of decapod larvae ranged from 0 to 200000/100 m^3; from 0 to 75000/100 m^3 and between 0 to 86300/100 m^3 along Mangalore section; Malpe section and Gangolli section respectively.

Oikopleura

Oikopleurans formed an important component of zooplankton. Major peak for three sections was noticed during February/March/April 1988. The occurrence of secondary peak varied from one station to the other. Near shore stations recorded higher numbers than those of off shore stations. Numerical abundance of this group varied from 0 to 24500/100 m^3; from 0 to 294727/100 m^3 and between 0 and 166000/100 m^3 along Mangalroe section, Malpe section and Gangolli section respectively.

Other Groups

The several other groups that contributed to the zooplankton population were ctenophores, siphnophores, polychate larvae, evadne sp., ostrocods, gastropod larvae, bivalve larvae, salpids, doliolids, fish eggs and fish larvae.

Discussion

In the present study zooplankton biomass distribution showed two peaks, the first peak occurring during September and the second peak occurring during February/November, thus exhibiting a bimodal seasonal fluctuation. Ramamurthy (1965) reported the occurrence zooplankton abundance from April to November along the North Kanara coast and attributed this abundance to the effect of upwelling. In the present study, along Dakshina Kannada coast the maximum abundance of zooplankton in general appears to coincide with phytoplankton, abundance which can be attributed to the occurrence of upwelling. Upwelling was found to occur during the southwest monsoon months along this coast. Suresh and Reddy (1978), Benkappa *et al.* (1979) and Pai (1980) also observed similar bimodal fluctuation in zooplankton abundance, along certain parts of Dakshina Kannada coast.

In the present study Medusae exhibited a bimodal oscillation in its seasonal variations, with the first peak during November/December and the second peak during February/April 1988. The higher values of hydromedusae during February/April 1988 in the present study may be attributed to clear waters prevailing during that period. The distribution of Chaetognaths exhibited a bimodal oscillation with a major peak during February and a minor peak during November/December. Ramamurthy (1965) observed abundance of Chaetognaths during October to February and reported that sea water temperature range of 26.5–28.2 C and salinity range of 30.16–35.41 per cent were optimum conditions for maximum occurrence of Chaetognaths along the west coast of India. The peak noticed in November/December could be attributed to breeding season of Chaetognaths as opined by Menon *et al.* (1977). During the present investigation the higher number of Evadne and Penilia occurred during November/December and February. Ramamurthy (1965) also reported occurrence of abundance of Cladocerons during September/October along North Kanara coast and he described both Evadne and Penilia as cold water forms. Purushan *et al.* (1970) recorded peak abundance of Cladocerons in the low saline waters off Kerala coast. In the present study also abundance of Cladocerons occurred during the months of moderate salinity conditions.

In the present study copepods were among the most important constituents of zooplankton. Seasonal abundance showed two peaks, one during February and other during November/December. Ramamurthy (1965) reported sudden fall in copepod number during south-west monsoon season when the salinity of coastal water decreased considerably, thus suggesting copepods generally require moderately higher saline conditions. In the present study also peak abundance of copepods occurred during moderately high saline conditions. The presence of Copepodites through out the period of investigation indicated the continuous breeding nature of copepods in the observation region as opined by Prasad (1958) along Palk Bay. Lucifers were observed in comparatively lesser numbers. The seasonal distribution showed two peaks, one during February/March and the other in November. Mukundan (1967) observed abundance of Lucifers from November to January in the inshore waters off Calicut. In the present study decapods larvae were recorded during both pre-monsoon period from February to April and also during post-monsoon period from November to January but their number widely varied from station to station. Ramamurthy (1965) recorded high number of decapods larvae during the period from October to April along the North Kanara coast. In the present investigation, Oikopleura was found in abundance during the months of February and March forming a major peak.

The minor peak was found in the month of December. This may be attributed to the fact that Oiklopleura thrives well in moderate saline conditions. Menon *et al.* (1977) observed Oikopleura through out the year off Mangalore with higher abundance occurring from November to February.

Acknowledgements

The authors are thankful to Prof. H.P.C. Shetty, Director of Instruction, College of Fisheries, UAS, Mangalore, for providing all the facilities for conducting the research. The research work carried out forms the part of the Ph.D. thesis of the first author.

References

Benakappa, S., M.P.M. Reddy and V. Hariharan (1979). Distribution of plankton in the Arabian Sea between Mukka and Kaup, South Kanara. *Mysore J. Agri. Sci.,* 13: 454–463.

Chidambaram, K. and M.D. Menon (1945). The correlation of the (Malabar and South Kanara) fisheries with plankton and certain oceanographical factors. *Proc. Indian Acad. Sci.,* 22: 355–357.

Kumar, R.S. (1984). Studies on the distribution of plankton in waters off Mangalore. *M.F.Sc Thesis,* University of Agricultural Sciences, Bangalore, pp. 231.

Menon, N.R., T.R.C. Gupta, V. Hariharan, R.J. Katti and H.P.C. Shetty (1977). Zooplankton of Mangalore waters: A population assessment. *Proc. Symp. Warm Water Zoopl. Publ.,* UNESCO/NIO, pp. 274–284.

Mukundan, C. (1969). Plankton of Calicut inshore waters and its relationship with coastal pelagic fisheries. *Indian J. Fish.,* 14(1–2): 271–292.

Pai, R. (1980). Studies on the oceanographic features in relation to fisheries off Malpe region, South Kanara. *M.F.Sc. Thesis,* University of Agricultural Sciences, Bangalore, pp. 237.

Prasad, R.R. (1958). Note on the occurrence and feeding habits of Noctiluca and their effects on the plankton community and fisheries. *Proc. Indian Acad. Sci.,* B, 47: 331–337.

Purushan, K.S., T. Balachandran and M. Sakthivel (1974). Zooplankton abundance off the Kerala coast during February and April, 1970. *Mahasagar,* 7: 165–175.

Ramamurthy, S. (1965). Studies on the plankton of the North Kanara coast in relation to pelagic fishery. *J. Mar. Biol. Assoc.,* India, 7(1): 127–149.

Subrahmanyan, R. (1959b). Studies on the phytoplankton of the west coast of India. Part II: Physical and chemical factors influencing production of phytoplankton with remarks on the cycle of nutrients and on the relationship of the phosphate content to fish landings. *Proc. Ind. Acad. Sci.,* B 50: 139–252.

Suresh, K. and M.P. Reddy (1978). Seasonal variation in plankton of the near shore waters off Mangalore. *J. Mar. Boil. Ass.,* India, 17(3): 664–674.

Chapter 19

Biochemical Composition and Calorific Potential of Zooplankton from the Mangrove Waters and the Bay Environment of Kakinada, South East Coast of India

N. V. Prasad

Division of Marine Biology, Department of Zoology, Andhra University, Visakhapatnam–530 003, A.P.

ABSTRACT

Zooplankton dry weight, biochemical constituents and calorific values have been determined from the mixed zooplankton collected from the mangrove-Bay environment of Kakinada. Zooplankton dry weight varied from 10.75 $mg.m^{-3}$ to 401.97 $mg.m^{-3}$. Of the principal biochemical constituents, protein formed the major component. Overall mean values reckoned as percentage of dry weight as 48.98 per cent protein, 10.55 per cent lipid, 5.68 per cent carbohydrate, 42.3 per cent organic carbon and 3.36 $k.cal.g^{-1}$ calorific value. The organisms in the Bay region had relatively less protein content than the plankton inhabiting in the mangrove habitat. All the three biochemical constituents showed seasonal variations in their composition. Significant positive correlation ($p < 0.01$) was observed between biochemical constituents indicate to certain extent act as a metabolic reserve in the zooplankton. The results revealed that the zooplankton of the mangrove-Bay environment of Kakinada, do not have extensive lipid and carbohydrate storage, suggesting that the protein in addition to lipid and carbohydrate serve as metabolic reserve. A strong correlation between protein content and calorific values was indicated that the calorific value greatly dependent on the protein composition of zooplankton in the dry weight.

Keywords: *Zooplankton, Biochemical composition, Calorific value, Mangrove-bay waters, Kakinada, Indian coast.*

Introduction

Studies on the biochemical aspects of zooplankton are very important in understanding their nutritive value and physiological functions which are relevant to the studies in the energy transfer and secondary production of any ecosystem (Madhupratap *et al.,* 1979). Zooplankton plays very significant role in marine food chain and in trophic dynamics. Zooplankton in the form of secondary producers acts as the connecting link between the primary producers like phytoplankton and higher carnivores (fish and large heterotrophic organisms). There should be always an equilibrium between the availability of nutrients, phytoplankton abundance and zooplankton grazing, since these three parameters control over all biological productivity of the system to a greater extent (Castro *et al.,* 1991; Buskey and Stockwell, 1993). The heterotrophic energy derived from the primary production at the primary consumer level is readily available to the higher trophic level in the form of three basic principle biochemical components, *i.e.* protein, lipid and carbohydrates. The study of these biochemical constituents of zooplankton community yields a picture regarding their feeding habits and also gives a comprehensive picture of the amount of potential energy available for the secondary producers to survive. Though there are many studies on the zooplankton composition and distribution, very few reports are available on the biochemical composition of zooplankton (Stephen *et al.,* 1979; Nandakumar *et al.,* 1988; Krishna Kumari and Achuthankutty, 1989; Bhat and Wagh, 1992; Krishna Kumari and Goswani, 1993; Maruthanayagam and Subramaniam, 1999; Goswami *et al.,* 2000; Nageswara Rao and Krupanidhi, 2001; Lau and Lane, 2002). The present paper deals with biomass, major biochemical constituents, organic carbon 'and calorific content of the mixed zooplankton from the mangrove-Bay environment of Kakinada.

Materials and Methods

Zooplankton dry weight and biochemical composition of mangrove-Bay environment of Kakinada including Kakinada Bay, Gaderu and Coringa rivers were studied during March 1997 to February 1998. Altogether 6 stations, two each in Kakinada Bay (Stations 1 and 2), Gaderu (Stations 3 and 4) and Coringa (Stations 5 and 6) were selected (Figure 19.1). Zooplankton samples were collected using 40 cm diameter net of 120 mμ mesh size. As soon as the plankton net was hauled the samples were deep frozen on the boat, later brought to the laboratory and washed with distilled water and dried at 60 C until constant weight was obtained. The protein content was determined by the method of Lowry *et al.* (1951). The lipid content was estimated by the method of Folch *et al.* (1957) and the total carbohydrate content by the method of Dubuois *et al.* (1956). The ogranic carbon content was determined by the method of Parsons *et al.* (1984). The calorific content was estimated by using the conversion factors of 5.7, 9.3 and 4.0 k.cal.g^{-1} (Wineberg, 1971).

Results and Discussion

Protein, lipid, carbohydrate, organic carbon and calorific value (per cent dry weight) of the mixed zooplankton in the mangrove-Bay environment of Kakinada is given in the Table 19.1. In the present study, the protein content ranged from a minimum of 17.76 (Station 3) to a maximum of 73.47 per cent (Station 6). Carbohydrate content varied from 1.55 (Station 1) to 12.73 per cent (Station 2) and lipid ranged from 4.61 (Station 1, near proximity of sea) to 15.05 per cent (Station 3, mangrove environment). Organic carbon in the mixed zooplankton varied from 20.16 (Station 6) to 49.00 per cent (Stations 4 and 5) and the calorific value varied from 2.06 to 5.66 k.cal.g^{-1} dry weight. Whereas the dry weight of the zooplankton varied from 10.75 to 401.97 mg.m^{-3}. During the present study, high values of dry weight, organic carbon values were observed in the Bay region (Stations 1 and 2) and lower values were observed in Coringa (Stations 5 and 6) (Figure 19.2). Whereas higher protein content and calorific values were observed in the mangrove habitat (Stations 3, 4, 5 and 6) relative to Bay stations (Stations 1 and 2) (Figure 19.2). Significant positive correlation ($p > 0.01$) were observed between total population, displacement volume, dry weight, biochemical constituents, organic carbon and calorific value implying that biochemical constituents play an important role in energy metabolism. A strong correlation ($r = 0.88$) between protein content and calorific value was indicated that the calorific value greatly dependent on the protein composition of zooplankton in the dry weight.

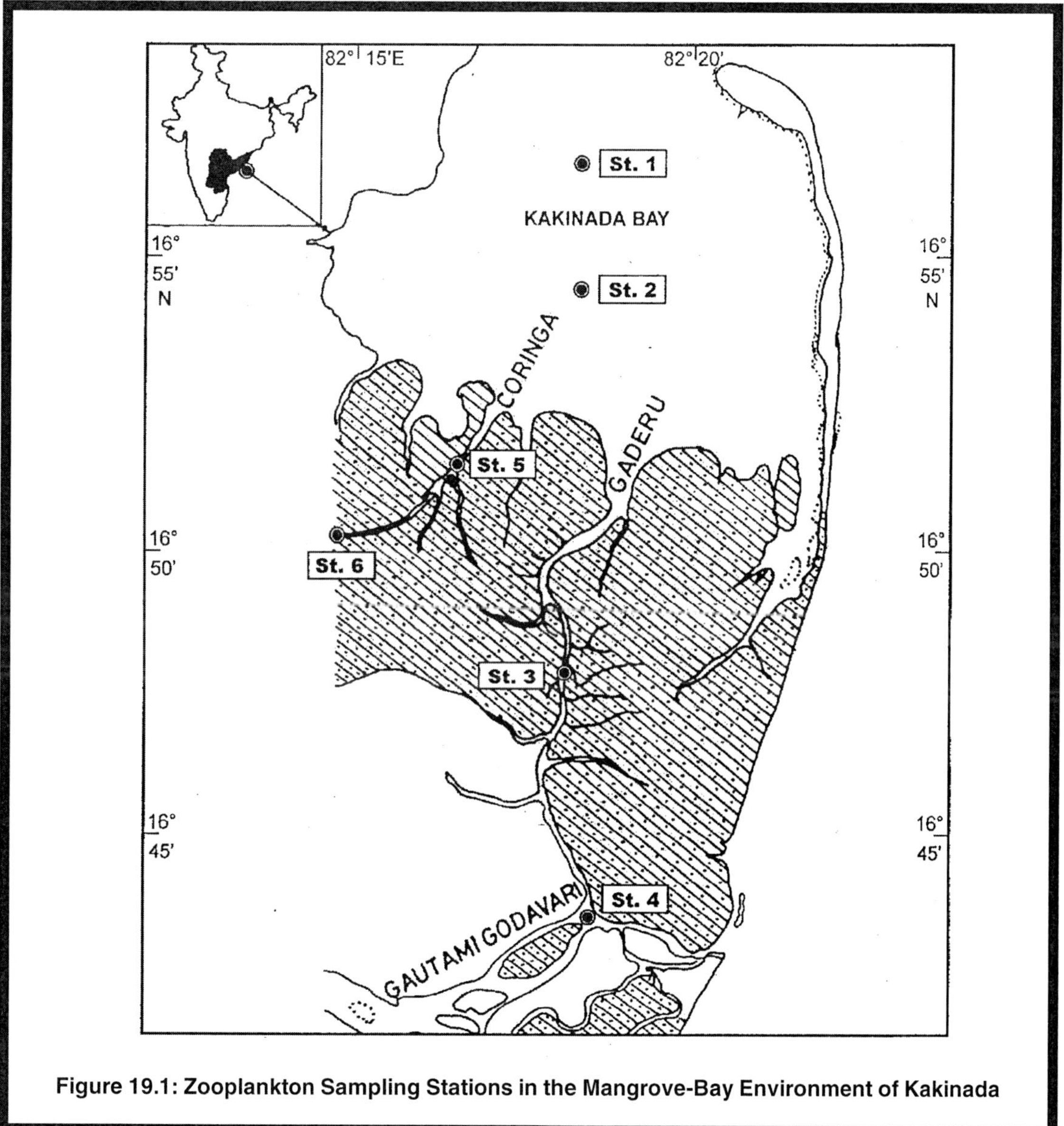

Figure 19.1: Zooplankton Sampling Stations in the Mangrove-Bay Environment of Kakinada

Table 19.1: Range of Biochemical Constituents and Calorific Values of Zooplankton in the Mangrove-Bay Environment of Kakinada (values in parentheses denote mean)

Component	*St. 1*	*St. 2*	*St. 3*	*St. 4*	*St. 5*	*St.6*
Dry Weight	24.3–306.4	33.88–401.97	27.51–288.0	15.69–112.36	10.75–126.33	11.46–122.34
(mg.m^{-3})	(130.05)	(156.18)	(124.73)	(107.21)	(67.0)	(71.26)
Organic Carbon	27.64–47.31	35.65–47.70	30.13–47.94	33.77–49.00	21.72–49.00	20.16–51.23
(% dry weight)	(43.34)	(44.70)	(43.10)	(42.29)	(40.72)	(40.02)
Protein	29.47–53.14	27.56–66.79	17.76–53.87	23.41–63.24	25.38–70.01	24.02–73.47
(% dry weight)	(43.55)	(45.19)	(48.43)	(46.27)	(56.33)	(54.12)
Carbohydrate	1.55–7.04	1.66–12.73	2.46–9.3	2.93–9.45	2.38–8.89	2.19–9.12
(% dry weight)	(4.10)	(6.70)	(5.26)	(6.07)	(5.96)	(6.01)
Lipid	4.61–12.7	8.5–14.6	8.0–15.05	8.15–12.62	7.1–13.0	7.09–12.86
(% dry weight)	(10.77)	(10.60)	(10.78)	(10.84)	(10.18)	(10.13)
Calorific value	2.96–4.84	2.81–5.05	2.37–5.12	2.81–5.66	2.92–5.09	2.06–5.11
(k.cal.g^{-1})	(3.16)	(3.01)	(3.57)	(3.69)	(3.31)	(3.44)

Among the biochemical constituents the protein formed the major component in the mixed zooplankton (Table 19.1). The organisms in the Bay region had relatively less protein content than the plankton inhabiting in the other areas like Gaderu and Coringa rivers, where mangrove vegetation is dense (Figure 19.2). The plankton population dwelling in mangrove area were dominated by detritus

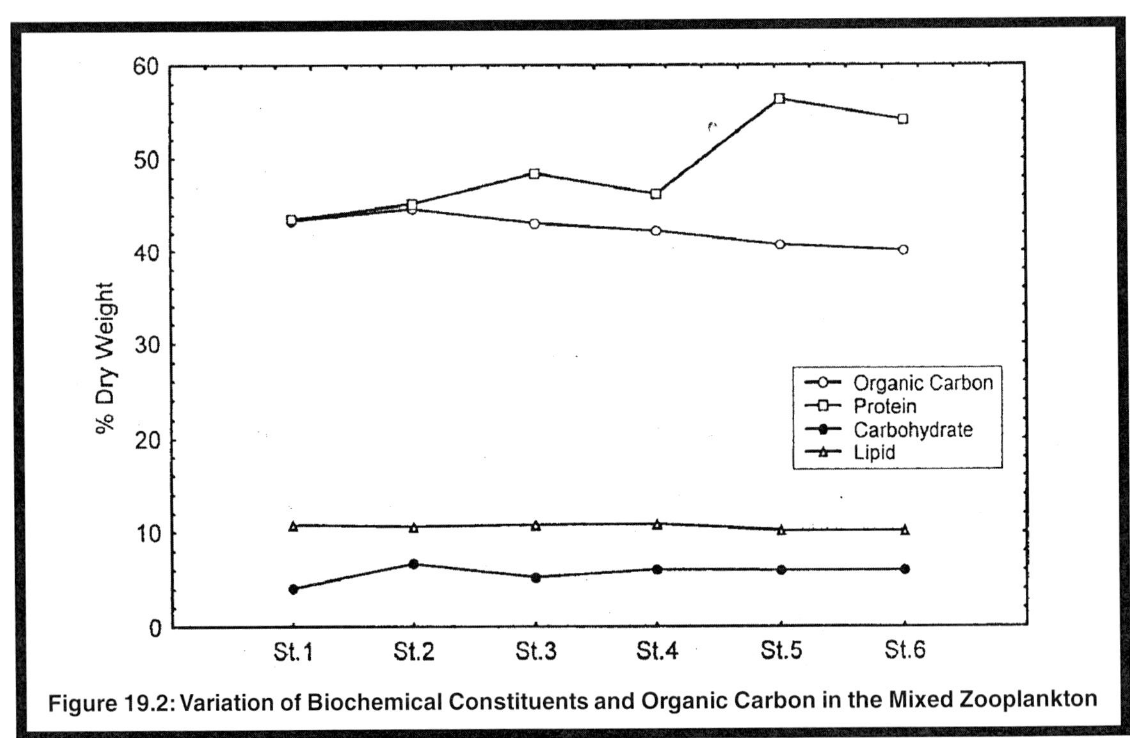

Figure 19.2: Variation of Biochemical Constituents and Organic Carbon in the Mixed Zooplankton

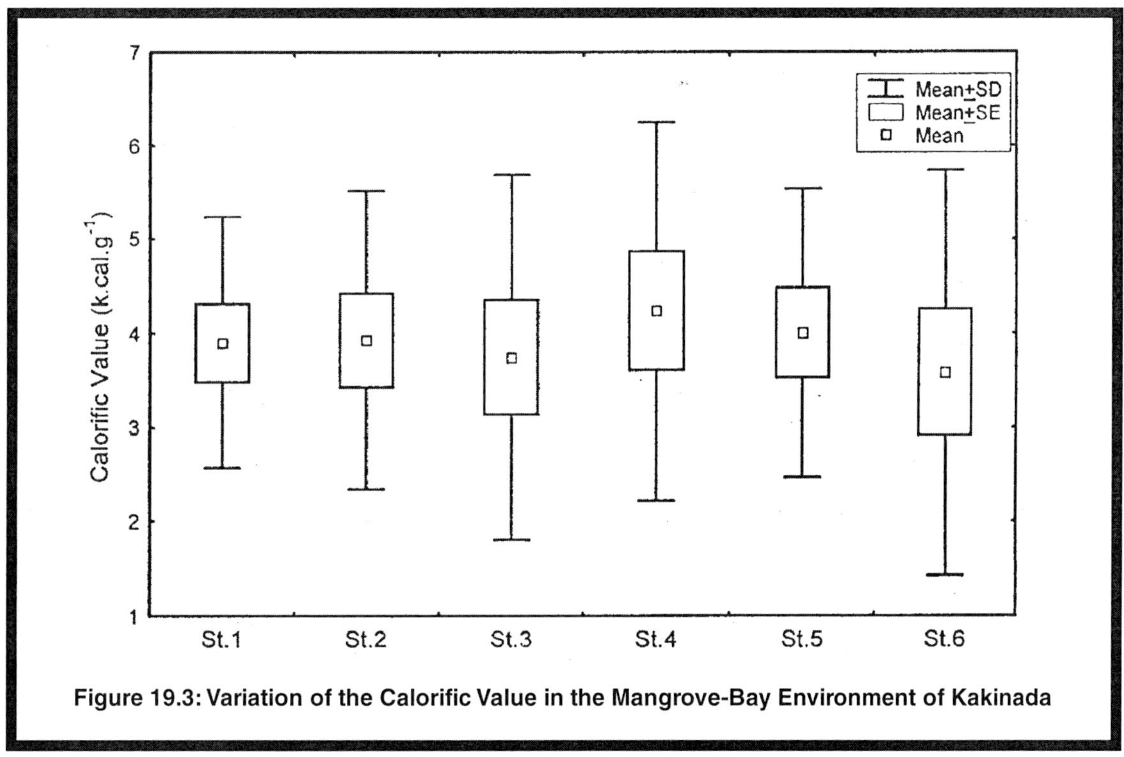

Figure 19.3: Variation of the Calorific Value in the Mangrove-Bay Environment of Kakinada

feeders mostly. They may be accumulating more of the protein than of the lipid and carbohydrates. Protein content was relatively high during the late monsoon months (August and September) and pre-monsoon period, when the hydrographical conditions were suitable for the procurement of food. Protein formed major fraction compared to lipid and carbohydrate, indicating its usefulness as energy reserve (Conover, 1964; Conover and Corner, 1968). Druing the present study, variation of protein may be either due to its utilization as a metabolic substate or due to seasonal change, age or organism at the time of collection and salinity of the waters (Raymont, 1972). Goswami *et al.,* (1993) suggested that those species exhibiting the large range of diurnal migration tend to show higher protein content, perhaps greater opportunities for increased feeding in richer upper water layers and greater muscular development.

The carbohydrate content in the mixed zooplankton during the present study was relatively low and comparable with earlier reports (Goswami *et al.,* 1981; Krishna Kumari and Goswami, 1993). Though zooplankton feed on phytoplankton directly, the low carbohydrate content suggest that glycogen, the usual storage form of carbohydrate does not form a substantial part of body reserve (Stephen *et al.,* 1979). The low carbohydrate content in zooplankton also considered as one of the important metabolic substrates and the high level of protein in addition to lipid may function as reserve food material (Raymont *et al.,* 1969; Reeve *et al.,* 1970).

During the present study, the lipid content showed both seasonal and spatial variation. Lipid content was more at Gaderu river (Stations 3 and 4) than the Bay and Coringa river. This could be due to the occurrence of high lipid containing groups like copepods, decapods, chaetognaths and lucifers (Krishna Kumari and Goswami, 1993). Variations in the lipid content attributed is to its storage and utilization during starved period when it serves as an effective energy reserve. Lipid values observed in the present study are lower when compared to the reported values from colder region (Fisher, 1962). In tropical environments, the rate of primary production far exceeds the rate of consumption of zooplankton and the continuous supply of phytoplankton food would render lipid reserve unnecessary which might account for the low lipid content in the tropical zooplankton (Qasim *et al.*, 1978). The lipid content in the zooplankton can also be related directly to temperature as it has been suggested by Krishna Kumari *et al.* (1993) that the higher temperature encountered by the tropical epipelagic species may inhibit lipid deposition. The higher lipid content recorded in the present study during the summer months in Kakinada Bay (11.0–14.6 per cent dry weight) suggests that it serve as an effective energy reserve to tide over the formidable monsoon period when the hydrographical conditions are not suitable for survival and procurement of food material. The lipids are known to play a major, often crucial role in the survival strategies of organisms in polar regions (Fahl and Katter, 1993).

Organic carbon content in the Kakinada Bay and mangrove regions of Gaderu and Coringa ranged between 20.16 per cent and 51.23 per cent dry weight of zooplankton. These values were higher than those reported in earlier studies of Arabian sea (Nandakumar *et al.*, 1988), Bay of Bengal (Sreepada *et al.*, 1992) and Andaman sea (Nageswara Rao and Krupanidhi, 2001). The monthly variation in the carbon content was not significant, and it was below 5–9 per cent of dry weight. Higher values of organic carbon observed in the Bay region (Stations 1 and 2) relative to Gaderu (Stations 3 and 4) and Coringa river (Stations 5 and 6) (Figure 19.2). This may be attributed to higher population densities of copepods, decapods, chaetognaths, bivalves, gastropods and lucifers in the total zooplankton. In general, the organic carbon of zooplankton is mainly dependent upon the species composition, the size of the different population, availability of food and physiological state of the individual organisms (Omari, 1969; Nair *et al.*, 1983).

Table 19.2: Correlation Matrix of Numerical Abundance, Biomass, Biochemical Constituents, Organic Carbon and Calorific Value of Zooplankton in the Mangrove-Bay Environment of Kakinada

	NA	*BM*	*DW*	*PRT*	*LP*	*CH*	*OC*
BM	0.94						
DW	0.92	0.92					
PRT	0.90	0.89	0.95				
LP	0.87	0.90	0.94	0.95			
CH	0.76*	0.72*	0.70*	0.62*	0.61*		
OC	0.81	0.73*	0.83	0.76*	0.72*	0.53*	
CV	0.90	0.90	0.96	0.96	0.98	0.71*	0.74*

* $P < 0.05$

NA: Numerical abundance; BM: Biomass; DW: Dry weight; PRT: Protein; CH: Carbohydrate; OC: Organic carbon; CV: Calorific value.

The mean calorific content at Kakinada Bay, Gaderu and Coringa were 3.08, 3.63, 3.37 k.cal.g^{-1}.The stations located in the mangrove dominant areas (Stations 3–6) had higher calorific content (Figure 19.3). The higher protein values in the percentage dry weight of zooplankton may be the reason for this high calorific values. Strong correlation ($r = 0.88$) between protein content and calorific content indicates that the calorific values is greatly dependent on the protein composition of dry weight of organism. The present observations agree with studies made by Bhat and Wagh (1992,

Bombay coastal waters) and Gupta (1977, Cochin backwaters). Variations in the calorific content were noticed between groups and species of zooplankton. The crustaceans particularly copepod species of zooplankton showed higher calorific values. Madhupratap *et al.* (1979) reported that the calorific content of zooplankton is limited by the protein content and also the presence of crustacean species. The reserve protein content determines the nutrient input into the digestive processes, thus the mangrove detritus feeding larval forms and also the pelagic meroplankton had higher protein content leading to the higher calorific values.

In the present study protein, lipid, carbohydrates, organic carbon and calorific values significantly correlated ($r = 0.74–0.92$) with copepods, decapods and chaetognaths, indicating that major fractions of biochemical constituents are derived from these group of plankton. Significant positive ($p < 0.01$) correlations (Table 19.2) among biochemical constituents, organic carbon and calorific values indicate that biochemical components play an important role in the energy metabolism in the mangrove-Bay environment of Kakinada. In conclusion it may be clearly stated that the variations in biomass and biochemical constituents are influenced by the species composition of zooplankton. Protein formed a major component and may serve as the main metabolic reserve as reported from the other areas. Since the Kakinada Bay and the surrounding mangrove regions are highly suitable environments for the zooplankton survival as the waters of the mangrove area and Kakinada Bay are nutrient rich due to the instantaneous input from the surrounding areas, the region supports high biomass of plankton 'population with varying biochemical constituents seasonally.

References

Bhat, K.L. and A.B. Wagh (1992). Biochemical composition of zooplankton of Bombay High area in the Arabian sea. *Indian J. Mar. Sci.,* 21: 220–223.

Buskey, E.J. and D.A. Stockwell (1993). Effects of a Persistent brown tide on zooplankton populations in the laguna Madre of South Texas. In: *Toxic phytoplankton Blooms in the Sea,* (Eds.) T.J. Smayda and Y. Shimizu, Elsevier Sci. Pub., p. 659–666.

Castro, L.R., P.A. Bernal and H.E. Gonzalez (1991). Vertical distribution of copepods and the utilization of the chlorophyll a rich layer within conception Bay, Chile. *Estuarine, Coastal and Shelf Science,* 32: 243–256.

Conover, R.J. and E.D.S. Corner (1968). Respiration and nitrogen excreation by some marine zooplankton in relation to their life cycles. *J. Mar. Biol. Ass.* U.K., 48: 47–49.

Conover, R.T. (1964). Food relations and nutrition of zooplankton. Proceedings of symposium on experimental marine ecology, Pub. No. 2, University of Rhode Island, p. 81–91.

Dubuois, M., K.A. Gilles, J.K. Hamilton and F. Smith (1956). Colorimetric method for determination of sugars and related substances. *Analyst. Chem.,* 28: 350–356.

Fahl, K. and G. Katter (1993). Seasonal variation in lipid content of two Antarctic Marine Crustacea. *Polar Biol.,* 13: 405–411.

Fisher, L.R. (1962). The total lipid material in some species of marine zooplankton. *Rapp. Cons. Explor. Mer.,* 153: 129–136.

Folch, J., M. Less and G.H. Sloane Stanley (1957). A simple method for the isolation and purification of total ;ipids from animal tissue. *J. Biol. Chem.,* 226: 497–509.

Goswami, S.C., T.S.S. Rao and S.G.P. Matondkar (1981). Biochemical composition of zooplankton from the Andaman sea. *Indian J. Mar. Sci.,* 10: 296–300.

Goswami, S.C., T.S.S. Rao and S.G.P. Matondkar (1993). Biochemical studies on some zooplankton off the west coast of India. *Mahasagar-Bull. Nat. Inst. Oceangr.,* 14: 313–316.

Goswami, S.C., L. Krishna Kumari and Y. Shrivastava (2000). Diel variation in zooplankton and their biochemical composition from Vengurla to Ratnagiri, West Coast of India. *Indian J. Mar. Sci.,* 29: 277–280.

Gupta, T.R.C. (1977). Studies on the chemical composition of zooplankton from coastal waters of Cochin. *Proc. Symp. Warm Water-Zooplankton,* UNESCO/NIO Spl. Pub., p. 511–514.

Krishna Kumari, L. and C.T. Achuthankutty (1989). Standing stock and biochemical composition of zooplankton in the northeastern Arabian sea. *Indian J. Mar. Sci.,* 18: 103–105.

Krishna Kumari, L. and S.C. Goswami (1993). Biomass and biochemical composition of zooplankton from the offshore oil fields of Bombay. *Proc. Nat. Acad. Sci.,* India., 63(B)II: 161–167.

Krishna Kumari, L., V.R. Nair and S.N. Gajbhiye (1993). Biochemical composition of zooplankton from the offshore oil fields of Bombay. *Proc. Nat. Acad. Sci.,* 63(B): 161–167.

Lau, S.S.S. and S.N. Lane (2002). Biological chemical factors influencing shallow lake eutrophication: A long-term study. *Science of Total Environment,* 288(3): 167–181.

Lowry, O.H., N.J. Rosenberg, A.L. Fare and R.J. Randall (1951). Protein measurement with the Follin-Phenol reagent. *J. Bio. Chem.,* 193: 265–275.

Madhupratap, M., P. Venugopal and P. Haridas (1979). Biochemical studies on some tropical estuaries zooplankton species. *Indian J. Mar. Sci.,* 4: 172–180.

Maruthanayagam, C. and P. Subrahmanian (1999). The zooplankton of the inshore waters of Palk Bay and Gulf of Mannar along the east coast of India. *J. Eco. Biol.,* 13(3): 205–211.

Nageswara Rao, I. and G. Krupanidhi (2001). Biochemical composition of zooplankton from the Andaman sea. *J. Mar. Biol. Ass.,* India, 43(1 and 2): 49–56.

Nair, V.R., C.T. Achutankutty and S.R.R. Nair (1983). Zooplankton variability in the Zuari estuary, Goa. *Mahasagar-Bull. Nat. Inst. Oceangr.,* 16(2): 235–242.

Nandakumar, K., L.K. Bhat and A.B. Wagh (1998). Biochemical composition and calorific value of zooplankton from Northern part of Central Arabian Sea. *Indian J. Mar. Sci.,* 17: 48–50.

Omari, M. (1969). Weight and chemical composition of some important oceanic zooplankton in the northern Pacific Ocean. *Marine Biology,* 3: 4–10.

Parsons, T.R., Y. Meritha and C.M. Lalli (1984). *A Manual of Chemical and Biological Methodology for Sea Water Analysis.* Pergamon Press, Oxford, pp. 173.

Qasim, S.Z., M.V.M. Wafar, S. Vijayaraghavan, I.P. Royan and L. Krishna Kumari (1978). Biological productivity of coastal waters of India from Dabhol to Tuticorin. *Indian I. Mar. Sci.,* 7: 84–93.

Raymont, J.E.G. (1972). *Essays in Hydrobiology.* (Eds.) R.B. Clark and R.J. Wootton, University of Siuthampton, p. 83–91.

Raymont, J.E.G., R.T. Srinivasagam and J.K.B. Raymont (1969). Biochemical studies on zooplankton, VII observation on certain deep sea zooplankton. *Int. Rev.Ges. Hydrobiol.,* 54: 357–365.

Reeve, M.R., J.E.B. Raymont and J.K.B. Raymont (1970). Seasonal biochemical composition and energy source of Sagitta hispida. *Marine Biology,* 6: 357–364.

Sreepada, R.A., C.V. Rivonker and A.H. Parulekhar (1992). Biochemical composition and calorific potential of zooplankton from Bay of Bengal. *Indian J. Mar. Sci.,* 21: 70–73.

Stephen, R., S.V. Panapunnayil, T.C. Gopala Krishnan and V.N. Sankaranarayanan (1979). Biochemical studies on zooplankton from the Laccadive sea. *Indian J. Mar. Sci.,* 8: 257–258.

Wineberg (1971). *Methods for the Estimation of Production of Aquatic Animals,* Academic Press, London, pp. 161.

Chapter 20

Seasonal Fluctuation of Different Zooplanktonic Groups of a Rainfed Wetland in Relation to Some Abiotic Factors

S.B. Patra and N.C. Datta

Fishery and Ecology Research Unit, Department of Zoology, University of Calcutta, 35, Ballygunge Circular Road, Kolkata – 700 019

ABSTRACT

The zooplanktonic community of the wetland was represented by Protozoa, Rotifera, Cladocera, Copepoda and Ostracoda. The protozoan population includes ciliates and flagellates and the maximum and the minimum abundance were 279 and 2 ind/lit respectively. Copepodans, the largest group, comprised a minimum of 1 ind/lit and a maximum of 708 ind/lit. Rotifers, the second largest group were represented by a minimum of 2 ind/lit and a maximum of 1425 ind/lit. The other two groups cladocerans and ostracods were found to be in lower profile which showed no definite pattern of their seasonal fluctuation. In this wetland water temperature varied from 19.75 C to 34 C, transparency from 21.15 to 88 cm, pH from 6.5 to 8.4, specific conductivity from 123.48 to 358.29 µmhos/cm, dissolved oxygen ranged from 1.2 to 11.2 mg/lit, total alkalinity from 62 to 82 mg/lit, DOM from 0.38 to 5.7 mg/lit, phosphate-phosphorus from 0.13 to 1.67 mg/lit, nitrate-nitrogen from 0.03 to 0.08 mg/lit and BOD from 1.2 to 10.2 mg/lit. Although seasonal fluctuation of zooplanktonic community is largely governed by abiotic factors, but pH exerted considerable influence.

Keywords: *Zooplankton, Seasonal fluctuation, Wetland, Abiotic factors.*

Introduction

A satisfactory understanding of the ecological processes in an aquatic system requires a thorough knowledge not only of the organisms but also of the environmental factors. Probably Prasad (1916) was the first to study the limnological characteristics of fresh water pond in India. Subsequently, several workers studied water bodies from limnological view point (Pruthi, 1933; Sewell, 1934; Sreenivasan, 1970; Jana, 1973; Datta *et al.*, 1983; Datta and Bandyopadhyay, 1985; Datta and Chaudhuri, 1986; Sharma and Rao, 1990; Banik, 1995; Michael and Sharma, 1998; Datta and Das, 1999; Biswas and Konar, 2000; Sharma *et al.*, 2000; Singh *et al.*, 2001). The present paper portrays the role of abiotic factors on the seasonal fluctuation of different zooplanktonic groups.

Materials and Methods

The chosen wetland is a rainfed one and situated by the side of Nazrul Mancha at Golpark, Kolkata (Lat 22 31′ N and Long 88 22′ E). The surface area of this wetland is about 0.4 ha and the average depth is 3.0 meter. Half of the wetland is covered with *Ipomea* sp., *Nelumbo* sp., *Azolla* sp., *Lemna* sp. etc. Neither fish culture nor any other domestic use has been noticed in this wetland.

Surface water samples and zooplankton were collected weekly for two years from January 1996 to December 1997 between 9 a.m. and 10 a.m. The data are presented as monthly mean. For the physico-chemical analysis of water Welch (1948), Jhingran *et al.* (1969), Michael (1990) and APHA (1995) were followed. The zooplankton were collected with a plankton-net made up of bolting silk No. 25 and their identification was done following Pennak (1955), Edmondson (1959), Ward and Whipple (1959), Battish (1992), and Michael and Sharma (1998).

Results and Discussion

Zooplankton is considered as one of the most important linkages in aquatic food chain and shows continuous seasonal variation. During the entire study period water temperature (WT) varied from 19.75 C to 34 C and showed its peaks in June (30.25 C) in 1st cycle whereas it was in May (34 C) in the second cycle (Table 20.1). On the other hand, transparency (TRN) ranged from 21.15 to 88.00 cm and showed its peaks in January in both the cycles (Table 20.1). pH ranged from slightly acidic (6.5) to alkaline (8.43) and showed its peaks in April (8.0) in the first cycle and March (8.43) in the second cycle (Table 20.1). Specific conductivity (Sp.CON) varied from 123.48 to 358.29 µmhos/cm and showed its peaks in August in 1st cylce and March in the second cycle (Table 20.1). Dissolved Oxygen (DO) content varied from 1.2 to 11.2 mgl^{-1} and showed its peak in February in the first cycle whereas March in the second cycle (Table 20.1). Total Alkalinity (TA) ranged from 62 to 82 mgl^{-1} and showed its peak in January in the first cycle and August in the second cycle (Table 20.1). The value of DOM ranged from 0.38 to 5.7 mgl^{-1} (Table 20.1). During study its peaks were in January in the first cycle and November in the second cycle. Phosphate-phosphorus (PO_4–P) varied from 0.13 to 1.67 mgl^{-1} and showed its peaks in January in both the cycles (Table 20.1). Nitrate-nitrogen (NO_3–N) was found in traces (Table 20.1) and BOD varied from 1.2 to 10.2 mgl^{-1} and its peaks were in September in both the cycles (Table 20.1).

In this wetland total zooplankton showed three peaks in the first cycle, the first peak (1031 unit/lit) was in February followed by a gradual decrease upto April (Table 20.2). The second and the highest peak (1607 unit/lit) was in July and the third peak (106 unit/lit), though small, was in September (Table 20.2). In the second cycle the highest peak (1815 unit/lit) was in April and other two small peaks were in July (181 unit/lit) and November (282 unit/lit) (Table 20.2). Das and Srivastava (1956), George (1966), Jana (1973), Vasisht and Sarma (1975), Nasar (1977), Ayyapan and Gupta (1980) and Sugunan (1980) also reported three peaks.

Table 20.1: Monthly Variations of Physico-chemical Characteristics of Water in a Rainfed Freshwater Wetland During January 1996 to December 1997

Year/Month	WT	TRN	pH	Sp.CON	DO	TA	DOM	PO_4	NO_3	BOD
1996										
January	20.00	88.00	6.90	246.96	10.40	82.00	4.13	1.67	0.03	8.60
February	22.50	80.30	7.55	267.54	11.20	66.00	2.55	0.77	0.03	3.60
March	29.00	77.70	7.85	329.28	10.40	68.00	0.61	0.21	0.05	5.60
April	28.00	68.80	8.00	305.27	7.60	70.00	2.44	0.29	0.04	1.20
May	30.00	50.20	7.10	219.52	2.00	66.00	0.98	0.25	0.08	5.00
June	30.25	45.60	7.18	234.96	2.70	63.50	0.87	0.52	0.04	1.30
July	29.63	40.78	6.88	236.67	2.10	62.00	2.38	0.47	0.06	3.60
August	29.30	31.83	6.80	353.29	2.10	61.50	2.23	0.28	0.04	1.13
September	29.75	21.15	6.77	310.42	4.30	63.50	2.63	0.23	0.05	1.20
October	27.50	28.95	6.67	343.00	1.50	65.50	2.57	0.47	0.06	3.47
November	26.38	38.5	7.02	327.57	1.80	66.00	1.79	0.62	0.05	8.17
December	20.38	46.00	6.74	301.84	1.20	68.50	0.80	1.36	0.04	6.34
1997										
January	19.75	82.50	6.93	243.53	3.60	69.00	1.54	1.67	0.03	5.24
February	23.25	74.00	7.60	264.11	10.20	69.00	2.01	0.74	0.05	4.60
March	30.50	63.25	8.43	327.57	10.60	71.00	1.77	0.36	0.05	3.80
April	29.00	53.50	8.13	315.56	8.60	66.00	2.14	0.73	0.05	3.40
May	34.00	40.50	7.80	144.06	7.40	66.00	0.75	0.38	0.06	3.20
June	33.00	38.30	7.35	322.42	8.20	60.00	1.81	0.39	0.04	3.40
July	30.50	34.00	7.08	260.68	6.20	69.00	1.65	0.32	0.05	2.94
August	30.00	31.30	6.50	192.08	2.40	76.00	3.23	0.31	3.05	3.33
September	29.50	37.00	6.63	154.45	1.30	68.00	1.73	0.15	0.03	6.67
October	28.00	45.60	6.65	123.48	2.00	62.00	0.38	0.13	0.03	4.67
November	26.00	52.30	6.69	123.48	2.80	66.00	5.70	0.13	0.04	2.00
December	23.00	61.40	6.75	130.34	2.60	69.00	4.95	0.36	0.03	2.00

According to Welch (1952) less zooplankton species are found in tropical water bodies than the temperate one. Contrary to above, the present study recorded large number of species. Altogether 76 species were observed and of them 57 were of Rotifera, 13 of Cladocera, 5 of Copepoda and 1 of Ostracoda. It is relevant to mention that Sewell (1934) recorded 10 species of Rotifera, 15 of Cladocera, 10 of Copepoda and 1 of Ostracoda from Indian Museum tank. George (1966) reported 32 species of Rotifera, 11 of Cladocera and 7 of Copepoda from five fresh water ponds in Delhi. Nasar (1977) reported 12 species of Rotifera, 8 of Cladocera, 3 of Copepoda and Ostracoda each from a pond of Bhagalpur.

The total number of protozoans was at its peak during February (27 units/lit), September (21 unit/lit) and June (27 unit/lit) in the first cycle. But in the second cycle there was only one large peak in April (223 unit/lit) and the other small peak in September (18 unit/lit) (Table 20.2). In accordance with Michael (1969) it may be stated that the nature and the amount of available food is the controlling factor in the distribution of fresh water protozoans, especially the ciliates.

Table 20.2: Monthly Numerical Abundance of Different Zooplanktonic Groups (ind/lit) of a Rainfed Freshwater Wetland from January 1996 to December 1997

Year/Month	*Protozoa*	*Rotifera*	*Cladocera*	*Copepoda*	*Zooplankton*
1996					
January	77	75	3	31	186
February	279	49	103	600	1031
March	25	54	11	174	264
April	3	36	0	51	90
May	7	34	3	152	196
June	27	790	101	689	1607
July	5	103	22	324	454
August	4	63	7	24	98
September	21	100	9	76	206
October	4	43	1	3	51
November	11	56	0	1	68
December	11	33	0	4	48
1997					
January	40	97	2	77	216
February	68	24	95	272	459
March	124	419	0	708	1251
April	223	1425	10	165	1823
May	21	109	2	103	235
June	2	16	2	28	48
July	5	65	8	103	183
August	3	90	2	53	148
September	18	22	2	34	76
October	6	61	0	1	68
November	9	14	3	256	282
December	15	9	7	183	214

Though the dominance of rotifers was observed by many workers, but in the present investigation copepodans constituted the largest group which is in conformity with Das and Srivastava (1959). So, the fluctuation pattern of the total community was influenced by the fluctuation of this group. In this wetland this group was more abundant in summer and monsoon and less abundant in winter months which corroborates the observation of Bandyopadhyay (1985).

In the present investigation rotifers constituted the second largest group and showed a number of peaks. In the first cycle it showed two peaks, one in June (790 unit/lit) and the other in September (100 unit/lit). In the second cycle there were also two peaks, one in summer (April–1425 unit/lit) and the other in monsoon (August–90 unit/lit) (Table 20.2). The irregular periodicity in the abundance of rotifer population has already been indicated by several authors (Pennak, 1955; Krishnamoorty and Visweswara, 1966; Vasisht and Sarma, 1976; Mukhopadhyay *et al.*, 1981; Datta *et al.*, 1984). It may be mentioned that according to Reid and Wood (1976) rotifers never follow any predictable population pattern in fresh water impoundment.

Cladocerans were in lower profile in both the annual cycles and as such no definite pattern of their variation was observed. However, they are mostly abundant in winter and summer months (Table 20.2). The present observation is in agreement with Hutchinson (1967), Prasadam (1977) and Datta *et al.* (1984).

Ostracods were found to be in less numbers. The population dynamics of this group is not clearly known though some species exhibit distinct seasonal periodicity (Wetzel, 1983). According to Pennak (1978) ostracods can tolerate wide range of ecological factors.

Generally in natural waters, an approximate biological equilibrium exists, although abundance of zooplankton varies from season to season. Such variations are mostly related to abiotic factors. In this study total zooplankton exhibited a positive correlation with pH ($P < 0.01$) which corroborates with Mandal (1985) and Bandyopadhyay (1985). Not only the total zooplankton but the Rotifera, Protozoa, Copepoda also show positive correlation separately with pH. However, Protozoa also showed positive correlation with DO ($P < 0.01$), transparency ($P < 0.05$) and oxygen saturation ($P < 0.01$) (Table 20.3).

Table 20.3: Simple Correlation Coefficient (r) Between Physico-chemical Parameters and Different Zooplanktonic Groups of a Freshwater Wetland

	Protozoa	*Rotifera*	*Cladocera*	*Copepoda*	*Ostracoda*	*Total Zooplankton*
WT	– 0.2451	0.1960	– 0.1664	0.0559	0.1619	0.1017
TRN	0.4723*	0.0095	0.3034	0.3064	– 0.1944	0.2205
Sp.CON	0.1937	– 0.0206	0.0284	0.0070	0.0212	0.1486
pH	0.5236**	0.4480*	0.2089	0.4558*	0.0339	0.5607**
TA	0.1241	– 0.0807	0.1224	– 0.0628	0.0799	– 0.0675
DO	0.6158**	0.1878	0.3217	0.3587	0.0627	0.3764
PO_4–P	0.2831	0.0788	0.0920	– 0.0665	– 0.1049	0.0674
NO_3–N	– 0.1335	0.1192	– 0.1680	– 0.0041	0.1094	0.0437
DOM	0.0558	– 0.1358	– 0.0897	– 0.0013	– 0.0812	– 0.0840
BOD	– 0.0093	– 0.2005	– 0.2051	– 0.3233	– 0.1293	– 0.2767

*: $P < 0.05$; **: $P < 0.01$.

It may be concluded that the growth of zooplankton is directly or indirectly influenced by the seasonal variation in the complexes of various abiotic factors. The annual changes in the community of zooplankton depend on the succession of its component species. Some plankton species increase slowly and more or less uniformly reaching to the maximum while others showed an almost starting burst of development rising from minimal population to a numerical dominance over the whole plankton within a very short period of time.

References

APHA (1995). *Standard Methods for the Examination of Water and Wastewater*, 19th edn. Washington D.C.

Ayyappan, S. and T.R.C. Gupta (1980). Limnology of Ramasundara tank. *J. Int. Soc.*, India, 12(2): 1–12.

Bandyopadhyay, B.K. (1985). Hydrobiology of some brackish and freshwater ecosystems of West Bengal, India. *Ph.D. Thesis*, University of Calcutta.

Banik, S. (1995). Zooplankton abundance in a freshwater fish farming pond in West Bengal in relation to some environmental factors. *Bangladesh J. Zool.*, 23: 3–5.

Battish, S.K. (1992). *Freshwater Zooplankton of India*. Oxford and IBH Publishing Co. Pvt. Ltd., Calcutta, pp. 232.

Biswas, B.K. and S.K. Konar (2000). Influences of Nunia Nullah (Canal) Discharge on Plankton Abundances and Diversity. In: *The River Damodar at Narankuri (Raniganj) in West Bengal. Indian J. Env. and Ecoplan.*, 3(2): 209–217.

Das, S.M. and V.K. Srivastava (1956). Some new observations on the plankton from freshwater ponds and tanks from Lucknow. *Sci. Cult.*, 21(8): 466–467.

Das, S.M. and V.K. Srivastava (1959). Studies of freshwater plankton III: Qualitative composition and seasonal fluctuations in plankton components. *Ibid.* 29B: 174–189.

Datta, N.C. and B.K. Bandyopadhyay (1985). Rotifers as Bioindicators of the State of Aquatic Environment. *Symp. Biomonitoring State Environ*, p. 159–166.

Datta, N.C. and S. Chaudhuri (1986). Effect of some physico-chemical variables on the population fluctuation of *Brachionus angularis angularis* Gosse. *Biol. Bull.*, India, 8(2): 155–158.

Datta, N.C. and N.K. Das (1999). A comparative study of primary productivity of two freshwater ponds-without and with sewage. In: *86th Ind. Sci. Congr.*, Chennai, January 3–7.

Datta, N.C., N. Mandal and B.K. Bandyopadhyay (1983). Diurnal variations in some physico-chemical factors and zooplankton population in the surface niche of two freshwater fish ponds. *Biol. Bull.*, India, 5(3): 236–243.

Datta, N.C., N. Mandal and B.K. Bandyopadhyay (1984). Seasonal variations of primary productivity in relation of some physico-chemical properties of a freshwater pond, Calcutta. *Int. J. Acad Ichthyol.* (*Proc. IV: AISI*), 5: 113–120.

Edmonson, W.I. (1959). *Freshwater Biology*. John Wiley and Sons, New York.

George, M.G. (1966). Comparative plankton ecology of five fish tanks in Delhi, India. *Hydrobiologia*, 27(1–4): 81–108.

Hutchinson, G.E. (1967). *A Treatise on Limnology, II: Introduction to Lake Biology and the Limnoplankton.* John Wiley and Sons, Inc., New York.

Jana, B.B. (1973). Seasonal periodicity of plankton in a freshwater pond in West Bengal, India. *Int. Rev. Ges. Hydrobiol.*, 58: 127–144.

Jhingran, V.G., A.V. Natarajan, S.M. Banerjee and A. David (1969). Methodology on reservoir fisheries investigations in India. CICFRI. Bull No. 12., Barrackpore, W.B., India.

Krishnamoorthi, K.P. and G. Visweswara (1966). Hydrobiological studies in Gandhi Sagar (Jumna Tank). Seasonal variations in plankton (1961–62). *Hydrobiol.*, 27: 501–514.

Mandal, N. (1985). Limnological studies of freshwater fish ponds with special reference to zooplankton. *Ph.D. Thesis,* University of Calcutta.

Michael, P. (1990). *Ecological Methods for Field and Laboratory Investigations.* Tata McGraw Hill Publishing Co. Ltd.

Michael, R.G. (1969). Seasonal trends in physico-chemical factors and plankton of a freshwater fish pond and their role in fish culture. *Hydrobiol.,* 33: 144–160.

Michael, R.G. and B.K. Sharma (1998). *Fauna of India.* Zoological Survey of India, Calcutta, pp. 262.

Mukhopadhyay, S.K., M. Babu Rao, S.V. Kuley and B.E. Yadan (1981). A study on the Rotiferan population Wagholi, Poona. *Proc. Symp. Ecol. Anim. Popul.,* Zool. Surv. India, 2: 47–62.

Nasar, S.A.K. (1977). Investigations on the seasonal periodicity of Zooplankton in a freshwater pond in Bhagalpur, India. *Acta. Hydrochim. Hydrobiol.,* 5(6): 577–584.

Pennak, R.W. (1955). Comparative limnology of eight Colorado Mountain lakes. IUNIV. Colo. Stud. Sr. Biol., 2: 75.

Pennak, R.W. (1978). *Freshwater Invertebrates of the United States,* 2nd edn. John Wiley and Sons, New York, pp. 803.

Prasad, B. (1916). The seasonal conditions governing the pond life in Punjab. *J. Asiat. Soc. Beng.,* 12: 142–145.

Prasadam, R.D. (1977). Observations on zooplankton population of some freshwater impoundments in Karnataka *Proc. Symp. Warmwat. Zooplankton. N.I.O., Goa,* p. 214–225.

Pruthi, H.S. (1933). Studies on the bionomics of freshwaters in India. I. Seasonal changes in the physical and chemical conditions of water of the tank in Indian Museum compound. *Int. Rev. Hydrobiol.,* 28: 46–67.

Reid, G.K. and R.D. Wood (1976). *Ecology of Inland Waters and Estuaries.* D. Van Nostrand Company, N.Y., pp. 485.

Sewell, R.B.S. (1934). A study of the fauna of the Saltlakes, Calcutta. *Rec. Indian Mus.,* 36: 45–121.

Sharma, L.L., S.K. Sharma, V.P. Saini and R.N. Vyas (2000). Some limnological aspects of seasonal pools in Ahar river, Udaipur with reference to human interference and organic waste management. *Eco. Env. Conserv.,* 6(1): 81–85.

Sharma, M.S., F. Liyaqual, D. Barbar and N. Chishty (2000). Biodiversity of freshwater zooplankton in relation to heavy metal pollution. *Poll. Res.,* 19(1): 147–157.

Sharma, S.S.S. and T.R. Rao (1990). Population dynamics of *Brachionus patulus* Müller (Rotifera) in relation to food and temperature. *Proc. Indian Acad. Sci. (Anim. Sci.),* 99: 335–343.

Singh, P.K. and S.P. Srivastava (2001). Studies on abiotic factors of Sher Shah Tomb Tank, Sasaram (Bihar) with special reference to the physical factors. *88th Ind. Sci. Congr.,* New Delhi, January 3–7.

Sreenivasan, A. (1970). Limnology of tropical impoundments: A comparative limnology of some South Indian reservoirs. *Hydrobiologia,* 36(3–4): 443–469.

Sugunan, V.V. (1980). Seasonal fluctuations of plankton of Nagarjunasagar Reservoir, A.P. India. *J. Inland Fish. Soc.,* Calcutta, 51(2): 1–12.

Vasisht, H.S. and B.K. Sharma (1975). Ecology of typical urban pond in Ambala City of Haryana State. *Indian J. Ecol.*, 2: 79–86.

Vasisht, H.S. and B.K. Sharma (1976). Seasonal abundance of rotifer population in a freshwater pond in Ambala City (Haryana), India. 28(1&2): 35–44.

Ward, H.B. and G.C. Whipple (1959). *Freshwater Biology.* (Ed.) W.T. Edmondson, 2nd edition. John Wiley and Sons., New York.

Welch, P.S. (1948). *Limnological Methods.* McGraw-Hill Book Co., Philadelphia.

Welch, P.S. (1952). *Limnology.* McGrow-Hill Book Co., New York.

Wetzel, R.G. (1983). *Limnology,* 2nd Edn. Saunders College Publ., pp. 767.

Chapter 21

Phytoplankton Dynamics in Anchar Lake, Kashmir

Shamim A. Bhat and Ashok K. Pandit

Centre of Research for Development, University of Kashmir, Srinagar – 190 006, Jammu and Kashmir

ABSTRACT

The present investigation on Anchar Lake was undertaken during 2000–01 and deals with the general ecology of phytoplankton in the lake ecosystem. Overall 143 species of algae were reported from the lake basin. In general, the phytoplankton exhibited its peak growth during summer and a very low growth during winter. The growth and abundance of phytoplankton are closely related to physico-chemical characteristics of water besides climatic conditions especially the temperature. On the basis of phytoplankton population the lake is at present undergoing accelerated cultural eutrophication.

Keywords: *Freshwaters, Phytoplankton, Nutrients, Eutrophication, Lake, Kashmir.*

Introduction

The valley lakes, indispensable aquatic ecosystems, are enveloped on all the sides by the Himalayan mountains. They are irreplaceable natural water-bodies aboding a rich and diverse gene pool. These water-bodies also provide food, fodder, green manure and vegetables besides being potential sources of water and recreation. Nevertheless, some of the lakes like Anchar has shown pronounced trophic evolution due to fast growing human populations and urbanisation in the catchment of the lake (Zutshi *et al.*, 1980; Pandit, 1996, 1999). Development of floating gardens for agricultural purposes and the utilisation of lake waters for navigation and disposal of sewage and sewerage from the surrounding human habitation, in addition to natural siltation and the toxic effluents coming out from Sheri-Kashmir Institute of Medical Sciences (SKIMS) complex (Bhat *et al.*, 2001) are greatly responsible for the deterioration of the lakes environment which subsequently have severely disturbed the ecology of both plant and animal communities.

Very few ecological investigations have been published on the plankton populations of freshwaters of Kashmir Himalaya (Kaul *et al.*, 1978; Pandit, 1980; Zutshi *et al.*, 1980; Kaul and Pandit, 1982; Yousuf *et al.*, 1986; Pandit, 1993, 1996, 1998, 2000 and Sarwar, 1999) but so far as the Anchar Lake is concerned not much information is available vis-à-vis phytoplankton dynamics. It is in this backdrop, Anchar Lake was selected and the present study, therefore, encompasses the study on species composition, seasonal variation and seasonal succession of phytoplankton communities in relation to the operative influence of physico-chemical environment. The study will help us in understanding the current trophic status of the lake.

Study Area

Anchar Lake, is situated 14 km to the northwest of Srinagar city at an altitude of 1584 m a.s.l. within the geographical coordinates of 34 20′–34 26′ N lat. and 74 82′–74 85′ E long. The lake is connected to the Khusalsar Lake which is in turn is connected to a famous Dal Lake through a small inflow channel Nalla Amir Khan. However, a network of channels resulting in a delta type formation from the cold water river Sind enter the lake on its western shore. The lake is also fed by a number of springs present in the basin itself and along its periphery. Towards the northeast of this water-basin is situated the complex of SKIMS draining its toxic effluents into the lake. The run-off from the surrounding paddy fields including floating gardens and sewage and sewerage from the surrounding human habitation are also drained into the lake thereby further enhancing the nutrient levels of the lake.

The lake is heavily infested with thick macrophytic growth and the littorals, constituting the major portion of the lake, are dominated by full growing emergents like *Phragmites australis, Typha angustata* and *Sparganium erectum*. However, dense patches of low growing emergents like *Myriophyllum verticillatum* are found at sites receiving waste waters. The growth of submerged plants is, in general, restricted due to silt-impregnated water and thick phytoplankton blooms resulting in algal mats. Deeper zones of the lake are mainly occupied by the sparse growth of submerged like *Ceratophyllum demersum* and *Potamogeton crispus*. A number of floating islands (Radhs) utilized for vegetable cultivation, being developed in the margins, result in a myriad of channels which are heavily infested with thick mats of obnoxious weed complexes like *Lemna-Salvinia*. In fact, Anchar Lake has a great diversity of plant communities wherein Zutshi (1975) reported twenty associations and Kak (1981) reported forty-two macrophytic species from this natural heritage of Kashmir.

The four sites (I, II, III, IV) selected differed on the basis on water depth, vegetation and other biotic variables (Figure 21.1).

Material and Methods

The sampling was carried out on monthly basis during April 2000–March 2001 and the samples were thoroughly mixed in equal proportions to form a representative sample of five litres, which were sieved through a plankton net of bolting silk (mesh size no. 20). The plankton sample thus collected was preserved in 4 per cent formaldehyde (APHA, 1998). The quantitative estimation of phytoplanktons was done under microscope, with the help of Sedgwick Rafter cell of 1 ml capacity. The unicellular algae were counted as individuals whereas in filamentous Cyanophyceae 100 μm length of the filaments were taken as the equivalents. Similarly, the filamentous Chlorophyceae were recorded as

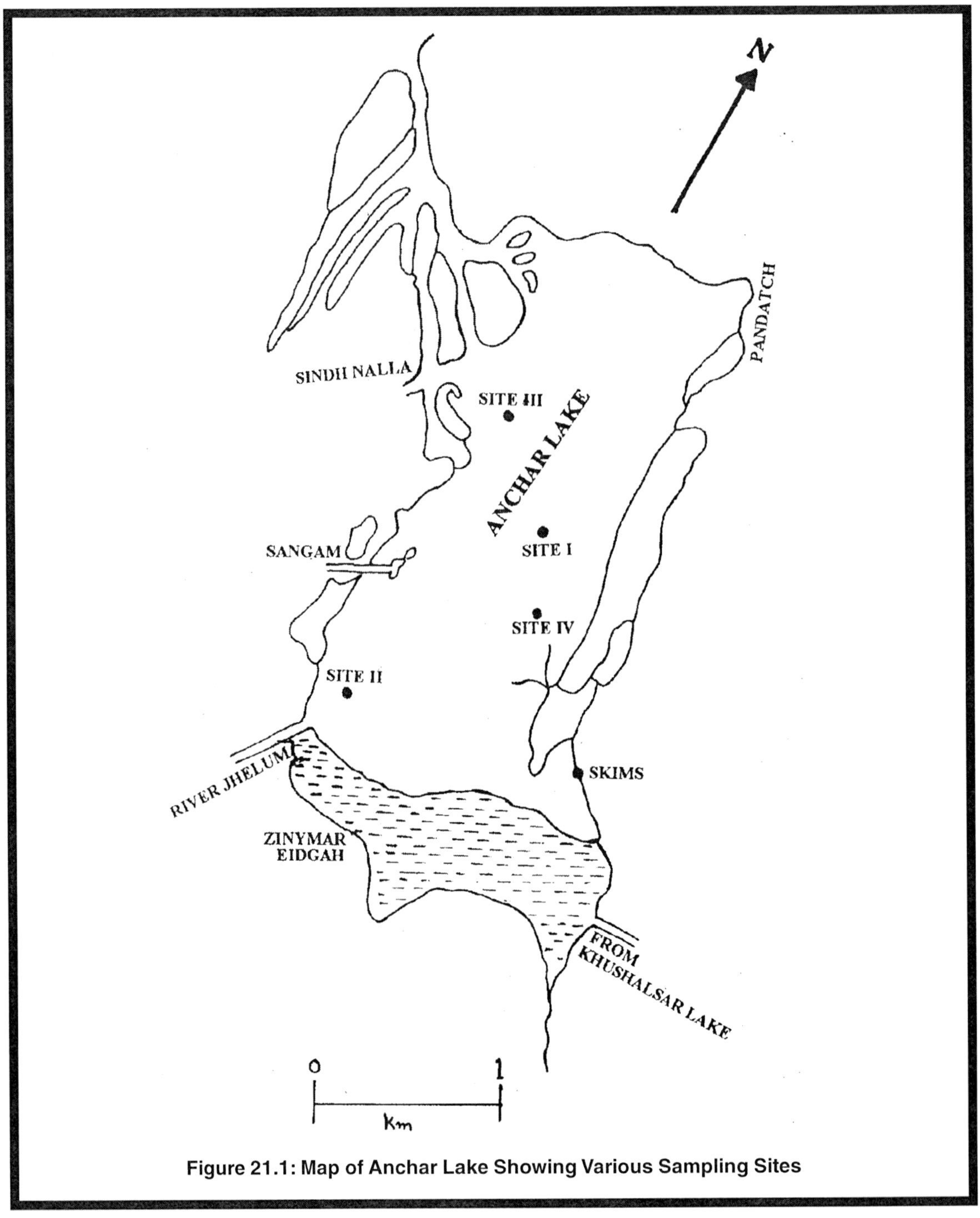

Figure 21.1: Map of Anchar Lake Showing Various Sampling Sites

cells while in the colonial forms like *Microcystis, Pandorina, Volvox, Gomphosphaera,* etc. the counting unit was the colony (Jumppanen, 1976).

Identification of the phytoplanktons was done with the help of standard works by Fritsch (1935), Desikachary (1959), Randhawa (1959), Prescott (1970), Palmer (1980), Edmondson (1992), Cox (1996) etc. The detailed chemical analysis of water samples followed the standard methods of Golterman and Clymo (1969), APHA (1998) and Gupta (2000).

Results

Physico-chemical Characteristics of Water

The average and mean values for various physico-chemical features of lake water at different study sites are summarised in Table 21.1. During the present investigation there were pronounced fluctuations in the depth of water-body with maximum depth (1.99 m) during autumn and the minimum (1.03 m) during summer. However, the secchi visibility remained high (0.64 m) during winter and low (0.43 m) during summer. The water temperature of the lake was highest (24.9°C) during summer as against the lowest (6.9 C) during winter. However, the dissolved oxygen values were more (4.4 mgl^{-1}) in spring and less (2.42 mgl^{-1}) in autumn. The pH of the water towards alkaline range fluctuated between 8.14 and 8.05 with maxima and minima during summer and autumn respectively. The conductivity values, an indication of total nutrient concentration varied from 323.2 μscm^{-1} to 408.4 μscm^{-1}, with two extremes being obtained during spring and autumn respectively. For calcium, the values fluctuated between 29.7 mgl^{-1} and 50.3 mgl^{-1}, with highest calcium content 50.3 mgl^{-1} during winter as against the lowest (29.7 mgl^{-1}) being recorded during summer. Magnesium followed a trend similar to that of calcium and ranged between 9.02 mgl^{-1} to 18.2 mgl^{-1}. The higher levels of sodium were found in the range from 4.8 to 15.7 mgl^{-1} as against the lower levels (2.6 to 7.3 mgl^{-1}) of potassium. Alkalinity is a measure of buffering capacity of the water with the values greatly fluctuating from 135.0 to 347.6 mgl^{-1}. The maximum value was, however, attained in autumn and minimum in summer. For the chloride concentration the values varied greatly at regular intervals with maximum (45.9 mgl^{-1}) during spring and minimum (21.6 mgl^{-1}) during autumn. Similarly, the maximum (3.98 mgl^{-1}) silicate level was obtained in winter as against the minimum (2.1 mgl^{-1}) in autumn. For ammonical nitrogen, the maximum content (491.2 $\mu g\ l^{-1}$) was obtained in autumn and the minimum (221.8 $\mu g\ l^{-1}$) in summer. The concentration of orthphosphate phosphorus was quite low (50.3 to 156.2 $\mu g\ l^{-1}$) as compared to total phosphate phosphorus (286.3 to 513.1 $\mu g\ l^{-1}$). In general, the maximum phosphorus concentration was obtained during hotter months and minimum during the colder months of the year.

Phytoplankton

Species Composition

The phytoplanktons exhibited a great diversity in species number at four different sampling sites of Anchar Lake. During the present investigation a total of 143 phytoplankton species and sub-species were recorded from the lake, out of which 31 belonged to Cyanophyceae, 49 to Chlorophyceae, 40 to Bacillariophyceae, 14 to Euglenophyceae, and 3 each to Chrysophyceae, Dinophyceae and Xanthophyceae respectively (Table 21.2).

Seasonal Variation in the Population Density

Pronounced seasonal variation occurred in the species composition and population density of phytoplankton (Table 21.3). There were variations in spatial distribution of various phytoplankters also.

Table 21.1: Physico-chemical Characteristics of Anchar Lake During Different Seasons, 2000–01

Sl.No.	*Parameters*		*Spring*	*Summer*	*Autumn*	*Winter*
1.	Depth (m)	A	1.38	1.03	1.99	1.07
		R	0.97–1.73	0.72–1.26	0.72–1.72	0.62–1.67
2.	Transparency (m)	A	0.54	0.43	0.55	0.64
		R	0.24–0.81	0.27–0.55	0.34–0.70	0.31–0.95
3.	Temperature (°C)	A	22.4	24.9	16.7	6.9
		R	22.0–22.8	24.3–25.6	16.5–16.9	6.7–7.4
4.	Dissolved oxygen (mgl^{-1})	A	4.4	3.9	2.42	3.6
		R	3.4–5.2	3.6–4.27	1.5–4.06	3.11–4.0
5.	pH	A	8.10	8.14	8.05	8.12
		R	7.90–8.19	7.97–8.41	7.94–8.15	8.02–8.21
6.	Conductivity (Scm^{-1})	A	408.4	332.2	323.2	330.7
		R	362.3–482.3	282.0–438.6	308.6–337.6	279.0–377.0
7.	Calcium (mgl^{-1})	A	33.8	29.7	49.3	50.3
		R	31.3–37.2	27.4–33.54	45.4–52.1	47.4–52.8
8.	Magnesium (mgl^{-1})	A	10.2	9.02	18.2	16.1
		R	9.16–12.9	6.62–12.1	16.06–19.8	14.6–17.9
9.	Sodium (mgl^{-1})	A	–	4.8	15.7	8.7
		R	–	3.1–6.55	14.7–17.9	8.0–9.0
10.	Potassium (mgl^{-1})	A	–	2.6	7.3	2.9
		R	–	1.6–3.15	6.6–7.9	2.8–3.0
11.	Alkalinity (mgl^{-1})	A	311.0	135.0	347.6	298.5
		R	256.6–358.3	110.6–160.6	320.0–381.3	292.0–310.0
12.	Chloride (mgl^{-1})	A	45.9	32.7	21.6	25.5
		R	40.4–52.5	18.6–40.6	20.0–23.3	19.0–29.0
13.	Silicates (mgl^{-1})	A	3.1	3.15	2.1	3.98
		R	2.96–3.19	3.06–3.25	1.68–2.48	3.20–4.47
14.	NH_3–N (g l^{-1})	A	240.7	221.8	491.2	407.1
		R	195.0–350.0	180.0–282.5	393.3–568.3	362.5–450.0
15.	NO_3–N (g l^{-1})	A	270.3	208.3	247.2	184.3
		R	237.6–304.0	140.6–247.6	199.0–321.6	135.0–280.0
16.	TOPP (g l^{-1})	A	50.3	80.3	156.2	145.1
		R	35.3–72.6	66.6–98.0	121.6–178.3	120.0–162.0
17.	TPP (g l^{-1})	A	286.3	513.1	403.2	401.8
		R	203.3–357.6	371.6–603.3	267.6–501.6	372.5–425.0

A: Average; R: Range.

Table 21.2: Total Number of Phytoplankton Species

Sl.No.	Taxonomic Groups	Total Number of Species
1.	Cyanophyceae	31
2.	Chlorophyceae	49
3.	Bacillariophyceae	40
4.	Xanthophyceae	03
5.	Chrysophyceae	03
6.	Dinophyceae	03
7.	Euglenophyceae	14
	Total	**143**

Cyanophyceae

The most common species *viz., Oscillatoria foreani, O. accuminata, Nodularia spumigena, Chrococcus turgidus* and *Rivularia minutula* occurred throughout the year and show their maximum density at the peak stages of growth in summer. However, some of the species showed only temporal variations and species like *Oscillatoria proboscidae, O. princeps, Lyngbya aestuari, Aphanizomenon holsaticum, Nodularia* sp. and *Astramoeba radiosa* were thus mostly restricted to spring and summer whereas species namely *Microcystis aeruginosa, Schizothrix* sp., *Oscillatoria nigra* and *Anabaena major* exhibited their mass occurrence in winter only. Of the study sites Cyanophyceae was the most dominant at Site–II during summer being mainly composed of *Oscillatori foreani* (800 ind/l).

Chlorophyceae

The most common species found throughout the year included: *Chlorella vulgaris, Chlorococcum* sp., *Microspora amoena, Ulothrix zonata, Uronenemalongatum, Dermatophyton radians, Oedocladium operculatum, Oedogonium* sp., *Spirogyra* sp., *Mougeotia* sp., *Zygnema cylindrospermum* and *Desmidium aptogonum.* However, species like *Gonium pectorale, Pediastrum biradiatum, Scenedesmus quadricauda, Sphaerocystis schroeteri, S. boryanum, Coleochaete scutata, Drapanaldiopsis* sp., *Closterium ehrenbergii, C. leiblenii* and *C. moniliferum* had a long growth period extending from spring to autumn while as forms like *Hormidium subtile, Hyalotheca* sp., *Pithophora* sp. and *Rhizoclonium* were thermophobic, loving to live in cold waters.

Among the study sites, the maximum density of green algae was registered during summer at Site–I with *Chlorella vulgaris, Pediastrum biradiatum* and *Spirogyra* sp., each making 800 ind/l; thus contributing the maximum towards the population density at the study site.

Bacillariophyceae

Most of the species occurring throughout the year included: *Cymbella cistula, Fragilaria capucina, Navicula subtillisma, Pinnularia nobilis, Eunotia minor* and *Synedra ulna.* Only one species (*Diatomella balfouriana*) was reported to occur in spring, summer and autumn. However, the species namely *Pinnularia cincta, Asterionella formosa* and *Cocconeis placentula* were restricted to winter season only.

Like the former group, the maximum density of diatoms was observed at Site–I during summer. *Navicula subtillisma* with 900 ind/l, however was the most dominant form at the study site during the season.

Table 21.3: Seasonal Variation in the Population Density (ind/l) of Phytoplankton at Four Different Study Sites of Anchar Lake

	Spring				Summer				Autumn				Winter			
	I	II	III	IV	I	II	III	IV	I	II	III	IV	I	II	III	IV
(A) CYANOPHYCEAE																
Anabaena major	–	–	–	–	–	–	–	–	–	–	–	–	–	100	–	–
A. solitaria	–	–	–	–	–	–	–	–	–	–	200	–	–	–	–	–
A. oscillatorioides	–	–	–	–	–	–	–	–	–	200	–	–	–	–	–	–
Aphanizomenon holsaticum	–	200	–	300	–	200	–	100	–	–	–	–	–	–	–	–
Coccochloris stagnina	200	–	–	–	400	–	–	–	400	–	200	–	–	–	–	–
Coccochloris sp.	40	–	–	20	200	–	–	100	400	200	200	300	–	–	–	–
Chrococcus turgidus	200	100	100	–	–	200	200	100	100	20	20	–	40	–	–	–
Cylindrospermum licheniforme	–	–	–	–	–	100	–	–	–	–	–	–	100	–	–	–
Gomphosphaeria wichurae	–	–	–	–	300	–	–	200	–	–	–	–	–	–	–	–
Lyngbya aestuarii	–	300	400	–	–	300	200	–	–	–	–	–	–	–	–	–
Nostoc linckia	600	–	–	–	400	–	–	–	300	200	–	–	–	–	–	–
Nodularia spumigena	300	–	–	–	200	–	–	–	100	–	–	–	–	300	100	–
Nodularia sp.	–	–	–	200	–	–	–	100	–	–	–	–	–	–	–	–
Oscillatoria acuminata	–	–	500	300	–	–	300	200	400	–	–	100	–	100	–	–
O. annae	–	–	–	–	–	100	–	–	–	–	–	–	–	–	–	–
O. foreani	600	300	–	400	700	800	–	300	100	200	400	100	800	300	100	100
O. proboscidea	–	–	–	300	–	–	–	200	–	–	–	–	–	–	–	–
O. princeps	–	–	–	200	–	–	–	100	–	–	–	–	200	300	100	100
O. nigra	–	–	–	–	–	–	–	–	–	–	–	–	800	400	300	200
O. tenuis	100	–	–	–	200	–	–	100	–	–	–	100	–	–	–	–
O. stagnina	40	–	–	–	200	–	–	100	100	–	–	–	–	–	–	–
Pediastrum tetras	–	–	–	–	200	100	100	–	–	–	–	–	–	–	–	–
Radiococcus nimbatus	200	–	–	–	100	–	–	40	200	–	–	300	–	–	–	–
Schizothrix sp.	–	–	–	–	–	–	–	–	–	–	–	–	400	–	–	–
Synechoccus ambiguus	–	–	–	–	–	–	–	–	–	–	–	–	–	–	–	100

Contd...

Table 21.3–Contd...

	Spring				*Summer*				*Autumn*				*Winter*			
	I	*II*	*III*	*IV*	*I*	*II*	*III*	*IV*	*I*	*II*	*III*	*IV*	*I*	*II*	*III*	*IV*
Trichodesmium lacustre	–	–	–	–	–	–	–	–	–	–	–	200	–	–	–	–
Spirulina major	500	300	40	–	300	100	200	–	–	–	–	–	–	–	–	200
Merismopoedia sp.	–	–	300	–	–	–	200	–	–	100	200	–	–	–	–	–
Astramoeba radiosa	40	–	–	–	200	–	–	–	–	–	–	–	–	–	–	–
Rivularia minutula	500	–	–	–	400	–	–	–	300	–	–	–	–	100	–	–
Microcystis aeruginosa	–	–	–	–	–	–	–	–	–	–	–	–	400	600	200	–
Total Cyanophyceae	**3320**	**1200**	**1340**	**1720**	**3800**	**1900**	**1200**	**1640**	**2400**	**920**	**1200**	**1100**	**2740**	**2200**	**800**	**700**
(B) CHLOROPHYCEAE																
(a) **Volvocales**																
Chlamydomonas sp.	40	–	–	–	–	–	–	–	–	–	–	–	–	–	–	–
Volvox sp.	–	–	–	–	–	–	–	–	200	–	–	–	–	–	–	–
Gonium pectorale	500	300	–	–	400	200	–	–	200	100	–	–	–	–	–	–
Pandorina chakourensis	–	–	–	–	–	–	–	–	–	200	–	–	–	600	200	–
P. major	–	–	–	–	40	–	–	–	–	–	–	–	–	–	–	–
(b) **Chlorococcales**																
Chlorella vulgaris	600	300	200	400	800	200	300	400	600	800	400	300	–	700	–	300
Chlorococcum sp.	500	–	300	–	300	–	–	–	200	–	–	–	–	100	20	20
Oocystis sp.	600	–	200	700	500	–	20	200	–	–	–	–	–	–	–	–
Pediastrum chodati	–	–	–	–	–	–	–	100	–	200	100	20	–	–	–	–
P. biradiatum	400	–	–	300	800	–	–	200	200	100	–	–	–	–	–	–
Scenedesmus quadricauda	500	300	300	200	300	200	200	100	700	500	400	–	–	–	–	–
S. acutus	–	–	–	–	–	20	–	–	20	–	–	–	500	–	–	–
Sphaerocystis schroeteri	400	–	–	–	600	–	–	–	20	–	–	–	–	–	–	–
S. boryanum	–	–	–	400	500	–	–	100	–	–	100	–	–	–	–	–
Radiococcus nimbatus	–	–	–	–	–	–	–	–	20	–	–	–	–	–	–	–

Contd...

Table 21.3–Contd...

	Spring				*Summer*				*Autumn*				*Winter*			
	I	*II*	*III*	*IV*	*I*	*II*	*III*	*IV*	*I*	*II*	*III*	*IV*	*I*	*II*	*III*	*IV*
Hydrodictyon sp.	–	–	–	–	–	–	–	–	20	–	–	–	–	–	–	100
Selanastrum gracile	–	–	–	–	–	20	–	–	–	200	300	–	–	–	–	–
Pediastrum ovatum	200	–	–	–	300	–	–	–	–	–	–	–	–	–	–	–
(c) **Ulotricales**																
Hormidium subtile	–	–	–	–	–	–	–	–	–	–	–	–	100	–	–	–
Microspora amoena	300	–	–	500	100	–	–	–	200	–	–	–	200	–	–	–
Binuclearia tetrana	20	–	–	–	–	–	–	20	–	–	–	–	–	–	–	–
Ulothrix zonata	200	–	–	100	200	100	20	200	200	200	400	300	–	300	200	–
Uronema elongatum	–	200	–	–	200	100	–	–	100	200	200	–	300	100	–	–
Enteromorpha intestinalis	500	300	–	200	300	200	100	–	–	–	–	–	–	–	400	–
Hyalotheca sp.	–	–	–	–	–	–	–	–	–	–	–	–	100	–	–	–
(d) **Chaetophorales**																
Chaetophora incrassata	–	–	–	–	–	–	–	–	–	–	–	100	–	–	–	–
Chaetonema irregulare	–	–	200	–	–	–	–	–	–	–	–	100	–	–	–	–
Coleochaete scutata	200	–	–	100	300	–	–	–	100	100	–	–	–	–	–	–
Dermatophyton radians	400	300	300	500	600	200	200	300	–	–	200	100	–	300	100	–
Draparnaldiopsis sp.	100	–	100	–	400	–	–	–	200	–	200	–	–	–	–	–
(e) **Oedogonales**																
Oedocladium operculatum	200	400	300	500	100	200	100	100	–	–	–	100	40	200	–	100
Oedogonium sp.	300	200	200	500	300	–	100	100	200	200	–	200	40	200	–	100
(f) **Cladophorales**																
Pithophora sp.	–	–	–	–	–	–	–	–	–	–	–	–	–	–	100	–
Rhizoclonium sp.	–	–	–	–	–	–	–	–	–	–	–	–	–	–	100	–
Cladophora glomerata	200	400	–	–	100	600	–	–	–	–	–	–	–	–	–	–

Contd...

Table 21.3–Contd...

	Spring				Summer				Autumn				Winter			
	I	*II*	*III*	*IV*	*I*	*II*	*III*	*IV*	*I*	*II*	*III*	*IV*	*I*	*II*	*III*	*IV*
(g) **Conjugales**																
(*i*) **Zygnemataceae**																
Spirogyra sp.	–	100	–	200	800	300	–	400	300	200	200	100	–	300	–	–
Mougeotia sp.	100	–	–	300	500	200	–	–	100	300	100	200	300	200	–	–
Zygnema cylindrospermum	–	–	100	–	–	–	–	100	200	100	–	–	100	–	–	–
(*ii*) **Desmidaceae**																
Desmidium aptogonum	400	300	200	200	300	200	100	100	300	100	–	–	100	200	100	100
Cosmarium granatum	400	200	200	–	200	100	100	–	–	–	–	–	–	–	–	–
C. monomarum	100	–	–	–	200	–	–	–	–	–	–	–	–	–	–	
C. monliforme	–	–	–	100	–	–	–	200	–	–	–	–	–	–	–	–
C. monomazum	–	–	–	–	–	–	–	–	–	–	100	–	100	100	–	–
Closterium ehrenbergii	400	–	–	–	300	–	–	–	300	–	100	–	–	–	–	–
Penium margaritaceum	–	–	–	–	–	–	–	–	100	100	–	–	–	–	–	–
Closterium granatum	–	–	–	300	–	–	–	100	–	–	–	–	–	–	–	–
C. leibleinii	700	–	300	400	600	–	200	–	400	–	–	–	–	–	–	–
C. moniliferum	300	–	–	–	300	–	–	–	–	–	200	–	–	–	–	–
Euastrum geminatum	–	–	–	–	–	200	–	–	–	100	–	–	–	–	–	–
Total Chlorophyceae	**8560**	**3300**	**2900**	**5900**	**10340**	**3040**	**1440**	**2720**	**4880**	**3700**	**3000**	**1520**	**1880**	**3300**	**1220**	**720**
(C) BACILLARIOPHYCEAE																
Bacillaria paradox	–	200	–	–	–	100	–	–	–	–	–	–	–	–	–	–
Asterionella Formosa	–	–	–	–	–	–	–	–	200	–	100	200	400	–	–	–
Amphora ovalis	–	–	–	200	–	100	–	100	–	–	–	–	200	100	–	–
Achanthes lanceolata	–	–	–	–	100	–	–	–	–	–	–	–	–	–	–	–
Cymbella cistula	400	–	–	300	600	100	–	100	–	100	400	–	–	100	200	300
Cyclotella sp.	100	–	–	–	100	–	–	–	–	–	–	–	–	–	–	–

Contd...

Table 21.3–Contd...

	Spring				Summer				Autumn				Winter			
	I	II	III	IV	I	II	III	IV	I	II	III	IV	I	II	III	IV
Cocconeis placentula	–	–	–	–	–	–	–	–	100	–	–	–	40	200	–	100
Doatomella balfouriana	300	–	300	–	600	100	200	–	100	–	300	300	–	–	–	–
Diatoma hemalis	–	–	–	–	300	–	–	–	–	–	100	–	–	–	–	–
Eunotia minor	100	–	–	–	300	–	–	–	–	–	300	200	100	–	–	–
Epjthenlia sp.	–	–	–	–	–	–	–	–	–	–	–	–	–	–	100	200
Fragilaria capucina	–	100	–	–	–	200	100	–	200	200	300	–	–	100	400	–
Gomphonema minutum	–	–	–	–	200	–	–	–	–	–	–	–	–	–	–	–
Meridion circulare	–	–	–	–	–	–	–	–	300	200	100	100	–	–	–	–
Nedium affinis	–	200	–	–	40	100	–	–	–	–	–	–	–	200	100	–
Nitzschia acicularis	–	–	–	–	100	100	–	–	–	–	–	–	–	–	–	–
N. diversa	100	200	–	–	200	–	–	–	–	–	–	–	–	–	–	–
N. radicula	–	–	–	100	–	–	–	–	–	–	–	–	100	100	–	–
N. capitellata	–	–	–	–	–	–	–	–	–	–	–	–	100	–	–	–
Navicula radiosa	–	100	–	–	–	200	–	–	–	–	–	–	–	–	100	200
N. subtillisma	200	–	100	–	900	–	–	–	–	200	–	100	–	40	400	–
N. sancta	–	–	–	–	–	–	–	–	300	–	200	–	–	–	–	–
N. phyllepta	–	–	–	–	–	–	300	–	100	–	–	–	–	–	–	–
N. cryptocephala	–	–	–	–	100	–	–	–	–	–	–	–	100	–	–	–
Pinnularia sudetica	–	–	–	–	–	–	–	–	100	–	–	–	–	–	–	–
P. appendiculata	–	–	–	–	–	–	–	–	100	–	–	–	100	–	–	–
P. nobilis	100	–	–	–	200	–	100	–	100	100	100	–	200	100	–	–
P. cincta	–	–	–	–	–	–	–	–	–	300	200	–	400	–	–	–
Stauroneis lauenburgiana	–	–	–	–	100	–	–	–	–	–	–	–	–	–	–	–
S. phoenicenteron	–	–	–	–	–	100	–	–	–	–	100	–	–	–	–	–
S. thermicola	–	–	–	–	–	–	–	–	–	–	100	200	–	–	–	–
Synedra ulna	100	–	100	300	3000	100	100	200	300	–	200	100	100	–	–	100

Contd...

Table 21.3–Contd...

	Spring				Summer				Autumn				Winter			
	I	II	III	IV	I	II	III	IV	I	II	III	IV	I	II	III	IV
S. famelica	–	–	–	–	–	–	–	–	–	200	200	40	–	–	–	–
Tabellaria flocculosa	–	–	200	–	–	–	300	–	–	–	–	–	500	300	100	–
Melosira granulata	200	–	–	–	400	100	100	–	–	–	–	–	–	–	–	–
Meridion sp.	–	–	–	200	–	–	–	–	–	–	–	300	–	–	–	–
Surirella robusta	–	–	–	–	300	–	–	200	–	–	–	–	–	–	–	–
Tabellaria sp.	–	–	100	–	–	–	100	–	–	–	–	–	–	–	–	–
Cymbella aequalis	–	–	–	–	–	–	–	–	–	300	–	200	–	–	–	–
Gomphonema geminatum	–	–	–	100	–	–	–	200	–	–	–	–	–	–	–	–
Total Bacillariophycae	**1600**	**800**	**800**	**1200**	**5200**	**1300**	**1300**	**800**	**1900**	**1600**	**2700**	**1740**	**2340**	**1240**	**1400**	**900**
(D) XANTHOPHYCEAE																
Tribonema bombycinum	–	–	–	–	–	–	–	–	100	–	100	–	–	–	–	–
Vaucheria aversa	100	–	–	–	40	100	200	–	40	–	–	–	–	–	–	–
Botryococcus braunii	–	–	–	–	–	–	–	–	–	–	–	–	200	300	100	–
Total Xanthophyceae	**100**	**–**	**–**	**–**	**40**	**100**	**200**	**–**	**140**	**–**	**100**	**–**	**200**	**300**	**100**	**–**
(E) CHRYSOPHYCEAE																
Dinobryon borgei	400	–	–	–	100	–	–	–	200	–	–	–	100	–	–	–
D. stipitatum	–	400	–	–	–	100	–	–	–	–	–	–	–	–	–	–
D. divergens	–	–	–	–	200	–	–	100	–	–	–	–	–	–	–	–
Total Chrysophyceae	**400**	**400**	**–**	**–**	**300**	**100**	**–**	**100**	**200**	**–**	**–**	**–**	**100**	**–**	**–**	**–**
(F) DINOPHYCEAE																
Glenodinium quadridens	300	100	–	200	100	–	–	–	–	–	–	–	–	–	–	–
Peridinium sp.	–	–	–	100	–	–	–	–	–	–	–	–	–	–	–	–
Gymnodinium aeruginosum	–	–	–	–	100	–	–	–	–	–	–	–	–	–	–	
Total Dinophyceae	**300**	**100**	**–**	**300**	**200**	**–**	**–**	**–**	**–**	**–**	**–**	**–**	**–**	**–**	**–**	**–**

Contd...

Table 21.3–Contd...

	Spring				Summer				Autumn				Winter			
	I	*II*	*III*	*IV*	*I*	*II*	*III*	*IV*	*I*	*II*	*III*	*IV*	*I*	*II*	*III*	*IV*
(G) EUGLENOPHYCEAE																
Astasia klebsii	–	200	100	–	–	–	–	–	–	200	300	100	100	–	100	100
Euglena deses	400	–	–	–	300	100	–	–	–	–	–	–	100	200	100	100
E. laepis	–	–	–	–	–	–	–	–	100	–	–	–	–	–	–	
E. acus	200	–	–	–	200	–	–	–	–	–	–	–	200	–	–	–
Chlorogonium sp.	–	–	–	100	–	–	–	–	–	–	–	–	–	–	–	200
Cryptoglena pigra	–	–	–	–	–	–	–	–	100	–	–	–	–	–	–	–
Phacus suecica	200	100	–	–	200	100	–	–	–	40	300	–	200	200	100	100
P. quinquemarginatus	100	40	–	–	40	–	–	–	–	–	400	–	–	–	–	100
Lepocynclis acicularis	–	–	–	–	–	–	–	–	–	–	600	100	100	100	100	–
L. texta	300	200	–	–	100	40	–	–	–	–	–	–	200	200	100	200
L. ovum	–	–	200	–	–	–	100	–	–	–	–	–	–	–	–	–
Trachelomonas hispida	–	–	–	200	–	–	–	–	–	–	–	–	–	–	–	–
T. intermedia	100	–	–	–	–	–	–	–	–	–	–	–	–	–	–	–
T. planctonica	200	–	–	–	–	–	–	–	–	–	–	–	–	–	–	–
Total Euglenophyceae	**1500**	**540**	**300**	**300**	**840**	**240**	**100**	**–**	**200**	**240**	**1600**	**200**	**900**	**700**	**500**	**800**

Euglenophyceae

Species like *Phacus suecica* and *P. quinquemarginatus* were regularly present throughout the year. However, species like *Trachelomonas hispida, T. intermedia* and *T. planctonica* were found in great abundance in spring while as forms like *Euglena laepis* and *Cryptoglena pigra* mostly occurred during autumn.

Among the study, Site–I again registered the maximum population with *Euglena deses* (400 ind/l), being the most dominant species during the spring.

Chrysophyceae

The most dominant species, *Dinobryon borgei,* occurred throughout the year while as *Dinobryon divergens* was restricted to summer only. Sites–I and II exhibited the higher density values, with species like *Dinobryon borgei* and *D. stipitatum* each with 400 ind/l making the decisive proportions of the group during spring.

Dinophyceae

Glenodinium quadridens was the only species found during both spring and summer as against the forms like *Peridinium* sp. and *Gymnodinium aeruginosum* having restricted growth in spring and summer respectively.

Site–I again depicted maximum density of the group during spring with *Glenodinium quadridens* (300 ind/l) contributing the maximum towards the population density of the group.

Xanthophyceae

Though none of the species could make its presence throughout the year yet species like *Vaucheria aversa* was found during spring, summer and autumn as against *Botryococcus braunii* found during the winter only. Of the study sites, Site–II had maximum population with *Botryococcus braunii* (300 ind/l); thus contributing significantly to the winter phytoplankton.

Total Phytoplankton

The phytoplankton, in general, exhibited a long growth period extending from spring to autumn, the growth being restricted during winter (cold water period). Site differences showed Site–I registering the highest population density of 20,720 ind/l during summer. In contrast, the minimum of the populations (3120 ind/l) were obtained during winter at Site–IV (Table 21.4).

Seasonal Succession

The annual seasonal succession of phytoplankton based on relative density is depicted in Table 21.5.

Spring

Features of spring phytoplankton were the dominance of Chlorophyceae (52.1–62.6 per cent), followed by Cyanophyceae (18.2–25.1 per cent), Bacillariophyceae (10.1–15.1 per cent) and Euglenophyceae (3.2–9.5 per cent) in a decreasing order. The order of the dominance was thus Chlorophyceae > Cyanophyceae > Bacillariophyceae > Euglenophyceae > Xanthophyceae. However, the site variations depicted Site–IV having maximum relative density (62.6 per cent) of Chlorophyceae and Site–I having the lowest contribution of Xanthophyceae being only 0.6 per cent.

Table 21.4: Total Phytoplankton Density (ind/l) at Four Different Study Sites of Anchar Lake

Taxonomic Groups	Spring				Summer				Autumn				Winter			
	I	*II*	*III*	*IV*	*I*	*II*	*III*	*IV*	*I*	*II*	*III*	*IV*	*I*	*II*	*III*	*IV*
(A) Cyanophyceae	3320	1200	1340	1720	3800	1900	1200	1640	2400	920	1200	1100	2740	2.200	800	700
(B) Chlorophyceae	8560	3300	2900	5900	10340	3040	1440	2720	4880	3700	3000	1520	1880	3300	1220	720
(C) Bacillariophyceae	1600	800	800	1200	5200	1300	1300	800	1900	1600	2700	1740	2340	1240	1400	900
(D) Xanthophyceae	100	–	–	–	40	100	200	–	140	–	100	–	200	300	100	–
(E) Chrysophyceae	400	400	–	–	300	100	–	100	200	–	–	–	100	–	–	–
(F) Dinophyceae	300	100	–	300	200	–	–	–	–	–	–	–	–	–	–	–
(G) Euglenophyceae	1500	540	300	300	840	240	100	–	200	240	1600	200	900	700	500	800
Total Phytoplankton	**15780**	**6340**	**5340**	**9420**	**20720**	**6680**	**4240**	**5260**	**9720**	**6460**	**8600**	**4560**	**8160**	**7740**	**4020**	**3120**

Table 21.5: Composition of Phytoplankton on the Basis of Percentage Density of Various Taxonomic Groups

Taxonomic Groups	Spring				Summer				Autumn				Winter			
	I	*II*	*III*	*IV*	*I*	*II*	*III*	*IV*	*I*	*II*	*III*	*IV*	*I*	*II*	*III*	*IV*
(A) Cyanophyceae	21.0	18.9	25.1	18.2	18.3	28.4	28.3	34.9	24.6	14.2	13.9	24.1	33.5	28.4	19.9	22.4
(B) Chlorophyceae	54.2	52.1	54.3	62.6	49.9	45.5	34.0	51.7	50.2	57.2	34.8	33.3	23.0	42.6	30.3	23.0
(C) Bacillariophyceae	10.1	12.6	15.1	12.7	25.0	19.5	30.6	15.2	19.5	24.7	31.3	38.1	28.6	16.0	34.8	28.8
(D) Xanthophyceae	0.6	–	–	–	0.1	1.5	4.7	–	1.4	–	1.1	–	2.4	3.8	2.4	–
(E) Chrysophyceae	2.5	6.3	–	–	1.4	1.5	–	1.9	2.0	–	–	–	1.2	–	–	–
(F) Dinophyceae	1.9	1.6	–	3.2	0.96	–	–	–	–	–	–	–	–	–	–	–
(G) Euglenophyceae	9.5	8.5	5.6	3.2	4.0	3.6	2.3	–	2.0	3.7	18.6	4.3	11.0	9.0	12.4	25

Summer

Among the various groups, the Chlorophyceae was the most dominant group with relative density values ranging from 34.0 to 51.7 per cent, followed by Cyanophyceae (18.3 to 34.9 per cent), Bacillariophyceae (15.2 to 30.6 per cent), Xanthophyceae (0.1 to 4.7 per cent), Euglenophyceae (2.3 to 4.0 per cent) and Chrysophyceae (1.4 to 1.9 per cent) in a declining order. Thus, the order of dominance of summer phytoplankton was Chlorophyceae > Cyanophyceae > Bacillariophyceae > Xanthophyceae >Euglenophyceae. Like spring phytoplankton, Site–IV registered maximum relative density for Chlorophyceae (51.7 per cent). In contrast, Site–I exhibited the minimum for Xanthophyceae (0.1 per cent).

Autumn

The autumn phytoplankton again is characterised by the dominance of Chlorophyceae ranging between 33.3 and 57.2 per cent, followed by Bacillariophyceae (19.5–38.1 per cent), Cyanophyceae (13.9–24.6 per cent), Euglenophyceae (2.0–18.6 per cent), Xanthophyceae (1.1–1.4 per cent) and Chrysophyceae (0–2.0 per cent) in a decreasing order. The sequence of autumn phytoplankton groups recorded were thus Chlorophyceae >Bacillariophyceae > Cyanophyceae > Euglenophyceae > Xanthophyceae. Unlike spring and summer phytoplankton, during autumn the Site–II registered the maximum proportions of Chlorophyceae (57.2 per cent) as against the least for Xanthophyceae (1.1 per cent) at Site–III during the same season.

Winter

During winter the proportion of Chlorophyceae was again maximum and that of Chrysophyceae the minimum. The order of dominance of various taxonomic groups was: Chlorophyceae (23.0–42.6 per cent) > Bacillariophyceae (16.0–34.8 per cent) > Cyanophyceae (19.9–33.5 per cent) > Euglenophyceae (9.0–25.6 per cent) > Xanthophyceae (2.4–3.8 per cent) > Chrysophyceae (0–1.2 per cent). There were again spatial variation with Site–II registering the maximum relative density (42.6 per cent) for Chlorophyceae and the Site–I showing minimum (1.2 per cent) for Chrysophyceae.

In general, the seasonal succession of phytoplankton thus showed the dominance of two major groups *i.e.,* Chlorophyceae and Cyanophyceae over other various taxonomic groups.

Discussion

The growth and abundance of phytoplankton are closely determined by the changes in the physico-chemical characteristics of water, besides biotic variables. The increasing concentrations of phosphorus and the optimal ratio between phosphorus and nitrogen have direct impact on the primary production and the development of phytoplankton community during the initial phase of eutrophication (Jumppanen, 1976). Similarly, Kaul *et al.* (1978) and Pandit (1980) reported that the freshwater bodies of Kashmir, rich in Calcium and Magnesium, have thick populations of plankton, especially Cyanophyceae. The authors opined that the higher amounts of Calcium and Magnesium help in the development of a rich population of the blue-greens as calcium in the form of calcium formate is required for the sheath formation of blue-greens while magnesium is an essential element for healthy pigmentation, particularly of chlorophyll-a which the blue-greens contain (Fogg, 1953). The development and abundance of blue-greens in Anchar Lake is, therefore, the outcome of higher levels of both the ions. Chlorophyceae grows vigorously at Site–I during the late summer when the nitrate content ranged low. Similar results have also been obtained by Pearsall (1923), Singh (1960), Zafar (1964), Munawar (1970), Kaul *et al.* (1978) and Pandit (1980, 1998). The comparatively higher transparency and temperature, however, associated with low water level seems to be conductive for the dominance of filamentous forms of green algae and the overall maximum phytoplankton density (10340 ind/l) at Site–I during summer. These findings are in consonance with the earlier findings of Kaul *et al.* (1978) and Pandit (1980) on Malgam Wetland in Kashmir. Dense populations of desmids were not usually observed in calcium rich water of Anchar Lake as according to Pearsall (1932) and Vass and Sachlan (1949) desmids form thick populations in water that are poor in calcium. These results are in agreement with the present study wherein the highest density of desmids (5200 ind/l) at Site–I were correlated with the lowest calcium level ($\overline{X}$ = 29.7 mgl^{-1}) during summer. The higher values of silicates as reported for Site–III during winter (4.47 mgl^{-1}) were also held responsible for the development of diatoms at the site. These findings are also in agreement with the earlier findings of Pearsall (1932), Bailey-Watts and Lund (1973) and Khan (1985), who believe that the diatom populations are well dependent on the silicate content of water and two have a positive correlation.

Eugleninae starts emerging during late winter and showed peak development in early spring. The group then start declining gradually during summer months. The peak development in group seemed to be a function of higher amounts of both phosphorus and nitrogen brought about by floods and effluents besides the availability of key elements after decomposition. This assumption gains further support from the earlier findings of Zutshi and Vass (1982) while working on Dal Lake, and other urban valley lakees of Kashmir. The more remarkable feature of phytoplankton of Anchar Lake is the thin population of *Dinobryon borgei* (Chrysophyceae) and *Peridinium* sp. (Dinophyceae) and their almost negligible contribution towards the total algal population. The extremely low population of these species are attributed to the high phosphrous level (513.1 μl^{-1}) and its inhibitory influence on the species as also reported by Rodhe (1948) and McMurry and Olive (1975). The present study revealed, however, peak growth of phytoplankton in summer with an average density per site 9225 ind/l, as a result of regeneration and availability of minerals due to the decomposition of organic matter in the sediments during autumn and spring seasons. Authors like Poltoracka (1963), Davis (1964), Spondniewska (1974), and Kaul *et al.* (1978) also support such a proposition. The high levels of P and N, therefore, justify the mass bursts of algal groups at their peak productive stages, especially when the light and temperature conditions are quite favourable for their growth.

In conclusion, the present nutrient status and phytoplankton populations justify the placement of the Anchar Lake in the advanced stages of eutrophication.

Acknowledgements

The authors wish to thank Prof. A.R Yousuf, Director CORD for providing laboratory facilities.

References

APHA (1998). *Standard Methods for the Examination of Water and Wastewater,* 20th Edition. American Public Health Assoc. Washington, D.C.

Bailey-Watts, A.E. and J.W.G. Lund (1973). Observation on a diatom bloom in Loch. Leven, Scotland. *J. Linn. Soc.,* London, 5: 235–250.

Bhat, S.A., S.A. Rather and A.K. Pandit (2001). Impact of effluents from Sher-i-Kashmir Institute of Medical Sciences (SKIMS), Soura on Anchar Lake. *J. Res. & Dev.,* 1: 30–37.

Cox, E.J. (1996). *Identification of Freshwater Diatoms from Live Material,* 1st Edition. Chapman and Hall, London, UK, pp. 158.

Davis, C.C. (1964). Evidence for the eutrophication of lake Erie from phytoplankton records. *Limnol. Oceanogr.,* 1: 47–53.

Desikachary, T.V. (1959). *Cyanophyta.* Indian Council of Agricultural Research, New Delhi, pp. 686.

Edmondson, W.T. (1992). *Ward and Whipple's Freshwater Biology,* 2nd ed. Intern. Books and Periodicals Supply Service, New Delhi.

Fogg, G. E. (1953). *The Metabolism of Algae.* Mathuen and Co. Ltd. London.

Fritsch, F.E. (1935). *The Structure and Reproduction of Algae.* Vol. I Cambridge Univ. Press, London.

Golterman, H.Z. and R.S. Clymo (1969). *Methods for Physical and Chemical Analysis of Freshwaters.* IBP Handbook No. 8. Blackwell Scientific Publication, Oxford, Edinburgh.

Gupta, P.K. (2000). *Methods in Environmental Analysis: Water, Soil and Air.* Agrobios Jodhpur, India, pp. 409.

Jumppanen, K. (1976). Effects of waste waters on a lake ecosystem. *Ann. Zool. Fennici.,* 13: 85–138.

Kak, A.M. (1981). Floristic composition, structure and ecology of some high altitude lakes of Kashmir Himalayas. In: *Proc. Abs. Silver Jubilee Symposium of Tropical Ecology. "Ecology and Resource Management in the Tropics",* (Eds.) R.S. Ambasht and H.N. Pandey, Banaras Hindu University, Varanasi, p. 106–108.

Kanth, T.A and S.M. Amin (1996). Drainage system in Kashmir Valley. *In: Ecology, Environment and Energy,* (Eds.) A.H. Khan and Ashok K. Pandit. The University of Kashmir, Srinagar–6, J &K, India, p. 185–197.

Kaul, V., D.N. Fotedar, A.K. Pandit and C.L. Trisal (1978). A comparative study of plankton populations of some typical fresh water-bodies of Jammu and Kashmir State. In: *Environmental Physiology, and Ecology of Plants,* (Eds.) D.N. Sen and R.P. Bansal. Bishen Singh, Mahendra Pal Singh, Dehra Dun, India, p. 249–269.

Kaul, V. and A.K. Pandit (1982). Biotic factors and food chain structure in some typical wetlands of Kashmir. *Poll. Res.,* 1(1–2): 49–54.

Khan, M.A. (1985). Spatio-temporal phytoplankton variability in two Himalayan lacustrine ecosystems. *Acta hydrochim et hydrobiol.,* 13: 249–259.

Mackereth, F.J.H. (1963). *Water Analysis for Limnologists.* Freshwater Biol. Assoc., 21: 1–70.

McMurry, G. and J.H. Olive (1975). Summer phytoplankton photosynthesis in a north eastern Ohio glacial lake. *Ohio J. Sci.,* 75: 238–250.

Munawar, M. (1970). Limnological studies on freshwater ponds of Hyderabad, India. II. The biocenose and distribution of unicellular and colonial phytoplankton in polluted and unpolluted environments. *Hydrobiologia,* 36: 105–128.

Palmer, C.M. (1980). *Algae and Water Pollution.* Castle House Publications Ltd., USA.

Pandit, A.K. (1980). Biotic factor and food chain structure in some typical wetlands of Kashmir. *Ph.D. Thesis.* University of Kashmir, Srinagar – 190 006 India.

Pandit, A.K. (1993). Dal Lake ecosystem in Kashmir Himalaya: Ecology and management. In: *Ecology and Pollution of Indian Lakes and Reservoirs,* (Eds.) P.C. Mishra and R.K. Trivedy. Ashish Publishing House, New Delhi, India, p. 131–202.

Pandit, A.K. (1996). Lakes in Kashmir Himalaya. In: *Ecology, Environment and Energy,* (Eds.) A.H. Khan and Ashok K. Pandit. The University of Kashmir, Srinagar – 190 006, J&K, India, p. 1–40.

Pandit, A.K. (1998). Plankton dynamics in freshwater wetlands of Kashmir. In: *Ecology of Polluted Waters and Toxicology,* (Ed.) K.D. Mishra. Technoscience Publications, Jaipur, India, p. 22–68.

Pandit, A.K. (1999). *Freshwater Ecosystems of the Himalaya.* Parthenon Publications, New York, London.

Pandit, A.K. (2000). Trophic evolution of lakes in Kashmir Himalaya. In: *Natural Resources of the Western Himalaya,* (Ed.) Ashok K. Pandit. Valley Book House, Srinagar, J&K, p. 213–242.

Pearsalll, W.H. (1923). A theory of diatom periodicity. *J. Ecol.,* 11–12: 165–183.

Pearsall, W.H. (1932). Plankton in English lakes. *J. Ecol.,* 20: 241–262.

Poltoracka, J. (1963). Phytoplankton in lakes near Wegorzewo in the light of peculiarities of environment. *Pol. Arch. Hydrobiol.,* 11: 189–217.

Prescott, G.W. (1939). Some relationships of phytoplankton to limnology and aquatic biology. *Publ. Amer. Assoc. Adv. Sci.,* 10: 65–78.

Prescott, G.W. (1970). How to know the freshwater algae. *Ibid,* pp. 348.

Rodhe, W. (1948). Environmental requirements of freshwater plankton algae. Experimental studies in the ecology of phytoplankton. *Symb. Bot. Upsal.,* 10: 1–149.

Sarwar, S.G. (1999). Water quality and periphytic algal component of Anchar Lake, Kashmir. In: *Inland Water Resources, India,* (Eds.) M.K. Durga Prasad and P. Sankara Pitchaich. Discovery Publishing House, New Delhi, p. 237–250.

Singh, V.P. (1960). Phytoplankton ecology of the inland waters of Uttar Pradesh. *Proc. Symp. Algology.,* ICAR, New Delhi, p. 243–271.

Spondniewska, I. (1974). The structure and production of phytoplankton in Mikolajskie Lake. *Ekol. Poll.,* 22: 68–105.

Vass, K.F. and M. Sachlan (1949). On the ecology of some lakes near Bittenzorg, Java. *Hydrobiologia,* 1: 238–250.

Yousuf, A.R., M.H. Balki and M.Y. Qadri (1986). Limnological features of a forest lake of Kashmir. *J. Zool. Soc.,* India, 38(1&2): 29–42.

Zafar, A.R. (1964). On the ecology of algae in certain fish ponds of Hyderabad, India. Physico-chemical complexes. *Hydrobiologia,* 23: 179–195.

Zutshi, D.P. (1975). Association of macrophytic vegetation in Kashmir lakes. *Vegetation,* 30: 61–65.

Zutshi, D.P. and K.K. Vass (1971). Ecology and production of *Salvinia natans* Hoffim in Kashmir. *Hydrobiologia,* 38: 303–320.

Zutshi, D.P., B.A. Subla, M.A. Khan and A. Wanganeo (1980). Comparative limnology of nine lakes of Jammu and Kashmir Himalayas. *Hydrobiologia,* 72(1–2): 101–112.

Chapter 22

Detoxification Efficiency of Four Fungal sp. on Dye Effluent

K.T.K. Anandapandian, S. Chandrasekarenthiran, S. Kirupaa and G. Ram Kumar
P.G. Unit of Microbiology, Thiagarajar College, Madurai – 625 009, Tamil Nadu

ABSTRACT

Dye effluent sample was collected from colour yarn limited and tested for parameters such as colour, pH, total dissolved solids, total suspended solids, total solids, chlorinity, hydrogen sulphide, dissolved oxygen absorbed oxygen, biological oxygen demand, chemical oxygen demand, phenol, oil and grease and toxicity determined by survival efficiency of fishes. Among seven fungal species isolated from dye amended soil, four isolates were used for detoxification process (Two different *Aspergillus* sp., *Rhizopus* sp. and *Geotrichum* sp.) and were named CRII, CRIII, CRIV and CRV. CRII shown best one for detoxification activity.

Keywords: *Dye effluent, Detoxification, Aspergillus sp., Rhizopus sp., Geotrichium sp.*

Introduction

Toxic chemicals discharged into the environment through various industrial wastes constitute one of the major causes of environment problem (Sudhabai *et al.,* 1998). Textile dye industries are one which contribute to this pollution than other industries. More than eight thousand chemically different types of dyes are being manufactured and they require volumes of water and generate equally volumes of wastes which is highly coloured and complex. For ex. Thirupur an industrial town, houses about five hundred textile dyeing units located on river bank of the Noyyal releasing effluents untreated into river (Namasivayam *et al.,* 1992). Microbiological cleanup of this types of pollution can be advantageous when compared to other chemical remediation techniques. *Phanerochaete chrysosporium* and other fungus caused 80 per cent detoxification in broth containing 2.5 per cent of effluent. There was reduction in BOD and COD values. *Fusarium* sp. causes 35 to 85 per cent of detoxification, fungi are good in the accumulation of heavy metal such a cadmium, copper, mercury, lead and zinc (Sullia, 2002). Out of eighteen commercially used textile dyes, eight were degraded by white rot fungus, *Phanerochaete chrysosporium* based on decrease of colour (Capalash *et al.,* 1992). *Aspergillus* are efficient in the detoxification of dye effluent due to the production of amylases, proteases and other enzymes. By carefully selecting the fungal isolate and cultural conditions used, degradation of dyes can be obtained. Detoxification of orange colour dye was highest in *Phanerochaete chrysosporium* and it was pre-cultured in phosphate buffered medium. (Chao *et al.,* 1994). In the present study an attempt has been made to detoxify the textile dye industry effluents by using four fungal species, and tested for physico-chemical analysis (pH, TDS, TSS), inorganic analysis (Chlorinity, Hydrogen Sulphide, Salinity), organic analysis (Dissolved oxygen, BOD, COD, Oil and Grease, Adsorbed oxygen, Phenol) and survival efficiency of fishes.

Materials and Methods

The effluent samples were collected from colour yarn limited Allampatty, Madurai at the work time of industry. The samples were taken to the laboratory immediately after collection. Four fungal species were used for treatment and were characterized and confirmed as per Alexopoulos classification of fungi. The freshly collected raw effluent was mixed with the broth culture of fungal species in the ratio of 90 : 10 and incubated for 8 days at 37 C. The parameters such as colour, pH, TDS, TSS, TS, Chlorinity, Hydrogen Sulphide, Dissolved Oxygen, Absorbed Oxygen, BOD, COD, Oil, Grease and Phenol were measured in the effluent before and after treatment (APHA, AWWA and WPCF, 1975).

The fresh raw effluent and fungal inoculated effluent were tested for its toxicity as survival efficiency of fishes. 30 fishes (*Lepidocephalicthys thermalis*) which were acclimatized to the lab condition, were exposed to five different types of sample by putting 10 in each type of sample.

Results and Discussion

The Tables 22.1 and 22.2 showed that the physico-chemical analysis, inorganic analysis, organic analysis and survival efficiency of fishes of the yellow colour dye effluent collected from Colour Yarn Limited.

Table 22.1: Physico-chemical Characteristics of Yellow Coloured Dye Effluent

Parameters	*Raw Effluent*	*CR II*	*CR III*	*CR IV*	*CR V*
Colour	Yellow	Light brown	Dark brown	Dark brown	Yellow
pH	8.6	6.2	6.4	6.5	6.7
TDS (mg/l)	2480	720	830	790	817
TS (mg/l)	2240	520	610	530	583
TSS (mg/l)	4720	1240	1440	1320	1400
Chlorinity	596.4	178.9	187.4	193.1	195.9
Salinity	32.2	9.68	10.01	10.4	10.6
Hydrogen sulphide	595	219	243	237	249
Dissolved oxygen	–	1.8	1.89	1.87	1.96
Absorbed oxygen	121	36	39	38.8	41
BOD	87.8	12.3	16.9	14.3	15.3
COD	3907	8d08	839	827	889
Oil and grease	178×10^3	54×10^3	61×10^3	$55 \times ^{103}$	60×10^3
Phenol	868	198	213.2	207.3	219.5

Table 22.2: Survival Efficiency of Fishes on Yellow Coloured Dye Effluent

Conditions Exposed	*Raw Effluent*	*CR II*	*CR III*	*CR IV*	*CR V*
100%	–	–	–	–	–
75%	–	1	–	1	–
50%	–	2	1	3	1
25%	–	5	2	4	2

Time exposed = 72 hrs

No. of fishes (*Lepidocephalicthys thermalis*) taken in each tank = 10.

The Tables 22.3 and 22.4 showed that the physico-chemical analysis, inorganic analysis, organic analysis and survival efficiency of fishes of the blue colour dye effluent collected from Colour Yarn Limited. The result also shows that the fungus species CR II was the best decolourizing activity of the yellow and blue colour dye effluent.

Table 22.3: Physico-chemical Characteristics of Blue Coloured Dye Effluent

Parameters	*Raw Effluent*	*CR II*	*CR III*	*CR IV*	*CR V*
Colour	Blue	Yellowish	Yellowish	Light brown	Yellowish
pH	8.3	6.2	6.3	6.3	6.4
TDS (mg/l)	2160	751	768	767	788
TS (mg/l)	1890	427	467	438	489
TSS (mg/l)	4050	1178	1235	1205	1277
Chlorinity	724.2	213	218.6	221	231.6
Salinity	39.2	11.5	11.8	11.9	12.5
Hydrogen sulphide	695	318	327	323	336
Dissolved oxygen	–	2.3	2.1	1.9	1.6
Absorbed oxygen	174.6	41	43.6	44.5	45.5
BOD	119.8	36.8	39.7	38.6	39.5
COD	3827	789	813	818	838.5
Oil and grease	188×10^3	39×10^3	46×10^3	47×10^3	49×10^3
Phenol	778	188.3	197.6	191.3	201.5

Table 22.4: Survival Efficiency of Fishes on Blue Coloured Dye Effluent

Conditions Exposed	*Raw Effluent*	*CR II*	*CR III*	*CR IV*	*CR V*
100%	–	–	–	–	–
75%	–	2	–	1	–
50%	–	3	1	2	1
25%	–	4	2	3	2

Time exposed = 72 hrs.

No. of fishes (*Lepidocephalicthys thermalis*) taken in each tank = 10.

The table indicated the difference between the raw effluent and treated effluent. It is noted that BOD and COD values were reduced and dissolved oxygen content was increased. The reduction in BOD and COD values shows that fungal can detoxify the textile dye effluent. Organic nutrients are often presented to fungi as large, insoluble macromolecule complex. These complex must be degraded to smaller substitution which is used as a source of nutrients. Elevated oxygen levels increase the rate of lignin degradation through the production of enzymes and induction of ligninolytic activity by *Phanerochaete chrysosporium* (Das *et al.,* 1999). The dissolved oxygen level is normally at least 5 mg/l. for the survival of fish and other aquatic life (Amudha *et al.,* 1997). Not upto 100 per cent, degradation and detoxification efficiency of fungi on dye effluent, but when compare with the other chemical remediation technique, fungi showed the best detoxifying efficiency of dye effluent. These type of treatment may be used for the development of detoxification of the textile dye effluents.

References

Alexopoulus, C.J. (1996). *Introductory Mycology.* p. 311–313, 228–219, 130.

Amudha, P., R. Nagendran and S. Mahalingam (1997). Studies on the effects of dairy effluents on the behaviour of *Cyprin carpio. J. Env. Biol.,* 18(4); 415–418.

APHA, AWWA and WPCF (1975). *Standard Methods for Examination of Water and Wastewater,* 14th ed. APHA, New York, USA.

Capalash, N. and Prince Sharma (1992). Biodegradation of textile azodyes by *Phanerochaete chrysosporium. World J. Microb. & Biotech.,* 8: 309–312.

Chao, W.L. and S.L. Lee (1994). Decolouration of azodyes by three white-rot fungi: Influence of carbon source. *World J. Microb. & Biotech.,* 10: 556–559.

Das, M. and S.K. Masud Hussain (1999). Biopulping of non-conventional lignocellulotic agro-waste material (Jute stick) by *Phanerochaete chrysosporium:* A pollution free pulping process. *Indian J. Env. Prot.,* 20: 284 –286.

Namasivayam, C. and R.T. Yamuna (1992). Removal of congo red from aqueous solutions by biogas waste slurry. *J. Chem. Tech. & Biotech.,* 53: 153–157.

Sudhabai, R. and T.K. Emilia Abraham (1998). Studies on biosorption of chromium (VI) by dead fungal, biomass. *J. Sci. Indus. Res.,* 57: 821–824.

Sullia, S.B. (2002). *Fungal Diversity and Bioremediation.* www.google.com.

Chapter 23

Stable Carbon Isotopic Studies on Zooplankton in Mangrove Waters and the Bay Environment of Kakinada, East Coast of India

N. V. Prasad and P. Chandramohan

Division of Marine Biology, Department of Zoology, Andhra University Visakhapamam – 530 003

ABSTRACT

To understanding the role of zooplankton as secondary producers and their role in transferring the mangrove carbon to estuary and Bay region, stable carbon isotopic studies were carried out in mangroves and the Bay environment of Kakinada. The overall $\delta^{13}C$ in zooplankton varied from –16.45 per cent (Station K1) and –13.13 per cent (Station C1). The carbon isotopic values were more negative in the zooplankton collected from the mangrove region (Stations G1 and C1), relative to Kakinada Bay (Station K1). The composition of zooplankton seems to have a direct bearing on the $\delta^{13}C$ values. The composition of zooplankton at the mangrove region was mostly detritivorus meroplankton, where as mostly holoplankton in the Kakinada Bay which appear to be mostly phytoplankton grazier. The SPOM had $\delta^{13}C$ higher values close to detritus at mangrove regions and gradually faded away in the SPOM of Bay region. The results indicated that mangrove detritus is an important source of carbon for zooplankton thriving in mangrove creeks of Gaderu and Coringa river.

Keywords: *Stable carbon, SPOM, Detritus, Zooplankton, Bay-mangroves.*

Introduction

Stable carbon isotope measurements ($\delta^{13}C$) are helpful tools in tracing carbon pathways in ecosystem (Fry and Sheer, 1984). Analysis of carbon isotopes in the biota of mangrove ecosystem indicated that number of consumer organisms are able to incorporate mangrove carbon in tropical coasts adjacent to mangroves (Rodelli *et al.*, 1984; Zieman *et al.*, 1984; Grey *et al.*, 2000). The stable isotope technique gives a time integrated objective measure of actual carbon assimilated by the organisms (Fleming *et al.*, 1990).

On account of physico-chemical and biological process that are involved in the leaf decay and the transient pools develop in the surrounding mangrove creeks. Here the continuous sequestering of the nutrients by the primary consumers takes place. Teal (1962) reported existence of an equilibrium. between these nutrient pools. Zooplankton plays an important role in the sequestering mechanism of the nutrients from the transient pool to the higher trophic levels. According to Rao (1997), the mangrove litter decay leads to the enrichment of the mangrove waters with nutrients. The source of carbon to higher trophic levels in the aquatic regime of mangrove environment may be through the SPOM (Suspended Particulate Organic Matter) and phytoplankton.

During the present study, modest attempt was made in understanding the role of zooplankton as secondary producers and their role in transferring the mangrove carbon to estuary and Bay regions. Similar studies on stable carbon isotopes in zooplankton were made in the mangrove regions of Godavari estuary (Chandra Mohan *et al.*, 1997), Equator mangroves (Cifuentes *et al.*, 1996) and mangrove habitat of Kenya (Marguillier *et al.*, 1997).

Materials and Methods

Samples of zooplankton were collected from 3 stations, one in Kakinada Bay (Station 1) located near the open sea, and two in the estuarine regions of Gaderu (Station 2) and Coringa (Station 3) surrounded by mangroves (Lat, 82 14′–82 22′ N and Long. 16 50′–17 00′E) (Figure 23.1). Monthly sampling was done during January 1995 to August 1996. Surface zooplankton was collected by oblique haul using 40 cm diameter of 120 mm mesh size. For the quantification of the zooplankton a T.S.K. flow meter was fixed at the opening of the net. Zooplankton samples after collection transported live or in Ice container (0–2 C) to the laboratory and processed immediately. Zooplankton was filtered through a gauze and rinsed with fresh water followed by distilled water and was oven dried at 60 C to a constant weight for 36 hr. All the samples were ground to fine powder with a pestle and mortal. 2 mg of zooplankton sample was taken for the determination of $\delta^{13}C$ ratios. Mass spectrometric measurements were performed using a Delta E Finnigan Mat isotope ratio Mass Spectrometer (at Vrije Universiteit Brussels, Belgium) (Rao, 1997). For carbon isotope ratios the organ materials was subjected to cumbustion in an elemental analyzer (Carlo Erbana, 1500). The CO_2 generated during the combustion was automatically trapped in an on Line Finnigan Mat trapping box for cryopurification before injection into the Mass Spectrometer. The normal working standard for carbon was CO_2 produced from carrare marble. A graphite reference material (USG–24) was used as a standard carbon isotopic ratio measurements values. Standard carbon ratios (Coplen, 1996) are presented as:

$$\delta^{13}C\,(\%) = \left\{ \frac{(\delta^{13}C/\delta^{12}C\ \text{sample})}{(\delta^{13}C/\delta^{12}C\ \text{standard PDP})} - 1 \right\} 1000$$

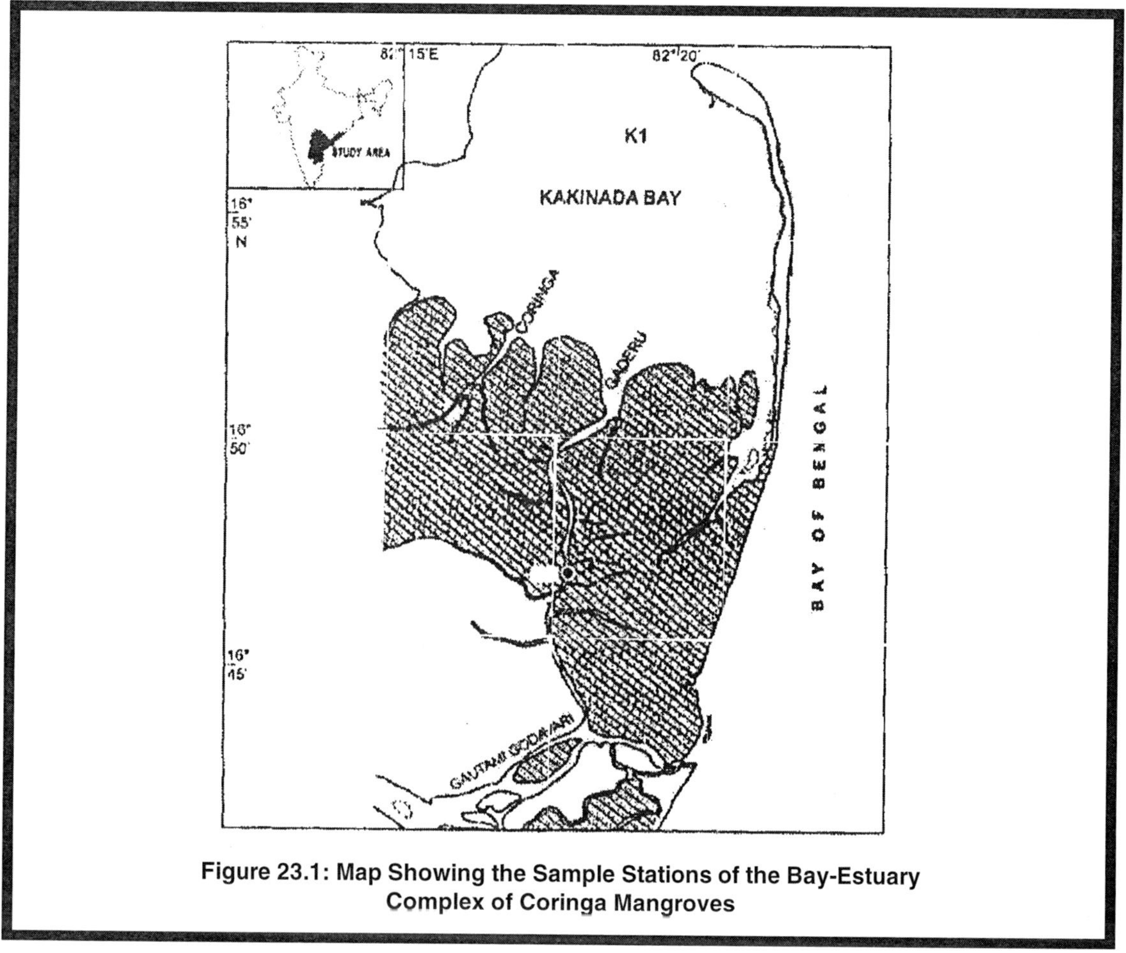

Figure 23.1: Map Showing the Sample Stations of the Bay-Estuary Complex of Coringa Mangroves

Results and Discussion

During the present study overall $\delta^{13}C$ in zooplankton varied from –16.45 per cent (June 1995; Station K1) and –30.13 per cent (January 1996; Station C1). In Kakinada Bay, $\delta^{13}C$ in zooplankton (Station K1) varied between –16.45 per cent (June 1995) and –28.19 per cent (October 1995), with a mean of –20.87 per cent. In Gaderu (Station G1) $\delta^{13}C$ values ranged from a minimum of –19.10 per cent (March 1995) to a maximum of –26.92 per cent (July 1996). The mean being –24.32 per cent. Whereas in the Coringa river (Station C1), $\delta^{13}C$ values ranged from a minimum of –20.62 per cent (April 1996) to a maximum of –30.13 per cent (January 1996), the mean being –25.74 per cent (Table 23.1). From the result it is obvious to state that $\delta^{13}C$ of zooplankton from Gaderu and Coringa rivers indicated more negative value than the Kakinada Bay (Table 23.2).

In the present study, the results revealed that distinctly two different groups of zooplankton organisms were seen. Those forms which showed more negative $\delta^{13}C$ were recorded in the Gaderu and Coringa river, indicating that the carbon source is from the detritus and leaf litter decay of the mangrove region. Here, composition of zooplankton was dominated by meroplankton forms *i.e.* decapod larvae, bivalve and gastropod veligers, which are mainly detritivorus. During the present study zooplankton collected from the Gaderu and Coringa rivers had $\delta^{13}C$ values of –24.32 and –25.74 per cent (mean) respectively. Chandra Mohan *et al.* (1997) from the Godavari estuarine mangrove ecosystem reported that the mangrove leaf, fruits and flowers had $\delta^{13}C$ values between –23.0 to –30.0 per cent. In contrast to the mangrove region (Gaderu and Coringa rivers) the carbon isotope signature were less negative in the zooplankton collected from the Bay region. Here, the zooplankton composition was dominated by Holoplankton population. The $\delta^{13}C$ was less negative in the zooplankton of this region (Bay), may be due to the reason that the carbon source for the zooplankton at this region may be other than mangrove detritus *i.e.* phytoplankton. In the present study, zooplankton from Kakinada Bay had less negative $\delta^{13}C$ than SPOM $\delta^{13}C$ value indicating that the zooplankton selectively feed on phytoplankton at this region (Bay). Earlier studies made by Fortugne and Duplessy (1981) from Bay of Bengal reported that phytoplankton had $\delta^{13}C$ value of –19.6 per cent (mean).

Table 23.1: Stable Carbon Isotopic Composition (%) of Zooplankton at Selected Stations in Kakinada Bay, Gaderu and Coringa

Months	*Kakinada Bay (Station K1)*	*Gaderu (Station G1)*	*Coringa (Station C1)*
January'95	– 25.43	– 26.19	N.S.
February'95	– 25.70	– 24.72	– 25.14
March'95	– 19.43	– 19.10	– 21.29
April'95	– 17.64	– 22.41	– 23.53
May'95	– 17.10	– 23.26	– 24.69
June'95	– 16.45	– 23.05	– 24.93
July'95	– 17.20	– 24.17	– 25.43
August'95	– 18.55	– 26.29	– 22.38
September'95	– 22.95	– 26.20	– 28.19
October'95	– 28.19	– 26.77	– 27.63
November'95	– 20.65	– 25.88	– 29.04
December'95	– 20.62	– 26.52	– 26.94
January'96	– 21.21	– 21.93	– 30.13
February'96	– 22.04	– 22.68	– 26.16
March'96	– 17.69	– 23.72	– 26.93
April'96	– 18.62	– 23.60	– 20.62
May'96	– 22.51	– 23.49	– 25.52
June'96	– 22.07	– 23.82	– 24.91
July'96	– 21.93	– 26.92	– 28.79
August'96	– 21.61	– 25.71	– 26.83
Minimum	– 16.45	– 19.10	– 20.62
Maximum	– 28.19	– 26.92	– 30.13
Mean	– 20.87	– 24.32	– 25.74

N.S.: No sampling.

Table 23.2: δ¹³C Isotopic (%) of Zooplankton and SPOM from the Kakinada Bay, Gaderu and Coringa

Station	$\delta^{13}C$ of Zooplankton (%)	$\delta^{13}C$ of SPOM (%)
K1	– 20.96	22.37
G1	– 21.37	– 23.67
C1	– 25.85	– 23.87

The role of zooplankton in the mangrove ecosystem is very significant because, the basic food chain that sustains in mangrove regions is the detritus food chain, commencing with mangrove leaf litter → Saprophyte community → detritus consumers → lower carnivores → higher carnivores (Odum and Heald, 1975). Later on it was shown that significant fraction of leaf and leaf litter reaching mangrove forest floor were consumed or carried under ground by crabs (Sasekumar and Loi, 1993; Robertson and Duke, 1987). In the mangrove regions the detritus consumers in the region of mangrove substrata are either crabs or molluscs, but the detritus which gets leached into the surrounding waters by the tidal action was probably consumed by the drifting zooplanktonic organisms, mostly meroplankton, like the larvae of prawns, crabs, bivalves, molluscs and fishes which are known to be obligatory feeders on assimable organic detritus. In view of this, the mangrove region with narrow creeks can be considered as good nursery grounds for many of the fin and shell fish.

Zooplankton and SPOM samples from Gaderu and Coringa rivers had negative $\delta^{13}C$ values. While least negative values were observed for the zooplankton from Kakinada Bay (Figure 23.2). This significant variation in the $\delta^{13}C$ and SPOM $\delta^{13}C$ values at the various locations may be due to selective grazing which inturn indicate the difference in the carbon source. In case of zooplankton from mangrove stations (Stations G1 and C1) marked depletion relative to suspended matter was observed which suggest their possible selective grazing on mangrove detritus. The zooplankton from the Kakinada Bay had less negative SPOM $\delta^{13}C$ indicating that the' zooplankton in this regions selectively feed on phytoplankton. Tackx *et al.* (1995) reported selective grazing of zooplankton from estuaries of temperate regions.

The composition of zooplankton seems to have a direct bearing on the $\delta^{13}C$ values. The composition at the mangrove regions, where mostly detritivours meroplankton dominates, greater magnitude of $\delta^{13}C$ values were obtained having close relation to available mangrove detritus. Similarly, zooplankton having mostly holoplanktonic forms in the Kakinada Bay which appears to be mostly phytoplankton grazers, relatively low values of $\delta^{13}C$ were obtained.

The magnitude and direction of the material flux between mangroves and adjacent coastal system depends on the geographical setting, hydrodynamics, soil and vegetation type (Robertson *et al.*, 1992). Newell *et al.* (1995) and Chandra Mohan *et al.* (1997) observed that mangrove carbon was ingested by Juvenile penaeid prawns living within tidal creeks but the signal of mangrove carbon faded away in the offshore regions. Whereas, Cifuentes *et al.* (1996) from the Equador mangrove estuarine system, reported that detritus from mangroves and other allochtonous sources had supplied much of the organic matter to the system. Marguillier *et al.* (1997) observed that gastropod *Terebralia* sp. from Kenya mangrove forest and adjacent creeks clearly had incorporated mangrove carbon, however the crustacean and fish sampled at greater distance from the mangroves, mangrove carbon did not contribute substantially to their diet.

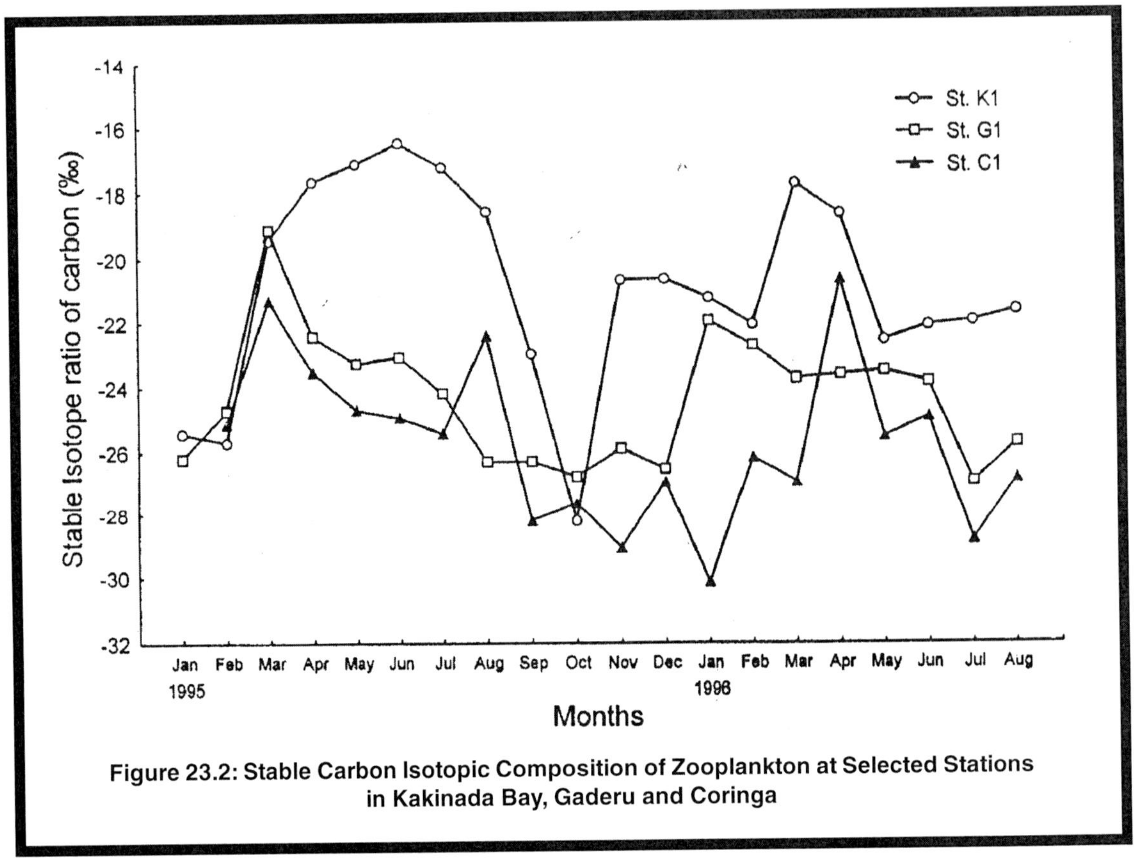

Figure 23.2: Stable Carbon Isotopic Composition of Zooplankton at Selected Stations in Kakinada Bay, Gaderu and Coringa

From the present study, it may be stated that the carbon source for the surrounding estuarine and Bay waters in Kakinada Bay may be from mangrove litter degradation. It may be seen that the SPOM had $\delta^{13}C$ higher values close to detritus at mangrove regions and gradually faded away in the SPOM of Bay region. The zooplankton plays an important role in assimilating the mangrove carbon and transferring to the next higher trophic level. Those zooplankton species which spend their larval and juvenile phase in the mangrove habitat but migrate to other habitats as adults may represent an important source of cabon input originating from primary production. The role of zooplankton at secondary producers needs to be emphasised as they seem to play a significant role in the food chain either through the mangrove detritus or through grazing as herbivores on available phytoplankton in the open waters. Thus, the mangrove vegetation compliments significantly to the uptake of carbon in certain groups of organisms both in pelagic and benthic assemblages. This particular aspect needs to be explored well so as to explain the possibility of mangroves contribution to the enrichment of organic matter in the nearshore water.

Acknowledgements

The present work was carried out as part of the project on the mangroves of the Kakinada Bay, sponsored by the Commission of European Communities (Brussels, Belgium). Due acknowledgements are given to the sponsors and to other investigator Dr. Frank Dehairs.

References

Chandra Mohan, P., R.G. Rao and F. Dehairs (1997). Role of mangroves (India) in production and survival of Prawn larvae. *Hydrobiologia*, 358: 317–320.

Cifuentes, L.A., R.B. Coffin, L. Solarzano, W. Cardenas, J. Espinoza and R.R. Twilley (1996). Isotopic and elemental variations of carbon and nitrogen in a mangrove estuary. *Estuar. Coastal and Shelf Sci.*, 43: 781–800.

Coplen, T.B. (1996). New guidelines for reporting stable nitrogen, carbon and oxygen isotope ratio data. *Geochemica et Cosmochemica Acta.*, 60: 3359–3360.

Fleming, M., G. Lin, S. Ldas and L. Sternberg (1990). Influence of mangrove detritus in an estuarine ecosystem. *Bull. of Mar Sci.*, 47: 663–669.

Fontugue, M.R. and J.C. Duplessy (1981). Organic carbon isotopic fractionation by marine plankton in the temperature range –1 to 31 C. *Oceanologica Acta.*, 4: 85–90.

Fry, B.M. and E.B. Sheer (1984). $\delta^{13}C$ measurements as indicators of carbon flow in marine and freshwater ecosystems. *Contribution to Marine Sci.*, 27: 13–47.

Grey, J., R.I. Jones and O. Sleep (2000). Stable isotope analysis in origins of zooplankton carbon in lakes of different trophic state. *Oecologia*. 123(2, 3): 232–240.

Marguillier, S., G. Vander Velde, F. Dehairs, M.A. Hemminga and S. Rajagopal (1997). Trophic relationships in an interlinked mangroves sea grass ecosystem as traced by $\delta^{13}C$ and $\delta^{15}N$. *Mar. Ecol. Frog. Ser.*, 151: 115–121.

Newell, R.I.E., N. Marshall, A. Sasekumar and V.C. Chong (1995). Relative importance of Benthic micro algae, phytoplankton and mangroves as source of nutrition for Panaeid prawns and coastal invertebrates from Malaysia. *Mar. Biol.*, 123: 595–606.

Odum, W.E. and E.J. Heald (1975). Detritus based food web of an estuarine mangrove community. In: *Estuarine Research*, (Ed.) L.E. Cronin. Academic Press Inc., New York, p. 265–286.

Rao, R.G. (1997). Tracing carbon flow in the mangrove ecosystem of Godavari delta, India. *Ph.D. Thesis*. Vrije University (Brussels), pp. 142.

Robertson, A.I. (1986). Leaf-burying crabs: Their influence on energy flow and export from mixed mangrove forest (*Rhizophora* sp.) in Northeastern Australia. *J. Expt. Mar. Biol. & Ecol.*, 102: 237–248.

Robertson, A.I. and N.C. Duke (1987). Mangroves as nursery sites: Comparison of abundance and species composition of fish and crustaceans in mangroves and other nearshore habitat in tropical Australia. *Mar. Biol.*, 96: 193–205.

Robertson, A.I., D.M. Alongi and K.G. Boto (1992). Food change and carbon fluxes. In: *Tropical Mangrove Ecosystems*, A.I. Robertson and D.M. Alongi. Am. Geophys. Union, Washington, p. 193–326.

Rodelli, M.R., J.N. Gearing, P.J. Gearing, N. Marshell and A. Sasekumar (1984). Stable isotope ratios a traces of mangrove carbon in Malaysian ecosystem. *Oecologia*, 61: 326–333.

Sasekumar, A. and J.J. Loi (1993). Litter production in three mangrove forest zones in the Malay peninsula. *Aquatic Botany*, 17: 283–290.

Tackx, M., W. DeCoster Zhul, R. Billones and M.H. Daro (1995). Measuring selectivity of feeding by estuarine copepods using image analysis combined with microscopy and coulter counting. *J. Mar. Sci.,* 52: 419–425.

Teal, J.M. (1962). Energy flow in a salt marshy ecosystem of Georgia. *Ecology,* 43: 614–624.

Zieman, J.C., S.A. Macko and A.I. Mills (1984). Role of sea grass and mangroves in estuarine food webs: temporal and spatial changes in stable isotope composition and aminoacids content during decomposition. *Bull. Mar Sci.,* 35: 380–387.

Chapter 24

Zooplankton Population Assessment in the Coastal Waters off Kakinada and Mumbai

V.W. Lande
National Environmental Engineering Research Institute, Nagpur – 440 020

ABSTRACT

Consequences of development of the oil and natural gas fields away from the shore in sediments along the coast monitored for zooplankton population at 12 sampling locations off Kakinada and Mumbai during pre-monsoon and post-monsoon seasons in the year 1993–1995. Frequency of general groups in coastal waters in relation to hydrological movements, upwelling areas indicate zooplankton enriched area of east coast. During the collections of zooplankton, representative groups comprise of 41 genera off Kakinada and 14 genera off Mumbai were encountered. Average zooplankton biomass varied in terms of volume 1.44-11.40 ml/m^3 and 0.32–9.2 ml/m^3 offVisakhapatnam and 0.003–1.15 ml/m^3 and 0.006–5.77 ml/m^3 off Mumbai. Copepoda appeared dominant group in the zooplankton. Indicator of water mass species of coelentrates madusae, siphonophora, chaetognatha were recorded off the east coast waters. The Shannon Wiener Index (SWI) ranges from 0.83 to 3.64 and 0.067 to 0.29 in the study area of these coastal waters.

***Keywords**: Coastal water, Zooplankton biomass, Shannon Wiener Index.*

Introduction

The activities of oil exploration along the coast have considerably affect the biological environment. Spillage of oil in transportation discharges from oil platforms and further refinery processes release many harmful hydrocarbons in the coastal waters. The environment which is divided into three main zones, the sea surface, water column and sediments, each of these zones possess plankton, fish and benthos as function of dynamic systems with interaction between organisms and environment as all these three zones interact freely (Dick, 1982).

The earlier work carried out about distribution of zooplankton biomass and conspicuous decrease in biomass during the peaks of south-west monsoon in the off-shores of Visakhapatnam was observed by Vijay Kumar and Sarma (1988).

Copepoda forming the predominant group except in Southern region of Bay of Bengal where a swarm of pelagic tunicate contributes 19 per cent of total catch (Achuthankutty *et al.*, 1980). Occurrence of lucifer, acetes, tunicate fritillaria, fish eggs and larvae was recorded in the coastal water -(Omori, 1977) shows some similarities of groups in the study area.

Considering the earlier studies, the present study has been carried out on zooplankton biomass using the samples collected from coastal waters of Kakinada (year 1995) and Mumbai (year 1993). Sinews to the present scenario, zooplankter of biological environment were assessed in the off-shore environments (latitude - 16 25′ N -17 02′ N to longitude 82 05′ E–82 25′ E) off Kakinada and (latitude 18 00′ N–20 50′ N to longitude 71 25′ E–72 50′ E) off Mumbai.

Materials and Methods

Zooplankton samples were collected using Heron tentor plankton net (200 μmesh size) from chartered vessel and MFV Saraswati (Research Vessel of C.I.F.E.) around offshore off Kakinada and offshore off Bombay High region of Mumbai. Zooplankton samples were preserved immediately in 5 per cent Formalin solution. The group and identification and enumeration of genera were carried out using Letiz Wetzler Microscope (Germany) under magnification of X100, X400 in the laboratory. Displacement volume measured as biomass ml/m^3 and diversity of zooplankton groups calculated as Shannon Wiener diversity index at each sampling location (APHA, 1992).

Results and Discussion

Zooplankton of the offshore off Kakinada were studied in terms of biomass in numbers/m^3, displacement volume, percentage compositions of identified groups and Shannon Wiener index in pre-monsoon and post-monsoon seasons. Population density fluctuated from 0.00–1693/m^3 in pre-monsoon and subsequently increased to 162–1982/m^3 in post in monsoon. Incisive observations of seasonal variations of per cent composition constituted 22.47–88.23 per cent and 0.00–85.08 per cent, copepoda; 0.00–17.40 per cent and 0.00–9.47 per cent, coelentrate medusae; 0.00–3.25 per cent and 0.00–10.85 per cent, Euphusiids; 2.10–27.23 per cent and 0.00–31.39 per cent, chaetognatha; 0.00–24.71 per cent and 0.00–29.00 per cent, fish eggs of zooplankton during pre-monsoon and post-monsoon respectively. Copepoda was dominated in the segregated samples. Zooplankton comprised of 41 genera identified in the east coast off-shore waters of Kakinada. Displacement volume varied 1.44–11.4 ml/m^3 to 0.32–9.2 ml/m^3 in pre-monsoon and post-monsoon. Coelentrates medusae, siphonophore, chaetognath, the indicator species of water masses formed large portion of percentage compositions in the samples. The displacement volume of biomass increases in the samples collected in pre-monsoon followed by decreasing volume during post-monsoon. Anthomedusae *Bougainvillea* sp. trachymedusae *liriope* sp. and *obelia* sp. and epipelagic medusae *Mnemiopsis* sp. were observed (Santhakumari and Vijayalakshmi Nair, 1999). Presence of coelentrate and chaetognatha were indicated greater diversity in groups of zooplankton next to group of copepoda in the east coast offshore (Terazaki, 1999). Copepoda organism belongs to groups of calanoida, cyclopoida and harpacticoida. Calanoida copepoda–*Paracalanus* sp., *Eucalanus* sp., *Acrocalanus* sp., *Acartia* sp., *Centropages* sp. followed by cyclopoida *Oithona* sp., *Oncaea* sp., *Corycaeus* sp., *Sapphirina* sp., *Copilia* sp. and harpacticoida-*Microsetella* sp. were shown prevalence in the coastal waters. Copepoda, crustacean larvae and euphausiids considered the inhabitant of surface and sub-surface layers in these coastal waters and amphipoda *Siriella* sp., *Hemisiriella* sp. and brachyuran larvae were recorded in pre-monsoon and lesser number post-monsoon. The *Lucifer* sp., *Acetes* sp. and *Sergestes* sp. swarming and pattern of diurnal migrations were observed in these shallow inshore areas (Table 24.1).

Sagitta sp., *Petrosagitta* sp., *Krohnitta* sp., inhabitant in low salinity coastal waters were seen abundant in pre-monsoon whereas *Petrosagitta* sp. observed in post-monsoon in the east coast offshore. The *Ophiopluteus* Larvae of starfish of benthic inhabitant was recorded in post-monsoon. Zooplankton production converted from displacement volume to dry weight varied from 117.65 mg/m^3 to 931.38 mg/m^3 in pre-monsoon to 24.51 mg/m^3 to 751.64 mg/m^3 in post-monsoon along the east coast off Visakhapatnam (Sumitra Vijaya Raghavan *et al.*, 1982).

SWI of diversity of species ranged from 1.00 to 2.85 at sampling locations and 2.74 to 3.64 around offshore platforms and from 2.35 to 3.12 at sampling locations and 0.83 to 2.20 around offshore platforms in pre-monsoon and post-monsoon (Hellawell, 1978). In offshore region of west coast off Mumbai, the zooplankton studies indicated lesser population of zooplankton as it forms the open ocean part of Arabian Sea (Table 24.2). The exploration activities are likely to disturb the biological environment in the offshore. During the studies, several species of zooplankton belonged to 14 genera were observed in the offshore off Mumbai. Copepoda was the dominant group in all the samples in pre-monsoon and post-monsoon. Foraminifera, radiolaria, ostracoda, polychaeta, mysidacea, chaetognatha, appendicularia were observed in pre-monsoon whereas chaetognatha thaliacea, salps, euphausiida, ctenophore, amphipoda–were recorded in post-monsoon. Biomass of zooplankton was ranged from 0.003–1.15 mg/m^3 in pre-monsoon 0.006–5.77 mg/m^3 in post-monsoon in the offshore waters.

Enumeration did not carry out due to meagre biomass. General trend showed zooplankton biomass increased in samples collected in post-monsoon. Zooplankton production in terms of dry weight in the coastal waters varied from 0.024 mg/m^3 to 94.36 mg/m^3 in pre-monsoon to 0.049 mg/m^3 to 471.40 mg/m^3 of the offshore waters in post-monsoon. SWI of diversity is found in the range of 0.088 to 0.103 at sampling locations and 0.067–0.207 and 0.094 to 0.29 around the platforms, subsequently increases in pre-monsoon, compared to post-monsoon.

Bay of Bengal is richer during the south west monsoon when the waters are generally less disturbed than during the warmer days. In the coastal waters of west coast, amphipoda are usually bathypelagic species and large dense swarms of Hyperiids monitoring the biological effects is primary concern in maintenance of commercial fisheries and marine life in the sea and to check where the effect of development of individual oil fields occurs which sufficiently cause threat to the marine life. Sources of impact around offshore platforms are localized thermal effect on water column inhabitant and discharge of drilling mud and produced water has been shown toxic to corals and range of marine organisms, but not detrimental and causes low order of toxicity incessantly on zooplankton of the coastal waters and fisheries (Koons *et al.*, 1975). It is found that inshore water showed greater richness in faunal composition and mean ratio of groups for recorded volumes of zooplankton. In thecosomata,

Table 24.1: Variation of Zooplankton at Different Sampling Locations in the Coastal Waters of Kakinada

Sl. No.	*Sampling Location*	*Percentage Composition of Species in Groups*																
		Bio-mass No./ml^{-1}	*Volume ml/m^{-3}*	*Shan-non Wiener Index*	*Coelen-trate Medu-sae*	*Siphon-ophore*	*Poly-chaeta*	*Gastro-trich*	*Cope-poda*	*Mysi-dacea*	*Amphi-poda*	*Euphau-siids*	*Crusta cean Larvea*	*Chae-tog-natha*	*The-coso-mata*	*Appen-dicu-laria*	*Echino-der-meta*	*Fish Eggs*
Pre-monsoon	**Average Vol. ml/m^3 1.44–11.4**																	
1.	1	1324	7.20	2.73	17.40	2.10	0.00	0.00	50.00	0.00	0.00	3.25	0.01	27.23	0.00	0.00	0.01	0.00
2.	3	1691	11.40	2.85	6.55	0.00	0.00	0.00	85.24	0.00	0.00	0.00	0.00	4.91	3.30	0.00	0.00	0.00
3.	7	518	7.00	2.82	5.40	8.10	0.00	0.00	56.75	0.00	0.00	2.70	0.00	27.05	0.00	0.00	0.00	0.00
4.	10	1584	9.00	1.00	0.00	2.24	0.00	0.00	22.47	28.08	4.50	1.15	0.00	16.85	0.00	0.00	0.00	24.71
Gas Exploration Areas																		
5.	1	1081	5.20	3.52	0.96	5.80	0,96	0.00	67.30	0.00	0.00	0.96	11.52	4.80	7.70	0.00	0.00	0.00
6.	2	387	1.90	3.38	6.80	4.90	2.94	0.98	59.84	0.00	0.98	3.04	6.80	10.78	1.96	0.00	0.98	0.00
7.	3	1047	3.40	2.74	0.00	1.68	0.00	0.00	88.23	1.68	0.84	1.68	2.95	2.10	0.00	0.00	0.00	0.84
8.	4	858	4.19	3.28	4.19	2.80	0.00	0.00	59.44	2.09	0.00	0.66	3.50	8.38	0.00	0.00	0.00	18.94
9.	5	0	0.00	0.00	0.00	0.00	0.00	0.00	0.00	0.00	0.00	0.00	0.00	0.00	0.00	0.00	0.00	0.00
10.	6	1110	1.44	3.64	14.41	1.80	2.70	7.20	43.24	9.00	0.00	1.85	1.80	9.00	0.00	0.00	0.00	9.00
11.	7	1693	6.25	3.37	6.25	0.00	0.00	0.00	68.47	5.34	0.00	2.17	4.74	9.78	0.00	2.17	0.00	1.08
12	8	932	4.20	3.55	6.30	4.50	0.00	0.00	69.36	0.00	0.90	0.00	5.40	3.60	0.00	0.00	0.00	9.94
Post-monsoon	**Average Vol. ml/m^{-3} 0.32-9.2**																	
1.	1	691	0.80	2.65	1.56	0.00	0.00	0.00	39.02	0.00	0.00	10.85	0.00	31.39	0.00	1.56	0.00	15.62
2.	3	342	1.35	2.97	1.28	11.53	0.00	0.00	67.23	0.00	0.00	1.28	8.41	6.41	3.86	0.00	0.00	0.00
3.	7	990	2.55	3.12	0.00	3.16	0.00	0.00	53.68	0.00	0.00	0.00	11.58	13.68	0.00	0.00	2.10	15.80
4.	10	1568	1.20	2.35	2.14	18.57	0.00	0.00	60.70	0.00	0.00	1.42	0.00	0.00	0.72	0.00	3.57	12.88
Gas Exploration Area																		
5.	1	1642	1.32	1.55	0.68	2.65	0.00	0.00	53.43	0.68	0.00	6.16	4.79	1.36	0.00	1.25	0.00	29.00
6.	2	1461	2.60	0.83	9.47	0.00	0.00	0.86	47.11	0.00	0.00	6.89	16.37	0.00	0.00	0.86	0.86	17.58
7.	3	1589	1.20	2.20	0.00	10.36	0.00	4.58	69.06	0.00	0.70	0.70	7.80	6.80	0.00	0.00	0.00	0.00
8.	4	162	0.60	1.70	0.00	0.00	85.72	14.28	0.00	0.00	0.00	0.00	0.00	0.00	0.00	0.00	0.00	0.00
9.	5	1982	9.20	1.60	0.00	0.00	0.00	0.00	85.08	0.00	2.76	1.65	3.31	1.10	4.45	1.65	0.00	0.00
10.	6	991	0.32	2.20	2.07	5.20	0.00	1.03	40.00	0.00	0.00	7.28	2.07	20.70	0.00	2.07	3.10	16.48
11.	7	1964	0.35	1.10	0.00	2.62	0.00	1.10	72.10	0.00	0.00	0.52	2.62	7.90	0.00	2.62	0.00	10.52
12.	8.	975	1.20	1.40	1.14	3.44	0.00	2.29	54.02	0.00	0.00	1.20	27.58	9.19	0.00	0.00	1.14	0.00

Table 24.2: Qualitative Abundance of Zooplankton and Biomass at Different Sampling Locations in the Coastal Waters of Mumbai

Sl. No.	Sampling Location	Biomass		Shannon Wiener Index	Radi-olaria	Coelen-trate	Siphono-phore	Cope-poda	Amphi-poda	Ostra-coda	Appen-dicularia	Thalia-cea	Chaeto-gnatha	Deca-poda	Fish Eggs
		Volume ml/m³	Poole Volume ml/m³												
Pre-monsoon															
1.	1	0.36		0.103	+	-	-	+	-	+	-	-	-	-	-
2.	2	0.02			-	-	-	+	-	-	-	-	-	-	-
3.	3	0.003			+	-	-	+	-	-	-	-	-	-	-
4.	4	0.15			-	-	-	-	-	+	-	-	-	-	-
Offshore Oil Exploration Area															
5.	10	0.006		0.207	+	-	-	+	-	-	-	-	-	-	-
6.	16	0.012			-	-	-	+	-	-	-	-	-	-	-
7.	33	0.006			-	-	-	+	-	-	-	-	-	-	-
8.	43	0.115			-	-	-	+	-	+	+	+	-	-	-
9.	12	0.115	0.003–1.15	0.290	-	+	-	+	-	-	-	-	-	-	-
10.	22	0.025			-	-	-	+	-	-	-	-	+	-	-
11.	36	0.006			+	-	-	+	-	-	-	-	-	-	-
12.	40	1.15			-	-	-	-	-	-	+	+	+	-	-
Post-monsoon															
1.	1	0.26		0.088	-	-	-	-	-	-	+	+	-	-	-
2.	2	0.06			-	+	-	+	+	-	+	+	+	-	-
3.	3	0.2			-	-	-	-	+	+	-	-	+	-	-
4.	4	0.006			+	-	-	+	-	+	-	-	+	-	-
Offshore Oil Exploration Area															
5.	10	0.13		0.067	-	+	-	+	-	-	-	-	+	-	+
6.	16	0.135			-	+	-	-	-	-	-	-	-	-	-
7.	33	0.31	0.006–5.77		-	+	-	+	-	-	-	-	+	-	-
8.	43	5.77			-	-	+	+	+	-	-	-	+	-	-
9.	12	0.412		0.094	-	+	-	+	-	-	-	-	+	-	-
10.	22	0.179			-	+	-	+	-	-	-	-	-	+	-
11.	36	0.12			-	-	-	+	-	+	-	-	+	-	-
12.	40	0.13			-	-	-	+	+	-	-	-	-	-	+

+: Presence; -: Absence.

warm and cold tolerant species *Creseis acicula,* is observed in the coastal waters off Visakhapatnam. Chaetognatha group occur from the Bay of Bengal to Singapore straits but its presence in the Bay of Bengal is extremely seasonal, usually in March and April (Table 24.3). Chaetognatha occurrence attributed to mainly availability of food and at the time when maximum copepoda density have been observed (Achuthankutty *et al.*, 1980). Seasonal distribution of fish eggs and larvae off the east coast of India were observed as recorded in the study area throughout the year (Vijayalakshmi Nair, Neelam Ramaiah and Gajbhiye, 1998). Indicator polychaete species *Sagitella* sp. appeared in one instance indicate the point source of pollution. Strong upwelling area of the east coast represented by the higher phytoplankton counts and increased number of zooplankton characterized by comparatively few dominant zooplankton species. In the offshore waters of Visakhapatnam, dendrogram of cluster of zooplankter indicated the correlation coefficients were positive and 47 per cent–52 per cent of the coefficients appear significant (Peter, Krishna Iyer, Simon John and Radha Krishnan, 1977). Correlation coefficients of three clusters (Figure 24.1). (I) mysidacea–echinoderma, siphonophora–euphausia. (II) gastrotricha–fish eggs, polychaeta–mysidacea. (III) copepoda–amphipoda, coelenterate medusae–copelata, crustacean larvae–thecosomata in pre-monsoon. Similarly three clusters (Figure 24.2) siphonophora–euphausiacea, copepod–amphipoda. (II) polychaeta–mysidacea, gastrotricha–fish eggs. (III) chaetognatha–copelata, polychaeta, echinodermata larvae, crustacea larvae–thecosomata were recorded in post-monsoon.

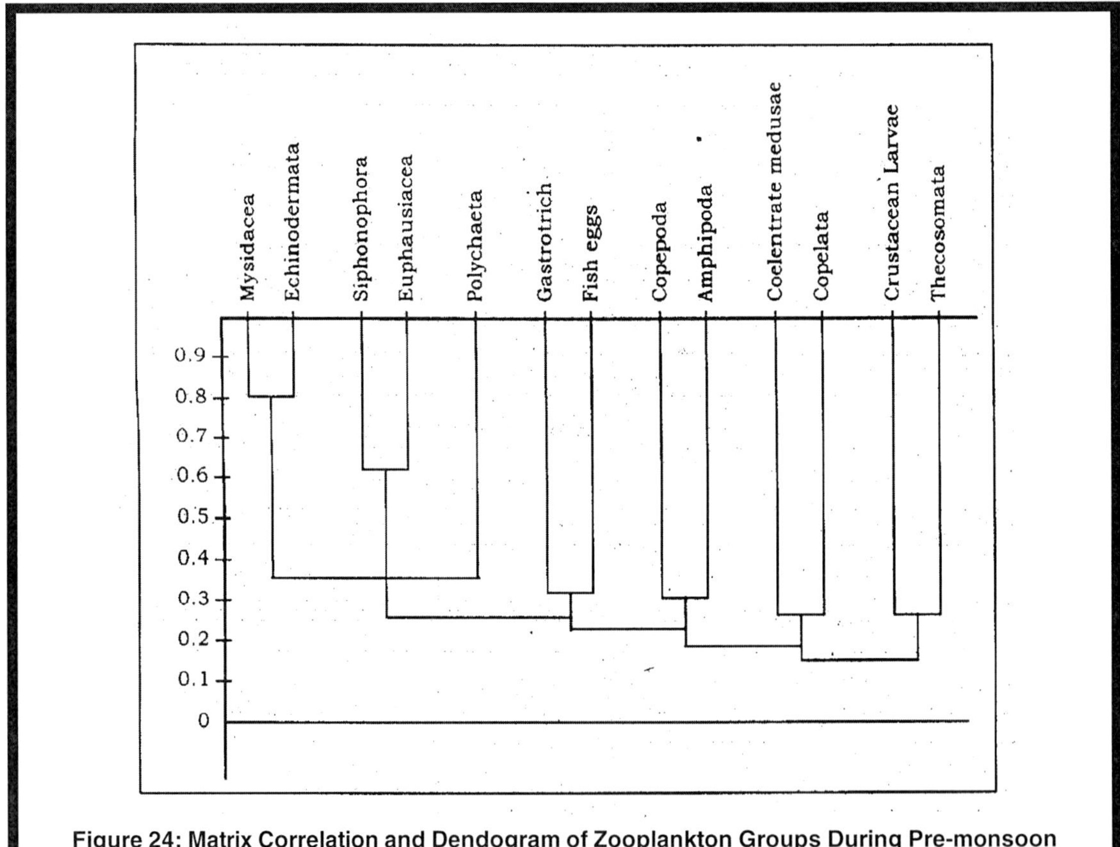

Figure 24: Matrix Correlation and Dendogram of Zooplankton Groups During Pre-monsoon

Table 24.3: List of Identified Zooplankton Species

Sl. No.	*Group*	*Species Identified*	
		Kakinada Coast	*Mumbai Coast*
1.	Coelentrata	*Obelia* sp.	1. Foraminifera
		Liriope sp.	2. Radiolaria
		Bougainvillea sp.	3. Coelentrata
		Mnemiopsis sp.	*Obelia* sp.
			Clytia sp.
2.	Siphonophore	*Diphyes* sp.	4. Siphonophore
		Lensia sp.	*Lensia* sp.
3.	Nematode	*Diplogaster* sp.	
4.	Polychaeta	*Neries* sp.	
		Terebellied sp.	
		Sagitella sp.	
5.	Gastrotrich	*Urodasys* sp.	
6.	Copepoda	*Paracalanus* sp.	5. Copepoda
		Acrocalanus sp.	(i) *Paracalanus* sp.
		Undinula sp.	
		Pontella sp.	(ii) *Acrocalanus* sp.
		Eucalanus sp.	(iii) *Centropages* sp.
		Corycaeus sp.	
		Centropages sp.	(iv) *Microsetella* sp.
		Oithona sp.	
		Copilia sp.	(v) *Copilia* sp.
		Acartia sp.	(vi) *Acartia* sp.
		Microsetella sp.	
		Sapphirina sp.	
		Miracia sp.	
7.	Amphipod	*Synopia* sp.	6. Amphipoda
		Ampelisea sp.	*Synopia* sp.
8.	Decapoda	*Lucifer* sp.	7. Ostracoda
		Acete sp.	*Cypris* sp.
9.	Crustacean larvae	*Lucifer* protozoa	8. Chaetognatha
		Brachyuran larvae	*Sagitta* sp.
		Anomuran galatheid larvae	
10.	Chaetognatha	*Sagitta* sp.	
		Petrosagitta sp.	
		Eukrohnia sp.	
		Krohnitta sp.	
11.	Theocosomata	*Creseis* sp.	9. Thaliacea
		Hyalocylis sp.	*Salpa* sp.
12.	Thaliacea	*Doliolum* sp.	
13.	Appendicularia	*Oikopleura* sp.	10. Appendicularia
			Olikopleura sp.
14.	Echinodermata	Ophiopluteus larvae	11. Fish eggs
15.	Pisces	Fish eggs	

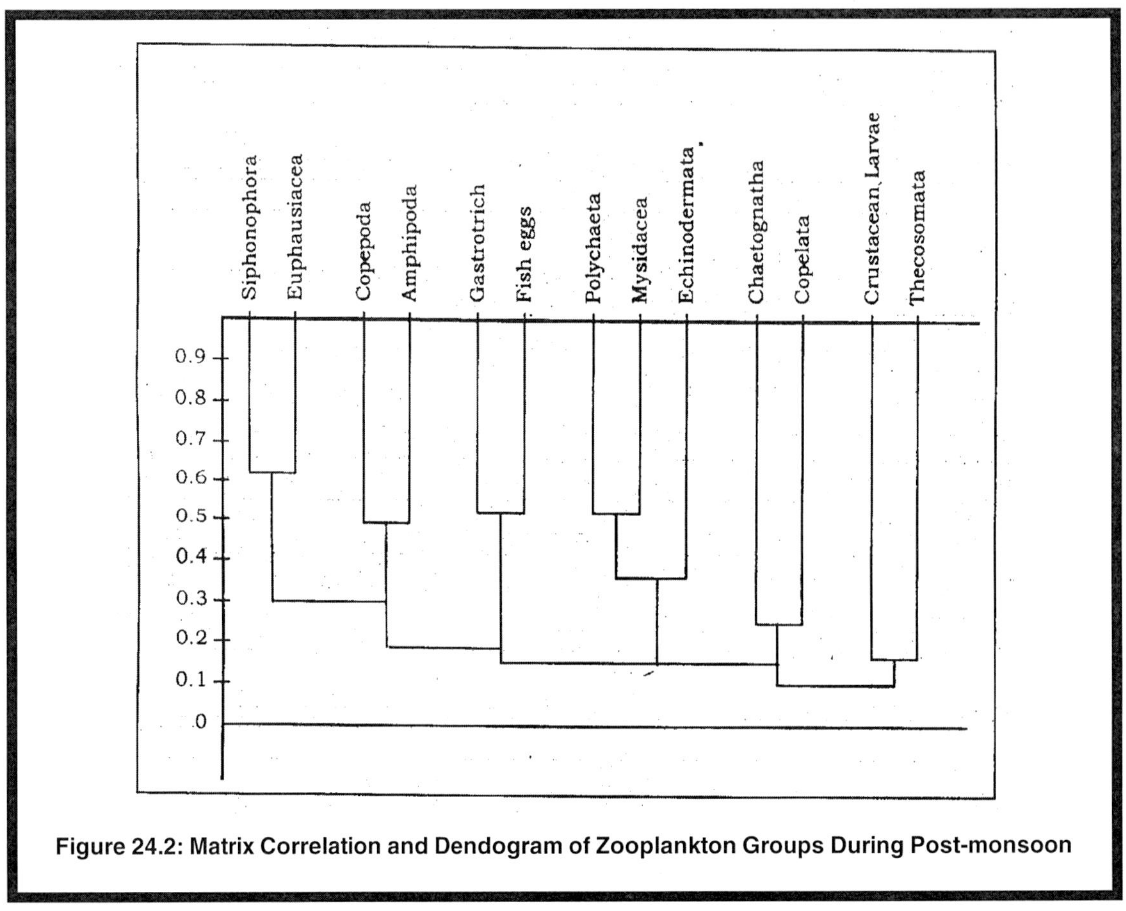

Figure 24.2: Matrix Correlation and Dendogram of Zooplankton Groups During Post-monsoon

Acknowledgements

Author gratefully acknowledges to Dr. S. Devota, Director, NEERI for encouragement and Dr. T. Chakrabarti, Scientist and Head, for providing laboratory facilities and Shri A.V.J. Rao and H.R. Dwivedi for carrying out offshore sampling.

References

Achuthankutty, C.T., M. Madhupratap, Vijayalakshmi R. Nair, S.R. Srikumaran Nair, T.S.S. Rao (1980). Zooplankton Biomass and composition in the western. Bay of Bengal during late south-west monsoon. *Ind. J. Mar. Sci.,* 9: 201–206.

APHA (1992). *Standard Methods for the Examination of Water and Wastewater,* 18th edition. APHA, AWWA, WPCF pp. 10–130, index 1–49. American Public Health Association, 1015, 15th street N.W. Washington DC 20005.

Dick, B. (1982). Monitoring the biological effects of north sea platforms. *Marine Poll. Bull.,* 23(7): 221–222.

Koons, C.B., C.C. Macantiffe and F.T. Weiss (1975). Environmental aspect of produced waters from oil and gas extraction operations in offshore and coastal waters. Report to sheer technical subcommittee offshore operation committee, New Orleans L.A.

Hellawell, J.M. (1978). Biological surveillance of river. Biological monitoring handbook. A collaborative production between natural environmental research council, water research centre, Regional water authorities. Stevanage, The Centre, pp. 331.

Nair, Vijayalakshmi, Neelam Ramaiah and S.N. Gajbhiye (1998). Zooplankton standing stock and diversity in the Gulf of Katchchh with special reference to larvae of decapoda and pisces. *Ind. J. Mar. Sci.,* 27(3–4): 340–345.

Omori, M. (1977). In: *Proc. Symp. Warmwater Zooplankton.* NIO-Unesco at Goa-PIR.

Peter, K.J., H. Krishna Iyer, Simon John and E.V. Radha Krishnan (1977). Distribution and coexistence of planktonic organisms along the southwest coast of India. In: *Proceeding of the Symposium on Warmwater Zooplankton,* Special Publication NIO, Goa, p. 80–86.

Santhakumari, V. and Vijayalakshmi R. Nair (1999). Distribution of hydromedusae from the exclusive economic zone of the west and east coast of India. *Ind. J. Mar. Sci.,* 28: 150–157.

Sumitra Vijaya Raghvan, R.A. Selvakumar, T.S.S. Rao (1982). Studies on zooplankton from the arabian sea off the south central west coast of India. *Ind. J. Mar. Sci.,* 11: 70–74.

Vijaya Kumar and V.V. Sarma (1988). Zooplankton studies in Visakhapatnam harbour and near shore water east coast of India. *Ind J. Mar. Sci.,* 17: 75–77.

Terazaki, M. (1999). Mass occurrence of bathypelagic chaetognath *Eukrohnia fowleri* from the arabian sea and Bay of Bengal. *Ind. J. Mar. Sci.,* 28: 163–168.

Chapter 25

Seasonal Variations in Freshwater Protozoans in Kali-Nadi, District Etah, U.P., India

A.K. Paliwal

Department of Zoology, Ganjdundwara, P.G. College, Ganjdundwara, District Etah, U.P.

ABSTRACT

This articles presents the data in regards to the seasonal variation in freshwater protozoans in Kali-Nadi, District Etah, U.P.

Introduction

Freshwater protozoans play an important role as bacterial and algal grazers and also in direct source of food for fish larvae and adults. Among previous workers Vasisht and Jindal (1980 and 1985), Rais and Sharma (1986) and Puri (1989) have made some contribution in the field of Indian protozoan. One year study (April 2003 to March 2004) was undertaken to understand seasonal distribution of freshwater protozoans in Kali-Nadi, District Etah. There is no earlier record of any such investigations of any such work in this field.

Materials and Methods

Monthly planktonic samples from four different locations of Kali-Nadi were collected by filtering two litres of water, at each experimental station, through a planktonic net *i.e.* 144 μ mesh size. Each samples were fixed at the spot in 5 per cent formaldehyde. In the laboratory protozoans were identified as Edmondson (1959), Kudo (1986), Nair *et al.* (1971) and Dutta (1983) and counted in Sedgewick refer (counting cell method).

Results and Discussion

Sixteen species of protozoans are recorded here which shows the predominance of *Difflugia* (5 sps.) followed by *Centrophyxis* (2 sps.), *Arcella* (2 sps.), *Loxodes* (2 sps.), *Paramecuim* (2 sps.), *Euplates* (2 sps.) *Unidentified Ciliate* (1 sps.).

Among various species genus *Difflugia* has shown its presence through out the year. Its maximum diversity is observed in November. Annual maximum count of *Centropyxis* sps. is observed in January. Genus *Arcella* recorded its annual maximum count in January. *Loxedes* Genus shows its annual maximum count in May. *Paramecuim* sps. indicates annual highest count in May.

Table 25.1: Monthly Variations in Protozoans in Kali-Nadi of District Eath, U.P.

Sl. No.	*Genus*	*Species*	*Monthly Variations*	*Presence*
1.	*Difflugia*	A–Lobostoma	May–June September–March	8 times
		B–Tubercula	April–June August–February	10 times
		C–Oblanga	May–July October–March	9 times
		D–Corona	April–May August–January, February	5 times
		E–Acuminata	May–July	3 times
2.	*Tentropyxis*	A–Ecornis	July–September November–March	9 times
		B–Aculeta	May–September	2 times
3.	*Arcella*	A–Dentola	August–September November–February	6 times
		B–Discoies	November–February	4 times
4.	*Loxodes*	A–Modesta	May–January	8 times
		B–Spiralis	January	9 times
5.	*Euplotis* sps.		May–June	2 times
6.	*Paramecium* sps.		May	1 time
7.	*Ciliates* sps.		May, June, August, January	4 times

An overall study of protozoans has revealed the seasonal presence of various sps. A similar type of observation was indicated earlier by Schonborn (1962), Dutta (1978), Jindal and Vasisht (1985), Rais and Sharma (1986) and Puri (1989).

Protozoans have shown May, November, and January as peaks with annual highest count in January. Their count remain low during monsoon period *i.e.* July–September and also in post-monsoon period *i.e.* October.

Dilution of water caused by monsoon rains *i.e.* July–Sept. may explain low protozoans count observed during July–October. Also fall in submerged macrophytic diversity, where some protozoans are known to grow as periphyton, appears as an additional factors explaining low diversity of protozoans.

Macrophytic decline contributing towards the enrichment of bacteria and detritus on which protozoans feed and also increase protozoans diversity *i.e.* May–June.

Detritus and bacterial enrichment contributed by decomposition of floating macrophytes which started decline after August which fall the water level and appearance of submerged macrophytes may also increase the number of protozoans.

Annual highest protozoans peak *i.e.* January–May be related to the optimum conditions of low temperatue, high dissolved oxygen, rise in pH and low bicarbonate etc.

May-June also indicated protozoans a positive relationship with temperature, bicarbonate, chloride, calcium, magnesuim etc.

References

Dutta, S.P.S. (1983). On some freshwater *Rhizopoda* from Jammu. *Jammu, Univ. Rev. Sc.,* 1: 91–94.

Edmondson, W.T. (1959). *Freshwater Biology.* John Wiley and Sons, U.S.A.

Jindal, R. and H.S. Vasisht (1985). Ecology of a freshwater pond at Nabha. *Zool. Ori.,* 2: 74–83.

Nair, K.N., A.K. Dass and B.N. Mukherjee (1971). On some freshwater rhizopoda and helizoa (Protozoa) from Calcutta and its environs. *Rec. Zool. Survey,* India, 65: 1–2–16.

Rai, D.N. and U.P. Sharma (1986). Studies on occurrence of periphytic Testaced Protozoa (Rhizopoda) in the weed infested swamps of Darbhanga (North Bihar). *India J. Ecol.,* 13: 334–339.

Trivedy, R.K., P.K. Goel and C.L. Trisal (1987). *Practical Methods in Ecology and Environmental Science.* Environment Publication Karad.

Vasisht, H.S. and R. Jindal (1980). Rheological survey of a Pucca Stream at Patiala, Punjab (India). *Limnologica,* Berlin, 12: 77–83.

Chapter 26

Ecologically Significant Species of Zooplankton in Coringa Mangrove Ecosystem, Andhra Pradesh

N. V. Prasad

Division of Marine Biology, Department of Zoology, Andhra University, Visakhapatnam – 530 003, India

ABSTRACT

Studies (1999–2000) on the community structure of zooplankton in Coringa mangrove environment on the east coast of India revealed altogether 12 Ecologically Significant Species (ESS) of zooplankton represented by 6 major groups. Description of zooplankton structure included species composition, density and spatial distribution patterns of ESS at 15 GPS fixed locations. *Acartia erythraea,* bivalve veligers, *Eucalanus elongatus, Paracalanus parvus,* zoea of *Scylla serrata,* gastropod veligers and *Acartia clausi* were the important ESS species. The distribution and density of ESS revealed that the majority of them are represented by typical brackish water forms. The ambient salinity and degree of stability achieved through neritic influence are prime influencing factors in limiting the BIS distribution in the mangrove environment of Coringa.

Keywords: *Zooplankton, Distribution, Ecologically significant species, Mangroves, Andhra Pradesh.*

Introduction

In marine biological investigations, species evaluation is important to find out association and assemblages, which usually characterise areas along a stress gradient. Lie (1968) noticed that since most communities are so complex and rich in species, it is necessary to elevate certain numerically important species and subject them to a detailed analysis. Such species have been variously termed as 'Characterising Species' (Thorson, 1957), 'Dominant Species' (Lie, 1968), 'Prevalent Species' (Jones, 1969), 'Biologically Important Species' (Feder *et al.*, 1973) and 'Ecologically Significant Species' (Ellis, 1969) etc. According to Boesch (1973), two approaches are usually followed to simplify and extract information from the very complex patterns of multi-species populations. The first is classification, in which species are classified on the basis of their distribution within a series of collections according to some mathematical criteria (*e.g.*, Thorrington-Smith, 1971). The second approach is the analysis of 'community structure' in which the distribution of importance (as measured by one of a number of criteria, usually abundance) among species is of primary interest (*e.g.*, Travers, 1971). The objective of the present investigation was to describe the Ecologically Significant Species and its distribution from the mangrove environment of Caringa, Andhra Pradesh.

Materials and Methods

The mangrove ecosystem of Coringa is situated near Kakinada, Andhra Pradesh (16 50′–17 00′ N; 82 14′– 82 22′E), along east coast of India. The mangrove ecosystem of Coringa including Kakinada Bay (north bay and south bay), Gaderu and Coringa rivers. Water and zooplankton samples were collected monthly from March 1998 to February 2000. Altogether 15 GPS fixed stations, 4 in north bay (Stations 1–4), 3 in south bay (Stations 5–7), 5 in Gaderu (Stations 8–12) and 3 in Coringa (Stations 13–15) were selected. Hydrographical parameters such as surface water temperature, salinity, dissolved oxygen and pH were measured by adopting standard methods (APHA, 1995). Zooplankton samples were collected by using a 120 μm mesh size net. For the enumeration of zooplankton, aliquot method was employed (Wickstead, 1965). Seasons were defined as monsoon (June–September), post-monsoon (October–January) and pre-monsoon (February–May) based on the seasonal monsoons. Based on the lines suggested by Feder *et al.* (1973), three criteria were evolved to designate a given species as 'Ecologically Significant Species' (ESS) during the present study. Each species classified as Ecologically Significant Species met at least one of the three conditions:

1. It was distributed in 50 per cent or more of the total stations sampled.
2. It comprised over 10 per cent of the composite population at anyone station.
3. Its population density was significant at any given station.

Results and Discussion

The estuarine mangrove environment of Coringa mangrove ecosystem received the maximum rainfall from June to October coinciding with seasonal monsoon. The two monsoons namely the southwest and northeast monsoons brings the maximum rainfall for the area. The hydrographical variables exhibited greater fluctuations both in time and space. The secchi depth was relatively high (mean 0.45–0.76 m) near the proximity of the sea (Stations 1 and 2) (Table 26.1). At the interior locations (Stations 11–15) water transparency was least (mean < 0.40 m) obviously due to the shallow nature of the region and with turbidity caused by suspension of bottom sediment. The turbidity is more influenced due to the prevailing ebb and flow tides in particular in the narrow creeks. Owing to the shallow nature (< 5 m) thermal conditions at most locations appeared largely determined by atmospheric temperature. Within the mangrove habitat, the waters are subjected to evaporation in the stagnated water, where higher temperatures were recorded. In the bay region, the temperature fluctuations can be attributed to the mixing process. The pH values (range 6.2–7.9) revealed that these waters are more alkaline in nature. Relatively high dissolved oxygen values were observed at mangrove locations of Gaderu (mean 4.66 ml/l) than north bay (mean 4.26 ml/l), south bay (mean 4.50 ml/l) and Coringa (mean, 4.26 ml/l). In Gaderu the relatively high range of dissolved oxygen can be attributed to mangrove/phytoplankton activity and community respiration (Prasad, 2003). Overall, salinity fluctuations during the study were very significant. Spatially, the Kakinada Bay is characterised by relatively high salinity (mean > 27.00 per cent) during most part of the year. In Gaderu subject to freshwater influx salinity was mostly 17.00 per cent (mean). In Coringa, where the influx of freshwater even more, salinity was very low (mean < 7.00 per cent). The distribution of salinity is under the influence of physical factors like the river water discharge, tides and the prevailing circulation pattern existing in the area.

Table 26.1: Hydrographical Characteristics (Range) of the Surface Water (Values in Parentheses Denote Mean)

Station	*Depth (m)*	*Sacchi Depth (m)*	*Temperature (°C)*	*Salinity (%)*	*pH*	*Dissolved Oxygen (ml/l)*
North bay	3.0–13.5	0.30–1.90	24.8–31.4	21.09–34.18	6.2–7.9	3.24–4.98
	(6.65)	(0.76)	(28.26)	(28.26)	(7.09)	(4.26)
South bay	2.6–8.5	0.24–0.70	25.4–34.8	18.95–34.06	6.7–7.9	3.19–7.94
	(5.76)	(0.45)	(27.92)	(27.04)	(7.13)	(4.50)
Gaderu	1.6–5.8	0.14–0.60	25.2–31.8	0.0–32.40	6.7–7.8	3.64–7.80
	(3.30)	(0.36)	(28.94)	(17.65)	(7.15)	(4.66)
Coringa	3.4–4.9	0.18–0.38	29.4–33.6	0.0–11.30	6.9–7.7	2.02–5.04
	(4.20)	(0.30)	(31.04)	(4.38)	(7.12)	(4.26)

Zooplankton numerical abundance fluctuated from 3,626 no.m^{-3} to 77,090 no.m^{-3}. Overall, copepods dominated the zooplankton during entire study period and constituted 67.04 per cent of the total zooplankton population (Table 26.2). Abundance of varies zooplankton groups in mangrove environment of Coringa fluctuated in accordance with salinity regime. Zooplankton groups such as copepods, decapod larvae, chaetognaths, bivalve veligers and gastropods were seen throughout the year in large numbers. Coelenterates, polychaete larvae and cladocerans were other major groups of zooplankton population, which showed a seasonal pattern, whereas ostracods, amphipods etc were obtained occasionally in sparse numbers. Altogether 12 Ecologically Significant Species (ESS) was observed during the study (Table 26.3). Among these the calanoid copepod, *Acartia erythraea,* was the most dominant species constituting numerically 13.88 per cent of the total zooplankton population followed by bivalve veligers (9.29 per cent), *Eucalanus elongatus* (9.09 per cent), *Paracalanus parvus* (7.84 per cent), zoea of *Scylla serrata* (7.54 per cent), gastropod veligers (6.56 per cent), *Acartia clausi* (6.36 per cent), *Acrocalanus gibber* (4.18 per cent), *Paracalanus aculeatus* (3.59 per cent), *Evadne* sp. (3.47 per cent), penaeid mysis (3.13 per cent) and fish eggs and larvae (2.61 per cent), which together formed 77.54 per cent of the total zooplankton population.

The distribution of ESS showed definite trends associated with environmental fluctuations. Based on the relative abundance and distribution, the ESS of zooplankton could be divided into the three categories:

1. Species characterising bay locations (neritic and estuarine).
2. Species which preferring mangrove habitat and found abundantly in mangrove channels.
3. Indifferent species (or) ubiquitous species being found almost everywhere in the bay-estuary waterways of Coringa mangroves.

Table 26.2: Percentage Composition (Overall) and Numerical Abundance (Range) of Major Zooplankton Groups

Group	Abundance (no.m^{-3})	Percentage
Copepods	8902–96586	67.04
Decapod larvae	3842–74418	7.56
Bivalves veligers	1203–87300	6.40
Gastropods	6371–12233	4.38
Chaetognaths	487–7820	1.96
Lucifers	187–8580	1.12
Appendicularians	54–1627	0.67
Polychaete larvae	66–568	0.52
Cladocerans	15–8217	0.28
Ostracods	22–415	0.19
Mysids	17–302	0.12
Medusae	5–308	0.09
Amphipods	26–171	0.04
Fish larvae and eggs	12–129	0.03
Miscellaneous groups	47–634	9.60

The ESS, which occurred mostly within the bay regions, was *Acartia erythraea,* bivalve veligers, *Eucalanus elongatus, Paracalanus parvus,* gastropod veligers, *Acrocalanus gibber, Paracalanus aculeatus.*

The *Acartia erythraea,* ranked first among ESS (Table 26.3) was occurred in overwhelming number in the bay region (mean 7,028 no.m^{-3}). The maximum density (13,661 no.m^{-3}) was observed at station 6 in south bay during pre-monsoon (May 1998), when recorded salinity was 31.56 per cent. In general, this species was found numerically rich in the south bay locations (mean > 5,000 no.m^{-3}) and preferring estuarine conditions (mean salinity 25.61 per cent). Similar observations regarding to distribution of *Acartia erythraea* was noticed by Santhanam *et al.* (1975) in Porto Novo waters, and Srinivasan and Santhanam (1991) in Pullavazhi brackish waters. *Acartia erythraea* is the most commonly found species in estuaries, coastal lagoons and embayments (Villate, 1991).

The bivalve veligers, were encountered predominantly in samples collected in the southern part of the Kakinada Bay (mean 5,679 no.m^{-3}) close to the confluence of mangroves. At Stations 5, 6 and 7, within the south bay near the proximity of mangroves, bivalve veligers were recorded highest number during post-monsoon (7,012 no.m^{-3}, 13,374 no.m^{-3} and 11,003 no.m^{-3} respectively). Earlier bivalve veligers were reported predominantly from the estuarine waters (Santhanam *et al.,* 1975).

The copepod, *Eucalanus elongatus* was relatively widespread in its distribution. This species was observed in large number (mean 1,946 no.m^{-3}) particularly towards the northern side of the Kakinada Bay (north bay), subject to much neritic influence (salinity > 25 per cent). Except in few numbers at Stations 8 and 9 (< 200 no.m^{-3}) in Gaderu, this species however remained absent in Gaderu and Coringa. In general, *Eucalanus elongates* found abundantly in high saline waters (Chandra Mohan, 1963). The ESS, *Paracalanus parvus* was encountered in large number (mean 3,765 no.m^{-3}) in the central and southern part of the Kakinada Bay (Stations 3, 6 and 7). The other locations in the bay recorded < 1000 no.m^{-3} on an average. This species preferred moderately saline waters (~25 per cent),

Achuthankutty and Nair (1980) described the species as euryhaline from Zuari estuary. In India, *Paracalanus parvus* was earlier reported from the southeast coast (Santhanam *et al.*, 1965) and west coast (Goswami, 1985) estuarine waters.

Table 26.3: List and Rank of Ecologically Significant Species (ESS) of Zooplankton in Coringa Mangrove Habitat

Rank	ESS Species/Group	Total Numerical Abundance (no.m^{-3})	Percentage by Number	Cumulative Percentage
1	*Acartia erythrae*	1,39,736	13.88	13.88
2	Bivalve veligers	93,589	9.29	23.17
3	*Eucalanus elongates*	91,500	9.09	32.26
4	*Paracalanus parvus*	78,901	7.84	40.10
5	*Scylla serrata* (zoea)	75,935	7.54	47.64
6	Gastropod veligers	66,021	6.56	54.20
7	*Acartia clausi*	64,073	6.36	60.56
8	*Acrocalanus gibber*	42,069	4.18	64.74
9	*Paracalanus aculeatus*	36,145	3.59	68.33
10	*Evadne* sp.	34,924	3.47	71.80
11	Penaeid mysis	31,534	3.13	74.93
12	Fish eggs and larvae	26,246	2.61	77.54
13	Others	1,79,954	22.46	100.00

The ESS *Acrocalanus gibber* was encountered in large number (mean 1,168 no.m^{-3}) in most bay locations, near confluence of the sea. Seasonally, this species was numerically high during post-monsoon period (December 1999) and was once (November 1998) recorded incredible number (10,642 no.m^{-3}) at Station 3 in north Bay. While this species was not found in Coringa, there were a few number (mean 316 no.m^{-3}) in Gaderu (Stations 8–11), when recorded salinity was > 21.48 per cent.

The gastropod veligers were observed in large numbers throughout the study period. These were seen in most parts of the bay-estuary waterways of Coringa mangroves, particularly at stations located in the southern region of the bay (south bay), close to the mangroves. Gastropod veligers were relatively low in north bay (< 2,000 no.m^{-3}) and Coringa (< 1.000 no.m^{-3}), while in Gaderu these were observed more at Stations 8, 9 and 10 (> 3.500 no.m^{-3}). Gastropod veligers preferring southern part of the bay, which receives nutrient rich waters from the surrounding mangroves. Earlier, high abundance of bivalve veligers was reported during post-monsoon and pre-monsoon periods (George, 1958; Nair and Tranter, 1971).

Last among bay ESS was *Paracalanus aculeatus.* This species appeared to prefer the comparatively high saline conditions and found throughout the study period. *Paracalanus aculeatus* was found in large number (mean 1,204 no.m^{-3}) particularly towards the northern side of the bay (north bay), subject to much neritic influence (salinity ~ 30 per cent). In general, this species preferring high saline conditions (Madhupratap, 1979) (salinity 25.5–35.5 per cent).

The ESS, which were found predominantly in the proximity of mangroves and channel habitat, were zoea of *Scylla serrata* and *Evadne* sp. The decapod larvae, zoea of *Scylla serrala* was found predominantly in the proximity of mangroves. Within mangrove locations of Gaderu, *Scylla serrata* zoea occurred in relatively high number (mean 4,162 no.m^{-3}) at Stations 9, 10 and 11, whereas at other locations the density was remained low (mean < 1.000 no,m^{-3}). Appreciably rich population (mean 6,276 no.m^{-3}) of *Scylla serrata* was recorded during monsoon period (August 1998). Nair and Tranter (1971) and Menon *et al.* (1971) reported the distribution of brachyuran zoeae in the plankton samples as more or less throughout the year.

Cladoceran *Evadne* sp. showed restricted distribution to mangrove locations, in particular at stations located in Coringa. During monsoon *Evadne* sp., attained its maximum abundance (mean 10,867 no.m^{-3}) in the proximity of Coringa, while it was not encountercd in bay, there were a few number in Gaderu (< 400 no.m^{-3}). Swarms of *Evadne* sp. was noticed in Coringa during monsoon months, when the freshwater conditions prevailed (0.42–6.23 per cent). During this period *Evadne* species was occurred in overwhelming number (> 8,000 no.m^{-3}) at Stations 14 and 15 (Coringa), constituted > 20 per cent of the total zooplankton populations (salinity 0.42–6.23 per cent). Within the Gaderu, the numerical abundance of *Evadne* sp. remained between 128 no.m^{-3}–359 no.m^{-3}. *Evadne* sp. preferring low saline or freshwater conditions (Rao *et al.,* 1975).

The ESS *Acartia clausi,* penaeid mysis and fish eggs and larvae were indifferent or ubiquitous in its distribution. The calanoid copepod *Acartia clausi,* was commonly found in relatively brackish waters in the middle reaches of the estuary. This species was observed maximum number during pre-monsoon (mean 1,428 no.m^{-3}), followed by post-monsoon (mean 869 no.m^{-3}) and monsoon (mean 823 no.m^{-3}). *Acartia clausi* was found relatively large number in Gaderu (mean 649 no.m^{-3}) followed by bay locations (mean 512 no.m^{-3}) and Coringa (mean 323 no.m^{-3}). In general, *Acartia clausi* preferring moderate saline conditions and exhibited wide range of salinity tolerance.

The ESS fish eggs ands larvae were relatively widespread in its distribution. Fish eggs and larvae were numerically high in bay locations relative to Gaderu and Coringa. This ESS attained maximum densities during post-monsoon (mean 589 no.m^{-3}), followed by pre-monsoon (mean 270 no.m^{-3}), and monsoon (mean 208 no.m^{-3}). Padmavathi and Goswami (1996) reported the high abundance of fish larvae from the upper reaches of the Mandovi-Zuari estuary, during post and pre-monsoon months.

The penaeid mysis was encountered everywhere in the entire Coringa mangroves waterways in relatively small numbers (mean 451 no.m^{-3}). The penaeid mysis was observed high densities in samples collected from the southern part of the bay and the mangrove creeks of Gaderu and Coringa. Penaeid mysis was rich in number during monsoon (mean 709 no.m^{-3}), relative to pre-monsoon (mean 482 no.m^{-3}) and post-monsoon seasons (mean 98 no.m^{-3}).

The distribution and abundance of ESS revealed that the majority of them is represented by typical brackish water forms, while some are neritic in nature. A few species were in particular favouring freshwater conditions. During the study, sporadic occurrence of some ESS indicated their preference to or specific effects of physical factors and atmospheric conditions. In this context, the cladoceran *Evadne* sp., usually favoured low saline conditions. For instance, in (monsoon) at Station 13, *Evadne* sp. was found in large number (14,613 no.m^{-3}), when freshwater condition was prevailed (salinity 6.23 per cent). In contrast, the copepods *Paracalanus aculeatus, Acrocalanus gibber* etc., preferred high saline conditions. They found in incredible numbers in the bay region, during monsoon and pre-monsoon months, when the salinity was > 30 per cent. ESS which occurred in large number in the proximity of mangroves appear to have entirely different requirement and favourable areas subject to considerable flows of detritus rich out welled water. The ESS species zoea of *Scylla serrata,* which was found predominantly in mangrove environment, due to the rich organic detritus and relatively small number of predation and other harmful organisms, provide necessary protection (Achuthankutty and Nair, 1980). The ESS characteristic of the bay locations are predominantly mixo-haline species such as *Paracalanus parvus,* gastropod veligers etc., while such as *Evadne* sp. etc., found in overwhelming number in the mangrove channels are mostly mixooligohalinc in nature. Based on the above, the following inferences can be drawn:

1. Among the ESS, there is considerable heterogeneity in distribution as a result of their specific preferences to varying hydro graphical conditions.
2. White some ESS (*Acartia erythraea, Paracalanus parvus, Eucalanus elongates* etc.) occurred more or less throughout the year a few species (*Evadne* sp. etc.) exhibited seasonality.
3. Some ESS *Evadne* sp. (14,613 no,m^{-3}, Station 15, July 1999), *Scylla serrata* (9,243 no.m^{-3}, Station 10, August 1998) were found in large numbers on specified months and locations attributable to the prevailing environmental conditions.
4. Based on species' individual preferences to the bay-estuary waterways of Coringa mangroves, *Scylla serrata* at Stations 9, 10 and 11 (Gaderu); *Evadne* sp. at Stations 14 and 15 (Coringa); gastropod and bivalve veligers at Stations 5, 6 and 7 (south bay); *Acrocalanus gibber* at Stations 1 and 2 (north bay) and *Acartia clausi* at Stations 5 and 6 (middle reaches of the estuary) were characteristic.
5. There were indifferent species as well. Examples include penaeid mysis etc., which exhibited cosmopolitan in their distribution, found in neritic, brackish water and mangrove environs.

References

Achuthankutty, C.T. and S.R.S. Nair (1980). Mangrove swamps as fry source for shrimp culture: A case study. *Mahasagar-Bull. Nat. Inst. Oceanogr.,* 13: 269–276.

APHA (1995). *Standard Methods for the Examination of Water and Wastwater,* 19th Edn. American Public Health Association, Washington D.C., USA. pp. 874.

Boesch, D.F. (1973). Classification and community structure of Macrobenthos in the Hampton roads area, Verginia. *Ibid,* 21: 226–244.

Chandra Mohan, P. (1963). Studies on zooplankton of the Godavari estuary. *Ph.D. Thesis,* Andhra University, Visakhapatnam, India, pp. 274.

Ellis, D.V. (1969). Ecologically significant species in coastal marine sediments of Southern British Columbia. *Syesis,* 2: 171–182.

Feder, H. M., G.J. Mveller, M.H. Dick and D.B. Hawkins (1973). Environmental studies of Port Valdez, (Eds.) D.W. Hood, W.E. Shields and E.J. Kelley. Institute of Marine Science and Oceanography, University of Alaska, USA, pp. 303–386.

Field, J.G. (1971). A numerical analysis of changes in the soft bottom fauna along a transect across the Bay, South Africa. *J. Exp. Mar. Biol. Ecol.,* 7: 215–253.

George, M.J. (1958). Observations on the plankton of the Cochin backwaters. *Indian J. Fish,* 5: 375–401.

Goswami, S.C. (1985). Zooplankton standing stock and composition in coastal waters of Goa, West coast of India. *Indian J. Mar. Sci.,* 14: 177–180.

Jones, G.F. (1969). *The Benthic Macro Fauna of the Main Land Shelf of Southern California.* Allan Hancock Monographs in Marine Biology, 4: 204–219.

Lie, U. (1968). A quality study of benthic infauna in Puget sound, Washington, USA in 1963–64. *Fisheries Directory Skr.,* 14: 223–356.

Madhupratap, M. (1979). Distribution, community structure and species succession of copepods from Cochin backwaters. *Indian J. Mar. Sci.,* 8: 1–8.

Menon, N.R., P. Venugopal and S.C. Goswami (1971). Total biomass and faunistic composition of the zooplankton in the Cochin backwaters. *J. Mar. Biol. Assoc.,* India, 13: 220–225.

Nair, K.K.C. and D.J. Tranter (1971). Zooplankton distribution along salinity gradient in the Cochin backwaters before and after monsoon. *J. Mar. Biol. Assoc.,* India, 13: 210–230.

Padmavathi, G. and S.C. Goswami (1996). Zooplankton ecology in the Madovi-zuari estuarine system of Goa, West coast of India. *Indian J. Mar. Sci.*, 25: 268–273.

Prasad, N.V. (2003). Composition and abundance of meroplankton in Coringa mangrove ecosystem with special reference to aquaculture. *J. Aqua. Biol.,* 18: 29–34.

Rao, T.S.S., M. Madhupratap and P. Haridas (1975). Distribution of zooplankton in space and time in a tropical estuary. *Bull. Dept. Mar. Sci.,* University of Cochin, 7: 695–704.

Santhanam, R., K. Krishnamurthy and R.C. Subbaraju (1975). Zooplankton of Porto Novo, South India. *Bull. Dept. Mar. Sci.,* University of Cochin, 4: 899–911.

Srinivasan, A. and R. Santhanam (1991). Tidal and seasonal variation in zooplankton of pullavazhi brackish water, South east coast of India. *Indian J. Mar. Sci.,* 20: 182–186.

Thorrington-Smith, M. (1971). West Indian ocean phytoplankton: A numerical investigation of phytohydrographic regions and their characteristic phytoplankton associations. *Mar. Biol.,* 9: 115–137.

Thorson, G. (1957). Bottom communities (sub-littoral or shallow shelf). *Mem. Soc. America,* 67: 461–534.

Travers, M. (1971). Diversity of microplankton of Golf of Marsellan, 1964. *Mar. Biol.,* 8: 308–343.

Villate, F. (1991). Annual cycle of zooplankton community in the Abra Harbour (Bay of Biscay): Abundance, composition and size spectra. *J. Plank. Res.,* 13: 691–706.

Wickstead, J.H. (1965). *An Introduction to the Study of Tropical Plankton.* Hutchinson Tropical Monograph, Hutchinson and Co. Publication, London, pp. 160.

Chapter 27

Feeding Modes and Associated Mechanisms in Zooplankton

Ram Kumar
Department of Zoology, Acharya Narendra Dev College, University of Delhi, Govindpuri, Kalkaji, New Delhi – 110 019
E mail: copepod65@rediffmail.com

Keywords: *Zooplankton, Feeding mode, Feeding mechanism, Herbivory, Omnivory, Bacterivory, Carnivory.*

Introduction

The zooplankton forms a vital link between the phytoplankton and the higher trophic levels in freshwater and marine ecosystems, comprise four major groups namely, cladocerans, copepods, rotifers and protozoa. Majority of zooplankton are filter feeding and suspension-feeding (on phytoplankton and bacteria), although predatory and facultative parasitic forms are also well represented. Cladocerans and copepods account for a greater proportion of the total zooplankton biomass than either protozoans or rotifers. Among zooplankton, all the four modes of feeding and associated mechanisms, depending on the type of food–bacteria, algae, other zooplankton and detritus–bacterivory, herbivory, omnivory and carnivory are common (Koste, 1978; Lampert, 1987; Williamson, 1991; González, 1998), because of which zooplankton as a group can exert considerable influence on the abundances and community organization of aquatic ecosystem. They playa vital role in the food web structure and trophic dynamics of all types of aquatic ecosystems being also the food of many planktivorous fishes. The various feeding modes in rotifers listed in Table 27.1, copepods and cladocerans show switching of the modes of feeding based on food size and type. The knowledge of feeding modes and associated mechanisms is warranted to understand the role of zooplankton in shaping and maintaining the community

Table 27.1: Species-wise Categorisation of Feeding Modes Among Rotifers (Koste, 1978)

Bacterivory	*Detritivory*	*Herbivory*		*Carnivory*	*Parasitism*	*Omnivory*
Brachionus rubens	*Brachionus patulus*	*Epiphanes clavulata*	*Kellicottia longispina*	*E. brachionus*	*Proales gigantea*	*E. brachionus*
B. dimidiatus	*B. quadridentatus*	*E. brachionus*	*L. ungulata*	*Eothinia lacustris*	*Proales parasita*	*E. senta*
Proales decipiens	*B. rubens*	*E. macrourus*	*Proales globulifera*	*Eosphora najas*	*Ptygura melicerta*	*C. gibba*
Encentrum putorius	*B. dimidiatus*	*E. senta*	*Tetrasiphon hydrocora*	*Notommata glyphura*	*Plerotrocha petromycon*	*A. brightwelli*
Conochilu unicornis	*B. angularis*	*Rhinoglena frontalis*	*Notommata copeus*	*C. gibba*	*Cephalodella catellina*	*S. vorax*
Filinia longiseta	*Keratella cochlearis*	*B. patulus*	*Cephalodella gibba*	*S. vorax*	*C. parasitica*	
	Anuraeopsis fissa	*B. quadridentatus*	*C. carina*	*Ploesoma hudsoni*	*A. volvocicola*	
	Lecane ungulata	*B. bidentata*	*Trichocerca collaris*	*Asplanchnopus multiceps*	*Dicranophorus uncinata*	
	Proales decipiens	*B. plicatilis*	*Synchaeta vorax*	*A. brightwelli*	*D. difflugiarum*	
	Cephalodella carina	*B. rubens*	*Asplanchna brightwelli*	*A. intermedia*	*Acyclus inquietus*	
	C. auriculata	*B. calyciflorus uncinata*	*Dicranophorus forcipatus*	*Dicranophorus*	*Balatro sp.*	
	E. putorius	*B. dimidiatus*	*Sinantherina socialis*	*E. putorius*	*Albertia sp.*	
	Hexarthra mira	*Keratella quadrata*	*Conochlius dossuaris*	*Collotheca cornuta*		
	F. longiseta	*K. cochlearis*	*Filinia longiseta*	*Cupelopagis vorax*		
	Lacinularia flosculosa	*Collotheca ornata*				

composition of aquatic ecosystem. This article discusses different feeding modes and associated mechanisms in different groups of zooplankton.

Bacterivory

The wide food size spectrum *i.e.* < 1–20 μm in rotifers (Pourriot, 1977), < 1–50 μm in cladocerans (Lampert and Sommer, 1997) and < 1–1,486 μm in copepods (Williamson, 1991; Kumar and Rao, 1999a; Kumar, 2003a; Rao and Kumar, 2002) includes the main microbial components also. Bacteria, directly or as part of detritus, form the principal diet of many cladocerans and rotifers. Recently, Work and Havens (2003) used fluorescently labelled bacteria in an *in situ* enclosure experiment in Lake Okeechobee, Fl., USA. They observed that all taxa of protozoa and metazoan zooplankton grazed on fluorescently labelled bacteria efficiently (Work and Havens, 2003). Field observations and experiments in the field, and under laboratory conditions, have conclusively shown that bacteria are an important food source for the rotifers and cladocerans (Starkweather, 1980; Porter, 1984; Boon and Shiel, 1990; Arndt, 1993), more so than probably for copepods. Copepods are able to utilize bacteria as a source of nutrition. However, bacterial diet does not support complete growth and reproduction in copepods (Kumar and Rao, 1998; Kumar, 2003a).

Pico and nanophagous rotifers such as *Keratella, Brachionus, Kellicottia, Filinia, Conochilus,* and many bdelloids effectively feed on bacteria (Bogdan *et al.*, 1980; Ricci, 1984; Ooms-Wilms, 1991). Some bdelloids and lecanids are adapted to live in wastewater, obviously feeding on the microbes in them. Microbial flocs have been found to be an effective source of food for *Brachionus plicatils* (Higashihara *et al.*, 1983).

Cladocerans are able to utilize bacteria as food source efficiently (Wylie and Currie, 1991; Jürgens, *et al.*, 1994). The bacterial diet supported reproduction at a higher rate than the algal diet in case of relatively smaller and medium sized cladocerans *Ceriodaphnia* and *Moina* (Kumar, unpublished data). Although many species differ in their ability to utilize bacteria as food, they may not be very efficient at collecting and ingesting such small particles (< 0.5 μm), which most natural bacteria appear to be (Vadstein *et al.*, 1993). Significant levels of ingestion may be possible only when the bacterial concentrations are very high (Starkweather *et al.*, 1979; Seaman *et al.*, 1986). Rotifers and cladocerans have a significant feedback effect on the microbial web by providing degraded algae, bacteria, and protozoans and may promote microbial activity (Arndt, 1993; Wickham, 1998). They thus form link between the classic food web (algae-zooplankon-fish) and the microbial web (bacteria-ciliates-zooplankton) (Starkweather, 1980; Bogdan and Gilbert, 1982; Walz, 1985; Heerkloss and Hlawa, 1995). When phytoplankton carry out photosynthesis, a substantial amount of the fixed carbon is excreted to the surrounding water as dissolved organic carbon (DOC) (Cole *et al.*, 1982). This carbon is taken up by bacteria, which in turn is grazed by protozoan and metazoan plankton. This path- way for carbon flow has been referred to as the microbial loop (Azam *et al.*, 1983). It can transport large amount of carbon in freshwater lakes (Weisse and Muller, 1990) and the open ocean (Cole *et al.*, 1988) and is the route whereby allochthonous DOC enters the food web (Jansson *et al.*, 1999). Koshikawa *et al.* (1996) directly compared the efficiency of carbon transfer in algal and bacteria-based food webs: they found that the transfer efficiencies were very low but nearly the same for both pathways. More detailed investigation of carbon dynamics indicates that the bacteria based pathway can transfer large quantities of carbon to zooplankton (Havens *et al.*, 2000).

Herbivory

Although herbivory is the predominant mode of feeding in zooplankton, recent studies indicate that exclusively herbivorous species might be more an exception than rule (Adrian and Frost, 1992,

1993; Burns and Gilbert, 1986; Wallace and Snell, 1991; Hopp *et al.*, 1997; Kumar and Rao, 1999b). Herbivorous zooplankton may be polyphagous, taking a wide variety of algal species, though often with a rank order of preference among the foods that they take, or they may be monophagous, specializing on a single algal species or a narrow range of closely related species. For example, some species of the genus *Notommata* are monophagous, specializing on a specific alga or a group with *N. copeus* showing a clear preference for the filamentous Zygnematales (*Zygnema, Mougeotia, Spirogyra*), *N. pachyura* for the solitary desmids (*Staurastrum, Closterium*), and *N. collaris* for the *Closterium* species (Pourriot, 1977; González, 1998). Many other monogonont rotifers are known to exhibit food selectivity based on particle size (Rothhaupt, 1990a; Hansen *et al.*, 1997).

Many species of cyclopoids and a few species of calanoids are omnivorous, with varying degrees of plant food in their diet. Even among the truly carnivorous cyclopoid species, the post-embryonic (nauplii and early copepodid) stages are known to be dependent on plant food (Santer and van den Bosch, 1994; Hansen and Santer, 1995; Kumar and Rao, 1998; Kumar, 2003).

The phytoplankton taxa with species that are actually or potentially food for herbivorous copepods include Cyanobacteria (blue-green algae) (unicellular species such as *Anacystis* and *Gleocapsa,* filamentous species such as *Anabaena, Oscillatoria* and *Nostoc,* and colonial species such as *Microcystis* and *Gomphosphaeria*), Chlorophyta (green algae), (*Chlorella, Chlamydomonas, Scenedesmus, Chlorococcum,* etc.) and diatoms (*Ceratium, Cyclotella,* etc.), spanning a particle size range of 2–300 µm.

Laboratory studies on the efficacy of different algal species for growth and reproduction in copepods brought to light the nutritional inadequacies of some food species. In a study on the nutritional adequacy of selected green algae and diatoms of comparable particle size, for *Metadiaptomus meridianus* and *Tropodiaptomus spectabilis,* it was observed that naupliar development was incomplete with *Scenedesmus* or *Selenastrum* (Kumar and Rao, 1998; Kumar, 2003) while naupliar stage duration was prolonged with the diatom *Cyclotella* (Hart and Santer, 1994). Xu and Burns (1991) tested the ability of nauplii and immature copepodids of *Boeckella triarticulata, B. hamata* and *B. dilatata* to survive and grow on monospecific diets of species of filamentous cyanobacteria, *Cyclotella* and *Cryptomonas.* Of these, only *Cryptomonas* supported the development from nauplius to adult stage in all the three calanoid species; *Cyclotella* and *Anabaena* supported complete development only in *B. dilatata.* The inadequacy of these algae for the calanoid post-embryonic developmental stages may be due to problems in capture and ingestion, indigestibility or some deficiencies in biochemical composition (Porter and Orcutt, 1980). The size, morphology, motility and physiological state of the algae are important factors influencing ingestibility (DeMott, 1986; DeMott and Watson, 1991; Kerfoot and Kirk, 1991). Given a choice of similar-sized algae the motile *Chlamydomonas reinhardti* and the non-motile *Monoraphidium minutum, Cyclops vicinus* ingested *C. reinhardtii* to a greater extent than *M. minutum* (Santer, 1993). Santer (1993) also found that *C. vicinus* population crashed if *Ceratium* or coccale algae were the only food available. Dinoflagellates are generally believed to be poor food because of their spiny shape. The green alga *Scenedesmus* is enveloped by a pellicle which may resist digestion sufficiently enough to preclude its adequate utilization as a sole food source, although the ingestion rates on this alga could be relatively high (Hart and Santer, 1994). Toxic cyanobacteria as a sole food are inefficient at supporting complete development of calanoid copepods, and their presence in the diet along with *Cryptomonas* sp. prolonged the duration of juvenile stages in *Eudiaptomus gracilis* (Santer, 1994). Nutritional adequacy also depends upon the physiological condition of the cells for example, fast growing *Thalassiosira weissflogii* cells contribute to higher fecundity in *Acartia tonsa* than the older cells as food (Jónasdóttir and Kiørboe, 1996).

Recent studies analysed the fatty acid composition of algae generally grazed upon by zooplankton and concluded that in general algal species containing high levels of long-chain polyunsatuated fatty acids (PUFA) are nutritionally the best foods for zooplankton (Gulati and DeMott, 1997). PUFA deficiency in algae could be one important reason why certain algae do not support optimal growth and reproduction in some copepods.

Two generalizations emerge from laboratory studies on the efficacy of algal foods for copepods:

1. The nutritional adequacy of an algal species is judged by the copepod species feeding on it. For example, *Chlamydomonas* is utilized reasonably well by cyclopoid species but poorly by calanoids (Chen and Folt, 1993; Hart and Santer, 1994).
2. Whether or not a given algal food is nutritionally adequate depends also on the developmental stages of the copepod; the naupliar and early copepodid development in *Eudiaptomus gracilis* could not be supported on a diet of *Chlamydomonas* although the adults were able to utilize it efficiently as food (Santer, 1994). Finally, it must be realized however, that in nature the diet of copepods is not necessarily made up of a single species, and therefore, any algal species considered nutritionally inadequate may still be acceptably adequate as part of a mixed algal diet (Vijverberg, 1989; DeMott, 1995a). Rather than categorising as nutritionally adequate or inadequate, the phytoplankton available to a copepod can be classified as either edible (*e.g.*, green flagellates, cryptomonads and diatoms-Tóth and Zankai, 1985; Knisely and Geller, 1986; Tóth *et al.*, 1987) or inedible blue-green and green algae other than flagellates phytolankton (Hansen and Jeppesen, 1992).

Cladocerans are predominantly herbivorous selectively feeding softer algal cells rejecting the harder one. Cladocerans have also been reported to have ability to select food on the basis of chemical composition, presence, or absence of toxins in different algal and cyanobacterial cells and morphology of algal cells (DeMott, 1995b). Bosmins longirostris eats algae preferring a size range of 1–5 µm, usually it does not filter particles larger than 20 µm but it can consume cells larger than 64 µm (Gliwicz, 1969). *Daphnia* can feed on algae ranging from 10–150 µm, 68 per cent of ingested algae being of small size. It has been estimated to consume daily upto 64 per cent of the algal biomass and 3–48 per cent of the bacteria in the epilimnion (Lair, 1991).

Carnivory

Inasmuch as plant foods constitute a portion, albeit small, of the total intake of even the acknowledged carnivorous zooplankton, it would be very difficult to name any zooplankton as exclusively carnivorous. It has been suggested (Fryer, 1957a) that a copepod species may be considered as carnivorous if at least 60 per cent of its gut contents are made up of animal remains. Thus, a number of omnivorous zooplankton can be classified as predominantly, but not exclusively, carnivorous. Calanoid copepods, by virtue of their suspension or filter feeding mechanisms have fewer representatives of predominantly carnivorous species. However, several species of freshwater calabnoids, which were previously thought to be herbivorous have recently been shown to have ability of utilizing animal food. In addition to phytoplankton calanoid copepods ingest micro-zooplankton such as ciliates and rotifers, which are presumed to increase their survival and reproduction (Schulze and Folt, 1990; Burns and Gilbert, 1993; Burns and Schallenberg, 1996; Sanders *et al.*, 1996, Couch *et al.*, 1999; Kumar, 2003a). Fryer (1957b), based on extensive studies on the evolution of feeding habits, proposed that cyclopoid copepods are primarily carnivorous species and that herbivory observed in smaller species of the group is evolutionarily a secondary development, having evolved in habitats with chronically low levels of suitable animal prey species. The food of carnivorous copepods includes a wide diversity of animals commonly found in plankton and representing a wide size range (10 – > 3000 μm): ciliates and flagellates (Burns and Gilbert, 1993; Hartmann *et al.*, 1993; Wickham, 1993, 1995), rotifers (Fryer, 1957a,b; Williamson and Gilbert, 1980; Williamson and Butler, 1986; Williamson, 1987; Greene, 1988), cladocerans (Fryer, 1957a,b; Brandl and Fernando, 1975; Stemberger, 1985; Krylov, 1988), nauplii of other copepods (Gabriel, 1985; Wiliamson, 1980; Stemberger, 1985; Krylov, 1988), mosquito larvae (Brown *et al.*, 1991; Kumar and Rao, 2003), and newly hatched fish larvae (Lillelund and Lasker, 1971). Cannibalism particularly on their own naupliar or copepodid stages has also been reported in some cyclopoid species (Gabriel, 1985; Krylov, 1988; Kumar, 2003b).

The cladoceran particularly *Leptodora kindti* is obligate carnivorous. In this species the setae of the first thoracic limb act as mechanoreceptors, the other thoracic limbs thorax and head together form the shape of an open basket in which after encounter the prey is pushed in by the aid of first thoracic limba and the furca. This cladoceran is able to utilize rotifers and small cladocerans as prey (Herzig and Auer, 1990). However, other cladocerans commonly found in our water-bodies have been shown to have ability of utilizing ciliates as food source. In a highly eutrophicated lake of Poland, the cladocera *Daphnia cucullata* has been observed to be the major user of ciliates (Wiackowski, *et al.*, 2001). Incidental carnivory is seen in case of certain *Daphnia* species, for instance, *D. similoides* has been observed to have ability to utilize the smaller rotifer *Brachionus angularis* as food source and the rotifer supported reproduction and survival in the cladoceran (Kak and Rao, 1998).

Many species of rotifers are carnivores, but from published literature it appears that very few are obligatory carnivores. Carnivorous species are found predominantly in the families Dicranophoridae, Asplanchnidae, Collothecidae and Notommatidae. Of these, species belonging to Asplanchnidae (such as *Asplanchna girodi, A. brightwelli,* and *A. intermedia*) have been studied extensively, particularly in terms of predation (Gilbert, 1980; Gilbert and Stemberger, 1985; Conde-Porcuna and Sarma, 1995; Iyer and Rao, 1996). The diet of carnivorous rotifers generally includes ciliates and smaller rotifers (Arndt, 1993), but *Asplanchnopus* has been found to include ostracods regularly in its diet (Koste, 1978). Facultative carnivores may 'capture' and ingest large algae such as *Phacus* and flagellates such as *Euglena* (Personal observation).

Omnivory

This mode of feeding refers to feeding at more than one trophic level (Pimm, 1982). Theoretical models of food webs (Pimm and Lawton, 1978; Pimm, 1980, 1982) suggest, and empirical observations indicate, that long food chains and extensive omnivory are rare in nature. The reasons are that food webs as such are dynamically unstable and that, omnivores rarely feed on species in non-adjacent trophic levels (Pimm and Lawton, 1978; Pimm, 1980, 1982). However, Saunders (1978) and Yodzis (1981) questioned the assumption that omnivory is rare because communities with many omnivores are dynamically unstable or because it is difficult for a species to evolve in a way to feed efficiently on, for instance both plants and animals. It appears that omnivory occurs among all major groups of planktonic organisms (protozoans, cladocerans, copepods, rotifers and aquatic insects) (Burns and Gilbert, 1986; Moore, 1988; Sherr and Sherr, 1988; Williamson and Butler, 1986; Tiselius and Jonsson, 1990; Adrian, 1991a; Taylor and Sanders, 1991; Wallace and Snell, 1991). In open water habitats, omnivory is more common because it is less costly in an evolutionary sense. Aquatic predators are typically larger than their prey, which they engulf whole, and their trophic position is largely a function of body size. Thus omnivory is achieved simply by becoming large, thereby broadening the size range of potential prey (Sprules and Bowerman, 1988; Taylor and Sanders, 1991).

Omnivory is common in the juvenile and adult stages of many planktonic copepods, both calanoid and cyclopoid (Paffenhöfer and Knowles, 1980; Landry, 1981; Williamson and Butler, 1986; Adrian, 1987; 1991b; Tóth *et al.*, 1987; Schulze and Folt, 1990; Verity and Paffenhöfer, 1996). The extent of herbivory in cyclopoids and carnivory in calanoids, shows wide variation. Cyclopoid copepods are more adapted to carnivorous mode of feeding as they are not efficient at utilizing cellulose invariably present in phytoplankton. Ingested phytoplankton too may pass through the cyclopoid gut undigested; undamaged, intact cells of *Scenedesmus* and *Cosmarium* have been recorded in the rectum of certain cyclopoid copepods (Fryer, 1957a). Any small green or blue-green alga with gelatinous sheath and swallowed without its cell wall intact, is most likely to emerge out of the copepod gut undamaged (Fryer, 1957a,b). However, recent studies have shown that different species of cyclopoids differ considerably in their ability to use algae as food. Larger species such as *Mesocyclops edax* tend to be more carnivorous and include very little of algae in their diet, while smaller cyclopoids do ingest plant food more regularly and utilize more efficiently too (Fryer, 1957a,b; Hopp *et al.*, 1997; Adrian and Frost, 1993; Hansen and Santer, 1995; Santer and van den Bosch, 1994). The smaller cyclopoids such as *Tropocyclops parsinus mexicanus* and *Cyclops vicinus* can complete their entire life cycle on exclusively algal diet (Adrian and Frost, 1993; Santer and van den Bosch, 1994). Among omnivorous cyclopoids, dependence on algal food may be more during juvenile stages (Fryer, 1957a).

Contrary to earlier views, Calanoida includes species, which are omnivorous (Mullin, 1966; Petipa *et al.*, 1970; Verity and Paffenhöfer, 1996). However, these calanoids feed on animal food types smaller than those fed upon by Cyclopoids. A recent study using video imaging analysis system, observed the feeding behaviour of *Eucalanus pileatus* under natural conditions (Verity and Paffenhöfer, 1996).

These calanoids were found to ingest a higher percentage of ciliates than diatoms and small heterotrophic dinoflagellates. The ingestion rates in *Eucalanus, Paracalanus* and *Temora* were higher on heterotrophic than on autotrophic food species (Verity and Paffenhöfer, 1996). Larger protozoa are ingested at high rates by various stages and species of calanoid copepods otherwise known to be herbivorous (Petipa, 1978; Gifford, 1993; Fessenden and Cowles, 1994; Ohman and Runge, 1994). For example, at similar abundances of larger phytoplankton and microzooplankton, the percentage of non-plant material (protozoa and metazoa) in the gut contents of *Calanus pacificus* ranged from 23 to 99 per cent, and of *Clausocalanus* sp. from 55 to 98 per cent (Kleppel *et al.*, 1988). In general, however, calanoids are more efficient in utilizing plant food; in case of freshwater calanoids, predation (carnivorous feeding) rates are reduced by 50 per cent in the presence of phytoplankton (Schulze and Folt, 1989; Kumar and Rao, 1999b). Similarly, in the marine species *Calanus pacificus,* predation rate declines as phytoplankton concentration is increased (Landry, 1981). However, in some freshwater calanoids, predation rates on ciliates exceed their grazing rates on algae (Hartmann *et al.*, 1993). Unlike in the case of cyclopoids, there is no information on the possible relation between body size and herbivory in calanoids.

The differences in the mouth-parts morphology of calanoids and cyclopoids have been used to explain the relative importance of herbivory and carnivory in their lives. The slender feeding appendages with setae in calanoids suggest for instance suspension feeding on small particles. Mouth part morphology alone however, is an unreliable predictor of diet (Wong, 1984). Enzymatic activity in the digestive tract may be a more reliable indicator of types of diet ingested and assimilated. The amylase/ trypsin activity ratio has been used as an indicator of herbivorous/carnivorous feeding. In the calanoid copepod *Acanthodiaptomus denticornis,* the amylase trypsin activity ratio is reported to be comparable with that in primarily carnivorous *Cyclops abyssorum,* pointing to the carnivorous potential of

Acanthodiaptomus (Hartmann *et al.*, 1993). The frequently observed incomplete digestion of phytoplankton with a high cellulose content in cyclopoid copepods suggests the absence of cellulase activity (Fryer, 1957a,b; Infante, 1973; Adrian, 1987; Adrian and Frost, 1993). On the other hand, in majority of calanoid copepods algal digestion is found to be complete (Schulze and Folt, 1989; Lair, 1992; Dagg, 1995).

In the copepod communities of many freshwater ecosystems it is often observed that the number of cyclopoid species is greater than the number of calanoid species (Dodson, 1991). One of the earlier explanations for this trend was the fundamental difference in their feeding modes. From recent studies it is clear that both cyclopoid and calanoids are capable of omnivory (Williamson and Butler, 1986; Schulze and Folt, 1990). Hence, factors other than feeding modes (such as life history patterns) may have to be invoked to explain the observed differences in the diversity of these two copepod groups.

Although the diets of some rotifers are highly specialized (Pourriot, 1977; Bogdan and Gilbert, 1987), many species consume regularly a wide variety of both plant and animal items and may be termed as generalists or omnivores (Rothhaupt, 1990a,b). For example, *Asplanchna* species are generally considered predatory, yet their main food could very often be large algae. In contrast, the generally herbivorous *Brachionus* may ingest small ciliates that may come along with the algae during suspension feeding (Arndt, 1993; Gilbert and Jack, 1993). Thus, the distinctions among carnivore, herbivore and omnivore are not usually sharp and clear-cut, which is a reflection of the opportunistic feeding strategies adopted by rotifers.

Many cladocera species consume regularly a wide variety of both plant and animal items and may be termed as generalists or omnivores (DeMott, 1995a; Vanni and Lampert, 1992). The smaller cladoceran for example *Ceriodaphnia* and *Scapholeberis* are predominantly herbivorous or bacterivorous whereas the larger *Daphnia* and *Daphnosoma* are efficient enough to utilize ciliates and flagellates as food source. Many studies have documented the high rate of predation by different species of *Daphnia* on ciliates (Jack and Gilbert, 1993; Wickham, 1998) and they are also able to utilize smaller rotifer *Brachionus angularis* as food source (incidental carnivory) in experimental condition (Kak and Rao, 1998).

Mechanisms of Food Gathering

Because of the wide size range of food particles ingested, zooplanktons exhibit all the three mechanisms of feeding, namely, filter feeding, suspension feeding, and raptorial feeding.

Filter Feeding Mechanism

This mechanism involves collection of particulate matter (algae, bacteria and detritus) by sieving ambient water through the fine meshes formed by filtering appendages of the animal (Jorgensen, 1966). The mechanisms of retaining particles smaller than the fibre mesh size include direct interception, inertial impaction, gravitational deposition, motile particle deposition and electrostatic attraction (Rubenstein and Koehl, 1977). For biological filters and particles, the direct interception mechanism (in which the particle is captured by the fibre of the mesh if the centre of massless comes within the range equal to one particle radius) seems to be the main mode of capture (Rubenstein and Koehl, 1977). The filtering animal can change the size of the particle on which it feeds by changing the pore size of its filter or by other means such as changing the velocity of water passing through the filter and altering the diameter of the adhesive filtering fibres (Rubenstein and Koehl, 1977). A filter feeding mechanism must consist of (1) a filter (2) a means of creating a current through the filter (3) some way of scrapping the food particles and transferring it to the mouth and (4) an exit for the filtered water

(Waterman, 1960). The main component of filtering apparatus is a set of the setae, each seta being provided with two rows of setules. These setules are set to its axis at a slight angle to the plane of the setal array and facing the current. Consequently when the current passes through, they spread out against the setules on the adjoining setae and form a meshwork or a filter. The size of the particles retained depends on the spaces in this mesh work.

Suspension Feeding

The term "suspension feeding" as used by Jorgensen (1966), includes a wider range of feeders. Whereas filter-feeding denotes feeding by passing the surrounding water through structures that retain particles mainly according to size and shape, the term 'suspension feeding' is used to denote other mechanisms for obtaining particulate food in addition to filter-feeding. Here the water with its content of suspended particles is not truly filtered, but is carried along surface which are capable of retaining particles, that contact the surfaces (Jorgensen, 1966). The term 'suspension feeding' covers all mechanisms of particle removal, regardless of the means. The mode of food gathering for a particular zooplankton may not be a single, specific mechanism. Instead, the animal may switch from one mode of feeding to another, depending upon the availability of food, size of food particles, their concentration, and its own developmental stage.

Mechanisms of Food Gathering in Copepods

Copepods occur in freshwater as well as in marine ecosystems. In contrast to cladocerans and rotifers copepods seem to play a more important role in marine systems. Their feeding process can not easily be generalized. However, adult cyclopoid copepods are commonly accepted to be predaceous (Gliwicz, 1994; Kumar, 2003b, Kumar and Rao, 1999b; Rao and Kumar, 2002). They exhibit strong food selectivity between food particle of different nutritional status (Cowles *et al.*, 1988), morphology and motility (Rao and Kumar, 2002), taste (DeMott, 1995a; Rao and Kumar, 2002), and toxicity (DeMott and Watson, 1991). A brief knowledge of the hydrodynamics of copepods movements in water is helpful for an appreciation of the peculiar problems faced by copepods and similar sized zooplankters in gathering food by any of these mechanisms, and avoiding predators (Zaret, 1980; Koehl and Strickler, 1981). Unlike that of the larger organisms in the medium (fish, dolphin, etc.), the physical world of a copepod is dominated by viscous rather than inertial forces. Their Reynolds Number (Re) (a parameter which is directly related to the density of the medium, body size of the organism and its velocity, and inversely to the viscosity of the medium) is ~ 1, which is the boundary between laminar ($Re < 1$) and turbulent ($Re > 1$) flows (Koehl and Strickler, 1981; Colinvaux, 1986). Under the laminar flow conditions, there is no mixing or stirring of surrounding water by the flapping of copepod appendages, which might confound the animal's chemoreception of food and predator signals. A copepod appendage cannot strain a food particle out of water, but must manoeuvre it by moving the water surrounding the particle (Alcaraz *et al.*, 1980). Copepods feed intermittently (Rosenberg, 1980) and when they are not flapping their appendages and passively sinking, the flow around the animal stops almost immediately. This allows the copepod to pick up the mechanical or chemical signals from the surroundings without any obstruction from any food particles (Koehl and Strickler, 1981).

Planktonic copepods filter the particles through the second maxillae which are modified as filtering appendages. In most copepods, the second antennae, the mandibular palps, and the first pair of maxillae are covered with setae. These appendages vibrate regularly between 600 and 2640 times a minute, creating two swirls of water passing along both sides of the body. The part of the swirl flowing toward the midline of the body turns anteriorly and is sucked forward into a median filter chamber. The feeding current is sucked out of the chamber anteriorly and laterally by the vibrations of the first maxillae. As feeding current passes, it is filtered by the dense screen of the setae on the second maxillae. The collected particles are cleaned off by the endites of the first maxillae and also by special long setae on the base of the maxillipeds, which eventually pass to the mandibles and mouth (Barnes, 1980).

Certain models of copepod foraging have been based on the assumption that when an animal is feeding, water is continuously passed through its maxillary filter (Lam and Frost, 1976; Lehman, 1976). However, recent use of high-speed micro-cinematography to study copepod feeding has revealed the complexity of appendage movements that create water currents which carry food toward the second maxillae (Alcaraz *et al.*, 1980). The study showed that second maxillae are not always held stationary, but rather periodically actively capture parcels of water containing algal cells. The high-speed films also revealed that algal cells are usually redirected without actually being touched by the feeding appendages (Koehl and Strickler, 1981). It was shown that there is an 'active space' of chemical around the algal cells which the copepods use as a means to track and capture them (Alcaraz *et al.*, 1980). Food particle type- and size-related differences in mechanisms of handling particles by the copepods have been observed (Price *et al.*, 1983).

When the particle size is large (> 10 μm), the cell never comes in contact with the appendages during oscillations but is pushed and pulled by the layer of water around the maxillipeds. The second maxillae are stationary until the cell is detected by the maxilliped, and then they flap outwardly, accompanied by coordinated motions of the other appendages to bring the cell closer and then the cell disappears inside the feeding basket formed by the maxillary setae (Price *et al.*, 1983). The individually captured cells may be rejected if warranted, by repeated sweeping motions of the second maxillae, second antennae and maxillipeds. In case of small food particles (< 8 μm). The second maxillae show continuous low amplitude movement which is periodically interrupted by "combings" of the appendages, presumably to remove small cells and pass them to the mouth. Combings, lasting upto 250 ms in each cycle, may also prevent any clogging of the mouth parts which would interfere with food detection, tasting, and handling (Price *et al.*, 1983). The most common type of combing motion consists of repeated scrapings of the setal of the second maxillae. The calanoid *Euchaeta pileatus* is shown to have a sensitivity threshold somewhere in the size range of 6 to 12 μm; cells larger than the sensitivity threshold are individually detected and handled and may be individually rejected, while cells smaller than the threshold size are captured by continuous low amplitude movement of the combing motions.

The interdigitation of the setae of the second maxillae and the setae of the endites probably serves to remove cells adhering to the setae and setules of the second maxillae. Cells below the sensitivity threshold thus appear to be accumulated and handled as a group. Different species of copepods show variations in this basic pattern of scanning and capture movements.

There was earlier a notion that copepods are 'filter feeders' with the underlying assumption that the animals were feeding with small passive sieves of a mesh size approximately equal to the setule spacing (Nival and Nival, 1976). However, Friedman and Strickler (1975) found numerous sensory structures in the setae of the mouthparts, suggesting that these animals have the tools to perceive chemical signals originating from the surrounding water as well as from physical contact with food items. Alcaraz *et al.* (1980) and Price *et al.* (1983) concluded that the viscous forces of the flow do not allow the calanoids to sieve water constantly through their mouthparts, and that algae are captured from the feeding current by special motion patterns. Micro-cinematographic observations indicate that they capture and handle the food particles not passively according to size and shape but in most cases, actively using sensory inputs for detection, capture, and ingestion. These observations support the notion that herbivorous calanoid copepods are suspension feeders.

Large particles (> 10 µm) are transported to the mouth in a body of water by late copepodid and female calanoids. The large algae are directed and oriented towards the mouth, and are even rejected at times. The mouthparts are not used passively as sieves but actively capture and handle algae. Larger cells are ingested immediately upon arrival at the mouth, which indicates sensory induced movements.

A different mechanism is adopted when a colony or chain of algae such as of the diatom *Laudaria borealis* are encountered which is likely to break in the process of being moved to the mouth, resulting in only a few cells being ingested in the first round. The mouthparts immediately start to recirculate the water, which contains the remainder of the chain and try to capture them (Paffenhöfer *et al.*, 1982). In diaptomids, particles (> 50 µm) are generally captured actively or by attack; while in intermediate size range the frequency of active captures increases with increasing particle size (Vanderploeg and Paffenhöfer, 1985; Williamson and Vanderploeg, 1988).

Raptorial feeding is the predominant mode in cyclopoids of which the majority are predatory. However, many rotifers and few cladocerans are raptorial feeders. The raptorially feeding rotifer species swim through the water propelled by the bands of cilia on their anterior end, they randomly encounter and capture the individual prey items ranging in size from 4 to 18 µm (Bogdan and Gilbert, 1987; Rothaupt, 1990a). Thus the food size niches are much narrower in rotifers than that of cladocerans and copepods and exclude particles in size range of suspended sediments. In addition rotifers feed selectively and preferentially ingest large algal cells, particularly flagellates, particularly over smaller algal cells or bacteria (Rothaupt, 1990b).

The act of predation, as observed in many cyclopoid species, can be broken down into several sequential components including encounter, attack, capture and ingestion.

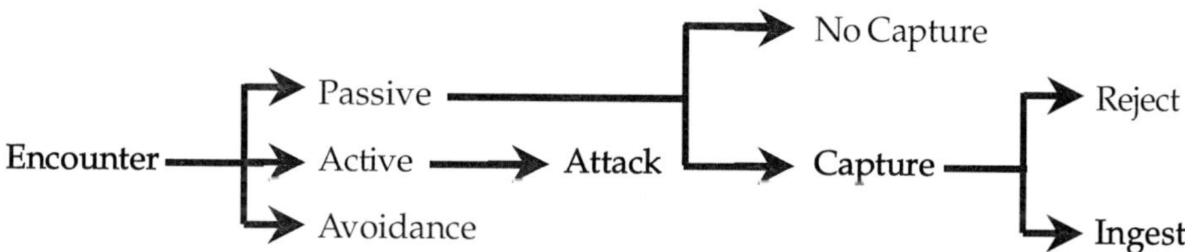

Encounter

The rate (E) at which a predatory copepod encounters a prey is a function of the concentration of the prey (n), volume of water swept clear by copepod in a time unit (v), such that E = ß n V, where ß is an encounter rate coefficient (Kiørboe and Saiz, 1995). The value of ß depends on the behavioural and sensory characteristics of specific predator and prey (Greene, 1983).

An encounter, in case of the predatory *Cyclops* has been defined as prey contact with any part of the body excluding the caudal setae. An extremely adept predatory cyclopoid generally responds to even the slightest prey contact over most parts of its body. Occasionally it responds to the prey even before any direct contact has been made. The reactive distance (the maximum distance from which an individual prey can be detected by the predator depends upon the behaviour of prey. For example, *Diaptomus* species respond to different prey species at distances that vary with prey type (Williamson and Vanderploeg, 1988). However, *Mesocyclops* is much less responsive and never initiates attack unless it directly contacts the prey and unless the contact is almost exclusively with the feeding chamber of the predator. Thus for *Mesocyclops* an encounter is a prey contact with any part of the second antennae, or with parts of the first antennae more proximal to the mouth (Williamson and Gilbert, 1980).

Copepods can detect the presence of prey by one or more of the three mechanisms of reception–mechano, tactile and chemical (Williamson, 1991; Atkinson, 1995). The body surfaces of calanoids and cyclopoids are covered with two types of integumental sensilla: small pegs (sensilla basiconica) and fine hairs (sensilla trichodea); the pegs are thought to be chemoreceptors and hairs serve as both chemoreceptor and mechanoreceptors. The finest antennae of many cyclopoids and calanoids have setae that are modified ciliary processes that may serve in mechanoreception (Williamson, 1991; Atkinson, 1995). Mechanoreceptors sense the differences in pressure created by a prey in the vicinity, and prepare the predator for attack. Laboratory studies have demonstrated that, motile food particles are cleared by the cyclopoid predators at a higher rate than non-motile particles (DeMott and Watson, 1991; Atkinson, 1995). *Diaptomus* responds at different reactive distances depending on the size and 'quality' of the prey type (Williamson and Vanderploeg, 1988). The range for chemoreception depends on both the sensitivity of the animal's sense organs and the 'signal-to-noise' ratio in the immediate surroundings (Atema, 1987). High levels of various algal metabolites in the medium could generate the chemical 'noise'. The chemoreceptors of copepods probably help also in distinguishing between debris and particles of nutritional value and in food type selection (Alcaraz *et al.*, 1980; Huntley, 1988; Kerfoot and Kirk, 1991).

Attack

Once the prey is encountered and recognised, the predator has to decide whether to pursue the prey or not. Predator's active choice at this step can influence the overall selection process (Greene, 1983). Calanoids and cyclopoids differ in certain ways in their method of attack and capture of prey.

The prey capture mechanism of suspension-feeding calanoids is similar in both marine and freshwater species. Majority of calanoids clear smaller ciliates and other small heterotrophs at rates 6–10 times faster than they do the autotrophs by suspension feeding mechanisms (Verity and Paffenhöfer, 1996). Food is brought in towards the copepod on microcurrents created by rapid vibrations of the second antennae, mandibular palps, first maxillae, and maxillipeds. Smaller food particles are captured passively and are funnelled into the mouth through the setae of the second maxillae (Vanderploeg and Paffenhöfer, 1985; Price and Paffenhöfer, 1986). Larger food particles are captured actively with an outward fling of the second maxillae followed by an inward squeeze to remove the excess water surrounding the food particle (Williamson, 1991). When microzooplankton such as rotifer and nauplii are brought in on the feeding currents of suspension-feeding diaptomids, the copepod will often attack these small animal prey from a distance with an active thrust response in which the antennae and swimming legs are used to orient and pounce towards the prey (Williamson and Butler, 1986; Williamson, 1987; Williamson and Vanderploeg, 1988). Many larger predatory calanoid species such as *Heterocope, Epischura, Limnocalanus,* and some of the large species of *Diaptomus* cruise through the water, attack their prey with a pounce, and grasp them with their first and second maxillae. If a capture attempt fails, the copepod may swim in a vertical loop to try again to capture the prey (Kerfoot, 1978). Attacks in *Diaptomus pallidus* are not as vigorous as the leaps and pounces of cyclopoid copepods (Williamson, 1987). The passive response to prey shown by copepod, has been incorporated by Williamson (1987) as an important additional component in the sequence of events in predation proposed earlier by Williamson and Gilbert (1980).

Unlike the calanoids, the cyclopoids have less complicated mouthparts, with fewer and shorter setae. They are more adapted to 'grab' their food. Cyclopoid copepods do not create currents to aid their feeding but instead grab their food directly with their first maxillae or, to a lesser extent, with their second maxillae and maxillipeds. These appendages push food between the mandibles by which they tear the prey into pieces and stuff them into the oesophagus (Fryer, 1957a; Williamson, 1991). But in many smaller cyclopoids which ingest relatively small sized algae, raptorial method of food gathering may not be the only method available to the copepod.

Capture

The holding of the prey by the copepod's mouth parts is defined as capture (Roche, 1987, 1990). Once the prey is pursued, the probability of its escaping capture is usually quite low. The probability of capture after attack, among prey species without specialized escape response, is closely related to body size; the rotifer *Polyarthra major* is less likely to be captured by *Acanthocyclops robustus* due to its skipping escape response (Roche, 1987, 1990). *Hexarthra* possesses lateral appendages which enables it to escape capture through a rapid 'flight' response (Gilbert and Williamson, 1978). Some of the cladocerans like *Chydorus* and *Daphnia* reduce their vulnerability to capture by possessing hard, rounded carapace while *Ceriodaphnia dubia* and *Bosmina longirostris* protect their soft bodied parts by closing their valves when attacked; this behavioural response further inhibits capture by the predator (Jamieson, 1980; Williamson, 1980).

As in many other predators, the attack rates and capture efficiencies of predatory copepods is influenced by its satiety level. Often observed is a reduction in feeding activity as the anterior two-thirds of the gut starts getting full (Williamson, 1980). The attack rate and capture success of a hungry copepod are significantly higher than those of a satiated one (Williamson, 1987; Hartmann *et al.*, 1993).

Ingestion

All captured prey may not be ingested. Small soft bodied prey such as nauplii and some rotifers, are generally ingested whole within a few seconds (Gilbert and Williamson, 1978). Loricate rotifers are less likely to be ingested than illoricate forms (Roche, 1987). The larger soft bodied rotifers like *Asplanchna* are handled for variable periods of time and may only be partially ingested depending on their size (Williamson, 1986). The incomplete consumption of a large prey is a common feature of cyclopoid predation (Brandl and Fernando, 1975a; Papinska, 1985; Krylov, 1988). Fryer (1957a,b) showed that suction of soft tissues and partial consumption of prey is the most frequent feeding mechanism in *Mesocyclops.* It is not known how prevalent in nature is such "wasteful killing" of larger prey like cladocerans by cyclopoids which has been well documented in laboratory experiments (Stemberger, 1985; Krylov, 1988).

Mechanisms of Food Gathering in Rotifers

Depending on the feeding mode, rotifers have evolved special mechanisms for gathering their food. The most important structure associated with feeding in rotifers is mastax, a complex calcareous structure situated in the pharyngeal region. Besides being of taxonomic importance (Koste, 1978), the mastax with its jaw apparatus called trophi is uniquely constructed for a specialized feeding mode and is described in detail here.

Structure and Function of the Feeding Apparatus

The size and nature of the food particles ingested by the rotifer is governed by the ciliary apparatus that forms the *corona,* the size of the mouth, and the size structure of the *mastax.* The trophi may process the food in various ways before it is passed on to the oesophagus and swallowed. The trophi of rotifers are unique structures evolved for a specific feeding mode. Food preferences exhibited by a rotifer are generally within the range and limits defined by its mastax. Bdelloidea (*e.g., Philodina*) feed on fine, suspended particulate matter brought to the jaws by ciliary currents produced by the corona. Alternatively, the corona is laid against the substratum so that the current releases particles which are then brought to the jaws by centrifugal action. The jaws are of 'ramate' type, which grind larger particles before digestion.

The 'forcipate' type of mastax found in the carnivorous *Dicranophorus* is modified for grasping the prey. The trophi are thrust through the mouth with a simple, direct motion, and small organisms (rotifers and protozoans) are dragged back by the tips of the teeth.

The 'incudate' mastax found in the predatory *Asplanchna,* is also specialized for grasping, but the modifications are quite different from those of the forcipate type. Here the mastax performs a sucking function by means of a dorsal, expansible sac. Thus prey (loricate and illoricate, ciliates, large euglenoids) are caught by the everted, sac. Species such as *Collotheca* and *Cupelopagis* possess a trapping mastax of the type called 'uncinate', where the ciliary apparatus is transformed into a trapping funnel. When the prey (flagellates, ciliates, rotifers) enters the funnel, it is trapped by the rigid bristles which close upon it.

Rotifers with a 'virgate' mastax show different feeding habits derived from the sucking or pumping function of this mastax type. Certain species pierce the integument of the food cell and suck out the cellular content: for example, *Ascomorpha ovalis* feeding on dinoflagellates, or various species of *Notommata* and *Trichocerca* feeding on algae, *Synchaeta, Eosphora* and *Ploesoma* preying upon other rotifers, and large *Cephalodella* eating small *Cephalodella.* Organisms small enough are eaten whole, others are pumped out empty.

Rotifers with a 'malleate' type of mastax (*Anuraeopsis, Keratella, Kellicottia, Brachionus, Rhinoglena, Notholca*) are generally suspension-feeders. Although majority of planktonic rotifers such as *Asplanchna, Brachionus, Polyarthra,* and *Rhinoglena* swim at similar rates through comparable areas of the water column, encounter possibilities and food capture methods may differ among them (Wallace and Snell, 1991). For example, the ciliary movements of *Brachionus* produce a current of water that carries suspended particles towards the mouth. After the current has been filtered by the cilia, particles of suitable size pass to the mouth where they are seized by the jaws and swallowed. This mode of feeding is called suspension-feeding. Thus, each suspension-feeding rotifer in a community may have a different food preference and consequently the impact on the ecosystem would be different, according to its feeding habits.

The lecanidae have a malleate mastax, but a slightly modified version of the more primitive Brachionid type. Besides chewing particles of food, the lecanids can also produce a certain amount of suction (Donner, 1966), augmenting the centrifugal action of the corona. Unlike the brachionid mastax, the lecanid mastax, is not at the base of a funnel-shaped buccal area. It can therefore be applied directly to large pieces of edible material, which are then held by suction against the jaws to be chewed before entering the stomach. *Lecane* is seldom seen upright and feeding by ciliary currents alone. More often the corona is applied to the substrate and the jaws clip off pieces of organic material. They are largely detritivores, feeding on suspended and flocculated matter.

Discussion

Does this article deviates from the existing views centered on the herbivory or carnivory mode of feeding in zooplankton? It would be very difficult to name any zooplankton as exclusively carnivorous or exclusively herbivorous. The prevalence of omnivory in food web of a community is viewed as an adaptive property that confers and promotes community stability (Fagan, 1997). Many theoretical models have explored the role of omnivores. One such model, complex but realistic, incorporates population age structure and life-history omnivory (Pimm and Rice, 1987). Life history omnivores are those that feed on different trophic levels as they mature. Model food webs which incorporate such omnivores are found to be more stable than those that do not, primarily because the predatory pressure brought to bear by a species is subdivided among a number of dynamically independent developmental stages in a manner that minimizes chances of extinction of any particular prey species. Examples of life history omnivores among zooplankton are cyclopoid copepods, which are herbivorous as nauplii but carnivorous as adults. There are life history omnivores among fishes also, which are planktivorous during larval stages and piscivorous during the adult stage. Analysis of aquatic plankton food webs has thus established the existence of structural complexity in freshwater communities that probably derives from specific life history characteristics of zooplankton species, and from a close coupling of omnivory with higher trophic levels. Because zooplankton communities are often confronted with limited food resources (Sterner, 1986), omnivory seems to be a favourable strategy, that may explain the success of some species within planktonic communities (Sprules and Bowerman, 1988). There is not much information however, on the relative importance of plant and animal prey in the diet of omnivorous zooplankton species (Williamson and Butler, 1986). Omnivory has important consequences for the interpretation of the forces influencing zooplankton community diversity and for the general food web theory. The occurrence of omnivory among all major groups of planktonic organisms (protozoans, cladocerans, copepods, rotifers, and aquatic insects) (Burns and Gilbert, 1986; Williamson and Butler, 1986; Moore, 1988; Wallace and Snell, 1991) contradicts suggestions from food web analyses that omnivory is rare and that omnivores rarely feed on species in non-adjacent trophic levels (Pimm and Lawton, 1978; Pimm, 1980). The above-mentioned taxonomic groups have converged on omnivory as a feeding habit in planktonic communities, suggesting that conditions in aquatic ecosystems generally favour a substantial degree of flexibility in feeding behaviour. The generalist feeders being bacterivorous, detritivorous, herbivorous and omnivorous form most important link between the microbial food web and the traditional food chain and has been in focus of interest recently. Zooplankton play a key role in determining the structure and functioning of aquatic communities and in the transfer of cascading effects to bacterioplankton. The planktonic food web structure, therefore, may be important in determining the qualitative and quantitative characteristics of bacterial and algal consumption and the transfer of bacterial and algal production to higher trophic levels.

Assessment of the specific conditions that favour such broad feeding strategies should provide an interesting arena for future research in zooplankton feeding. There is not much information however, on the relative importance of bacterial, algal and animal prey in the diet of zooplankton species having wider food niche. For any type of policy decision we need to have a clear idea about the width of the food niche of different zooplankton species. In addition following questions are yet to be solved to have an understanding of the feeding strategy of zooplankton: How does the suspended clay affect the zooplankton feeding in tropical and subtropical water bodies? Like terrestrial insects and vertebrates (Taghon, 1981; Calow, 1982) does zooplankton show any compensatory feeding in the presence of low quality food in the ambient medium? To what extent does the filamentous and colonial algae interfere the feeding process of zooplankton? Are the patterns of food selectivity reported in the temperate oliogotrophic water bodies similar to that in eutrophicated and hypertrophicated water bodies of tropical and subtropical areas?

Acknowledgements

I am grateful to Prof. T. Ramakrishna Rao for meaningful suggestions while preparing the review. I appreciate the assistance extended by my wife Dr. Poonam during the preparation of the manuscript. The bibliographic assistance provided by Priyanesh Prasad, School of Environmental Sciences, JNU is gratefully acknowledged. Financial support provided by the Department of Science and Technology, Government of India under young scientist scheme (SR/OY/LS–09/2001) is gratefully acknowledged.

References

Adrian, R. (1987). Viability of phytoplankton in fecal pellets of two cyclopoid copepods. *Arch. Hydrobiol.,* 110: 321–330.

Adrian, R. (1991). The feeding behaviour of *Cyclops kolensis* and *C. vicinus* (Crustacea, Copepoda). *Verh. Internat. Verein. Limnol.,* 24: 2852–2863.

Adrian, R. and T.M. Frost (1992). Comparative feeding ecology of *Tropocyclops parsinus mexicanu.* (Copepoda, Cyclopoida). *J. Plankton Res.,* 14: 1369–1382.

Adrian, R., and T.M. Frost (1993). Omnivory in cyclopoid copepods: Comparisons of algae and invertebrates as food for three differently sized species. *J. Plankton Res.,* 15: 643–658.

Alcaraz, M., G.-A. Paffenhöfer and J.R. Strickler (1980). Catching the algae: A first account of visual observations on filter-feeding calanoids. *Am. Soc. Limnol. Oceanogr. Spec. Symp.,* 3: 241–248.

Arndt, H. (1993). Rotifers as predators on components of microbial web (bacteria, heterotrophic flagellates, ciliates): review. *Hydrobiologia,* 255/256: 231–246.

Atema, J. (1987). Chemoreception in the sea: Adaptations of chemoreceptors and behaviour aquatic stimulus conditions. In: *Sensory Biology of Aquatic Animals,* (Eds.) J. Atema, and R.R. Ray. Springer, New York, pp. 387–402.

Atkinson, A. (1995). Omnivory and feeding selectivity in five copepod species during spring in the Bellingshausen Sea, Antarctica ICES. *J. Mar. Sci.,* 52: 385–396.

Azam, F., T. Fenchel, J.G. Field, J.S. Gray, L.A. Meyer-Reil and F. Thingstad (1983). The ecological role of water-column microbes in the sea. *Mar. Ecol. Prog. Ser.,* 10: 257–263.

Bogdan, K.G. and J.J. Gilbert (1982). Seasonal patterns of feeding by natural populations of *Keratella, Polyarthra and Bosmina:* clearance rates, selectivities and contributions to community grazing. *Limnol. Oceanogr.,* 27: 918–934.

Bogdan, K.G. and J.J. Gilbert (1987). Quantitative comparison of food niches in some freshwater zooplankton. A multi-tracer-cell approach. *Oecologia,* 72: 331–340.

Boon, P.I. and R.J. Sheil (1990). Grazing on bacteria by zooplankton in Australian billabongs. *Aust. J. Mar. Freshwat. Res.,* 41: 247–257.

Brandl, Z. and C.H. Fernando (1975). Investigations on the feeding of carnivorous cyclopoids. *Verh. Internat. Verein. Limnol.,* 19: 2959–2965.

Brown, M.D., B.H. Kay and J.G. Greenwood (1991). The predation efficiency of north-eastern Australian *Mesocyclops* (Copepoda: Cyclopoida) on mosquito larvae. *Bull. Plankton Soc. Japan. Spec., p.* 329–338.

Burns and M. Schallenberg (1996). Relative impacts of copepods, cladocerans and nutrients on the microbial food web of a mesotrophic lake. *J. Plankton. Res.,* 18: 683–714.

Burns, C.W. and J.J. Gilbert (1993). Predation on ciliates by freshwater calanoid copepods: Rates of predation and relative vulnerabilities of prey. *Freshwater Biology*, 30: 377–393.

Burns, C.W. and J.J. Gilbert (1986). Effects of *Daphnia* size and density on interference between *Daphnia* and *Keratella cochlearis*. *Limnol. Oceanogr.*, 31: 848–858.

Calow, P. (1982). Homeostasis and fitness. *Am. Nat.*, 120: 416–419.

Chen, C.Y. and C.L. Folt (1993). Measures of food quality as demographic predictors in freshwater copepods. *J. Plankton Res.*, 15: 1247–1261.

Cole, J.J., G.E. Likens and D.L. Strayer (1982). Photosynthetically produced dissolved organic carbon: An important carbon source for planktonic bacteria. *Limnol. Oceanogr.*, 27: 1080–1090.

Cole, J.J., S. Findlay and M.L. Pace (1988). Bacterial production in fresh and salt water: A cross system overview. *Mar. Ecol. Prog. Ser.*, 43: 1–10.

Colinvaux, P. (1986). *Ecology*. John Wiley and Sons, Hong Kong, pp. 725.

Conde-Porcuna, J.M. and S.S.S. Sarma (1995). Prey selection by *Asplanchna girodi* (Rotifer): The importance of prey defence mechanism. *Freshwater Biology*, 33: 101–108.

Couch, K.M., C.W. Burns and J.J. Gilbert (1999). Contribution of rotifers to the diet and fitness of *Boeckella* (Copepoda: Calanoida). *Freshwater Biology*, 41: 107–118.

Cowles, T.J., R.J. Olason and S.W. Chisholm (1988). Food selection by copepods discrimination on the basis of food quality. *Mar. Biol.*, 100: 41–50.

Dagg, M.J. (1995). Copepod grazing and the fate of phytoplankton in the northern Gulf of Mexico. *Continental Shelf Res.*, 15: 1303–1317.

De Bernardi, R.G. and D.C. McNaught (1975). Selective feeding of zooplankton with special reference to blue green algae in enclosure experiments. *Mem. Ist. Ital. Idrobiol.*, 40: 113–128.

DeMott, W.R. (1986). The role of taste in food selection by freshwater zooplankton. *Oecologia*, Berlin, 69: 334–340.

DeMott, W.R. (1995a). Food selection by calanoid copepods in response to between lake variation in food abundance. *Freshwater Biology*, 33: 171–180.

DeMott, W.R. (1995b). The influence of prey hardness on Daphnia's selectivity for large prey. *Hydrobiologia*, 307: 127–138.

DeMott, W.R. and M.D. Watson (1991). Remote detection of algae by copepods: Responses to algal size, odours and motility. *J. Plankton Res.* 13: 1203–1222.

Dodson, S. (1991). Species richness of crustacean zooplankton in European lakes of different sizes. *Verh. Inter. Verein. Limnol.*, 24: 1223–1229.

Donner, J. (1966). *Rotifera*. Frederick Warne, London.

Fagan, W.F. (1997). Omnivory as a stabilising feature of natural communities. *Am. Nat.*, 150: 554–567.

Fessenden, L. and T.J. Cowles (1994).. Copepod predation on phagotrophic ciliates in Oregon coastal waters. *Mar. Ecol. Prog. Ser.*, 107: 103–111.

Fryer, G. (1957a). The food of some freshwater cyclopoid copepods and its ecological significance. *J. Anim. Ecol.*, 26: 263–286.

Fryer, G. (1957b). The feeding mechanism of some freshwater cyclopoid copepods. *Proc. Zool. Soc. London,* 129: 1–25.

Gabriel, W. (1985). Overcoming food limitation by cannibalism. Arch. Hydrobiol. Beih Ergebn. *Limnol.,* 21: 373–381.

Gifford, D.J. (1991). The protozoan-metazoan trophic link in pelagic ecosystems. *J. Protozool.,* 38: 81–86.

Gifford, D.J. (1993). Protozoa in the diets of *Neocalanus* spp. in the oceanic subartic pacific Ocean. *Prog. Oceanogr.,* 32: 223–237.

Gilbert, J.J. (1980). Observations on the susceptibility of some protists and rotifers to predation by *Asplanchna girodi. Hydrobiologia,* 73: 87–91.

Gilbert, J.J. and C.E. Williamson (1978). Predator-prey behavior and its effect on rotifer survival in associations of *Mesocyclops edax, Asplanchna girodi, Polyarthra vulgaris* and *Keratella cochlearis. Oecologia,* 37: 13–32.

Gilbert, J.J. and J.D. Jack (1993). Rotifers as predators on small ciliates. *Hydrobiologia,* 255/256: 247–253.

Gilbert, J.J. and P.L. Starkweather (1977). Feeding in the rotifer *Brachionus calyciflorus.* I. Regulatory mechanisms. *Oecologia,* 28: 125–131.

Gilbert, J.J. and R.S. Stemberger (1985). Prey capture in the rotifer *Asplanchna girodi. Verh. Internat. Verein. Limnol.,* 22: 2997–3000.

Gliwicz, Z.M. (1969). Studies on the feeding of pelagic zooplankton in lakes with varying trophy. *Ekol. Pol. Ser. A.,* 17(36): 663–708.

Gliwicz, Z.M. (1970). Calculation of food ration of zooplankton community as an example of using laboratory data for field conditions. *Pol. Arch. Hydrobiol.,* 17: 169–175.

Gliwicz, Z.M. (1994). Retarded growth of cladouran zooplankton in the presence of a copepod predation. *Oecologia,* p. 548–461.

González, E.J. (1998). Natural diet of zooplankton in a tropical reservoir (Embalse El Andino, Venezuela). *Verh. Internat. Verein. Limnol.* 26: 1930–1934.

Greene, C.H. (1983). Selective predation in freshwater zooplankton communities. *Int. Revue. Ges. Hydrobiol.,* 68: 296–315.

Greene, C.H. (1988). Foraging tactics and prey-selection patterns of omnivorous and carnivorous calanoid copepods. *Hydrobiologia,* 167/168: 295–302.

Gulati, R.D. and W.R. DeMott (1997). The role of food quality for zooplankton: Remarks on the state-of-the-art perspectives and priorities. *Freshwater Biol.,* 38: 753–768.

Hansen, A.M. and B. Santer (1995). The influence of food resources on the development, survival and reproduction of the two cyclopoid copepods: *Cyclops vicinus* and *Mesocyclops Kukkarti. J. Plankton. Res.,* 17: 631–646.

Hansen, A.M. and E. Jeppesen (1992). Life cycle of *Cyclops vicinus* inrelaion to food availability, predation, diapause and temperature. *J. Plankton. Res.,* 14: 519–605.

Hansen, B. (1991). Feeding bahaviour in the opisthobranch *Philine aperta.* II. Food size spectra and particle selectivity in relation to larval bahaviour and morphology of the velar structures. *Mar. Biol.,* 111: 263–270.

Hansen, B., T. Wemberg-Moller and L. Wittrup (1997). Particle grazing efficiency and specific growth efficiency of the rotifer *Brachionus plicatilis* (Muller). *J. Expt. Mar. Biol. Ecol.,* 215: 217–233.

Hart, R.C. and B. Santer (1994). Nutritional suitability of some unialgal diets for freshwater calanoids: unexpected inadequacies of commonly used edible greens and others. *Freshwater Biol.,* 31: 109–116.

Hartmann, H.J., L. Taleb Aleya and N. Lair (1993). Predation on ciliates by the suspension-feeding calanoid copepod *Acanthodiaptomus denticornis. Can. J. Fish. Aguat. Sci.,* 50: 1382–1393.

Havens, K.E. (1998). Size structure and energetics in a plankton food web. *Oikos,* 81: 346–358.

Havens, K.E., K.A. Work and T.L. East (2000). Relative efficiencies of carbon transfer from bacteria and algae to zooplankton in a subtropical lake. *J. Planton. Res.,* 22: 1801–1809.

Heerkloss, R. and S. Hlawa (1995). Feeding biology of two brachionid rotifers: *Brachionus quadridentatus* and *B. plicatilis. Hydrobiologia,* 313/314: 219–221.

Herzig, A. and B. Auer 1i990). The feeding behaviour of *Leptodora kindti* and its impact on the zooplankton community of Neusiedler See (Austria). *Hydrobiologia,* 198: 107–117.

Higashihara, T., S. Fukuoka, T. Abe, I. Mizuhara, O. Imado and R. Hirano (1983). Culture of the rotifer *Brachionus plicatilis* using a microbial flock produced from alcohol fermentation slop. *Bull. Jap. Soc. Sci. Fish.,* 49: 1001–1014.

Hopp, U., G. Maier and R. Bleher (1997). Reproduction and adult longevity of five species of planktonic cyclopoid copepods reared on different diets: A comparative study. *Freshwater Biol.,* 38: 289–300.

Huntley, M. (1988). Feeding biology of *Calanus:* A new perspective. *Hydrobiologia,* 167/168: 83–99.

Infante, A.G. De (1973). Untersuchungen über die Ausnutzbarkeit verschiedener Algen durch das zooplankton. *Arch. Hydrobiol.,* 42: 340–405.

Iyer, N. and T.R. Rao (1996). Response of the predatory rotifer *Asplanchna intermedia* to prey species differing in vulnerability: Laboratory and field studies. *Freshwater Biol.,* 6: 521–533.

Jack, J.D. and J.J. Gilbert (1997). Effects of metazoan predators on ciliates in freshwater plankton communities. *J. Euk. Microbiol.,* 43: 194–199.

Jamieson, C.D. (1980). The predatory feeding of copepodid stages III–adult *Mesocyclops* leuckarti (Claus). In: *Evolution and Ecology of Zooplankton Communities,* (Ed.) W.C. Kerfoot, Univ. Press of New England, Hanover, New Hampshire, pp. 518–537.

Jansson, M., A.K. Bergstrorn, P. Blomqvist, A. Issaksson and A. Jansson (1999). Impact of allochthonous organic carbon on microbial food web carbon dynamics and structure in Lake Ortrasket. *Arch. Hydrobiol.,* 144: 409–428.

Jónasdóttir, S.H. and T. Kiørboe (1996). Copepod recruitment and food composition: Do diatoms affect hatching success? *Mar. Bioi.,* 125: 743–750.

Jorgensen, C.B. (1966). Biology of suspension feeding. *Ann. Rev. Physiol.,* 37: 57–79.

Jürgens, B.K., J.M. Gasol, R. Massana and C. Pedrós-Alió (1994). Control of heterotrophic bacteria and protozoana by Daphnia pulex in the epilimnion of Lake Ciso. *Arch. Hydrobiol.,* 131: 55–78.

Kak, A. and T.R. Rao (1998). Does evasive behaviour of Hexarthra influence its competition with cladocerans? *Hydrobiologia,* 387/388: 409–419.

Kerfoot, W.C. (1978). Combat between predatory copepods and their prey: *Cyclops, Epischura* and *Bosmina*. Limnol. *Oceanogr.,* 23: 1089–1102.

Kerfoot, W.C. and K.L. Kirk (1991). Degree for taste discrimination among suspension feeding cladocerans and copepods: Implications for detrivory and hernivory. *Limnol. Oceanogs.,* 36(6): 1107–1123.

Kiørboe, T. and E. Saiz (1995). Planktivorous feeding in calm and turbulent environments with emphasis on copepod. *Mar. Ecol. Prog. Ser.,* 122: 135–145.

Kleppel, G.S., D. Frezel, R.E. Pieper and D.V. Holliday (1988). Natural diets of zooplankton of southern California. *Mar. Ecol. Prog. Ser.,* 49: 231–241.

Knisely, K. and W. Geller (1986). Selective feeding of four zooplankton species on natural lake phytoplankton. *Oecologia,* 69: 86–94.

Koehl, M.A.R. and J.R. Strickler (1981). Copepod feeding currents: Food capture at low Reynolds number. *Limnol. Oceanogr.,* 26: 1062–1073.

Korstad, J., Y. Olsen and O. Vadstein (1989). Life history characteristics of *Brachionus plicatilis* (Rotifera) fed different algae. *Hydrobiologia,* 186/187: 43–50.

Koshikava, H., H. Shigeki, M. Watnabe, K. Soto and K. Akehatu (1996). Relative contribution of bacterial and photosynthetic production to metazooplankton as carbon sources. *J. Plankton. Res.,* 18: 2269–2281.

Koste, W. (1978). Rotatoria. Die Radertiere Mittereuropas Geruder Borntraeger, Berlin and Stuttgart.

Krylov, P.I. (1988). Predation of the freshwater cyclopoid copepod *Megacyclops gigas* on lake zooplankton: Functional response and prey selection. *Arch. Hydrobiol.,* 113: 231–250.

Kumar, R. (2003a). Effect of different food types on the post embryonic developmental rates and demographic parameters of *Phyllodiaptomus blanci* (Copepoda: Calanoida). *Arch. Hydrobiol.,* 157(3): 351–377.

Kumar, R. (2003b). Effect of *Mesocyclops thermocyclopoides* (Copepoda, Cyclopoida) predation on population dynamics of different prey: A laboratory study. *J. Freshwater Ecol.,* 18(3): 383–393.

Kumar, R. (2004). Ecological perspectives in cyanobactrial blooms. Oikos, Publishers, Inc. USA., pp. 136.

Kumar, R. and T.R. Rao (1998). Postembryonic Developmental rates as a function of food type in the cyclopoid copepods *Mesocyclops thermocyclopoides* Harada. *J. Plankton Res.,* 20: 271–287.

Kumar, R. and T.R. Rao (1999a). Demographic responses of adult *Mesocyclops thermocyclopoides* (Copepoda, Cyclopoida) to different plant and animal diets. *Freshwater Biol.,* 42: 487–501.

Kumar, R. and T.R. Rao (1999b). Effect of algal food on animal prey consumption rates in the omnivorous copepod, *Mesocyclops thermocyclopoides. Internat. Review Hydrobiol.* 84: 419–426.

Kumar, R. and T.R. Rao (2001). Effect of the cyclopoid copepod *Mesocyclops thermocyclopoides* on the interactions between the predatory rotifer *Asplanchna intermedia* and its prey *Brachionus calyciflorus* and *B. angularis. Hydrobiologia,* 453/454: 261–268.

Kumar, R. and T.R. Rao (2003). Predation on mosquito (*Anopheles stephensi* and *Culex quinquefasciatus*) larvae by *Mesocyclops thermocyclopoides* (copepoda: cyclopoida) in the presence of alternate prey. *Internat. Review Hydrobiol.,* 88(6): 570–581.

Lair, N. (1991). Grazing and assimilation rates of natural populations of planktonic cladocerans in a eutrophic lake. *Hydrobiologia,* 215: 51–61.

Lair, N. (1992). Daytime grazing and assimilation rates of planktonic copepods *Acantho diaptomus denticornis* and *Cyclops vicinus vicinus.* Comparison of spatial and resource utilisation by rotifers and cladoceran communities in a eutrophic lake. *Hydrobiologia,* 231: 107–117.

Lam, R.K. and B. Frost (1976). Model of copepod filtering response to changes in size and concentration of food. *Limnol. Oceanogr.,* 21: 490–500.

Lampert, W. (1987). Feeding and nutrition in *Daphnia.* In: *Daphnia,* (Eds.) R.H. Peters and R. De Bernardi. *Mem Ist. Ital. Idrobiol.,* 45: 143–192.

Lampert, W. and Sommer, U. (1997). *Limnoecology: The Ecology of Lakes and Streams.* Oxford University Press, New York, Oxford, pp. 382.

Lampert, W., W. Fleckner, H. Rai and B.E. Taylor (1986). Phytoplankton control by grazing zooplankton: A study on the spring clear-water phase. *Limnol. Oceanogr.,* 31: 478–490.

Landry, M.R. (1981). Switching between herbivory and carnivory by the planktonic marine copepod *Calanus pacificus. Mar. Biol.,* 65: 77–82.

Lawton. J.H., J. Beddington and R. Bonser (1974). Switching in invertebrate predators. In: *Ecological Stability,* (Eds.) M.B. Usher, M.A. Williamson. Chapman and Hall, London. p. 141–158.

Lehman, J.T. (1976). The filter feeder as an optional forager and the predicted shapes of feeding curves. *Limnol. Oceanogr.,* 21: 501–516.

Lillelund, K. and R. Lasker (1971). Laboratory studies of predation by marine copepods on fish larvae. *Fishery Bull.,* 69: 655–667.

Moore, M.V. (1988). Differential use of food resources by the instars of *Chaoborus* punctipennis. *Freshwater Biol.,* 19: 249–269.

Mullin, M.M. (1966). Selective feeding by calanoid copepods from the Indian ocean. In: *Some Contemporary Studies in Marine Sciences,* (Ed.) H. Barnes. After and Unwin. London, p. 545–554.

Nival, P. and S. Nival (1976). Particle retention efficiency of a herbivorous copepods *Acartia clausi* (Adult and copepodid stages) Effects on grazing. *Limnol.Oceanogr.,* 21:24–38.

Ohman, M.D. and J.A. Runge (1994). Sustained fecundity when phytoplankton resources are in short supply: omnovory by *Calanus finmarchians* in the Gulf of St. Lawrence. *Limnol. Oceanogr.,* 39: 21–36.

Ooms-Wilms, A.L. (1991). Ingestion of flourescently labeled bacteria by rotifers and cladocerans in Lake Loosdrecht as measure of bacterivory. Memorie dell' Institute italiano di idrobiologia, 48: 269–278.

Paffenhöfer, G.A. and S.C. Knowles (1978). Feeding of marine planktonic copepods in mixed phytoplankton. *Mar. Biol.,* 48: 143–157.

Papinska, K. (1985). Carnivorous and detritivorous feeding of *Mesocyclops leuckrti* Claus (Cyclopoida, Copepoda). *Hydrobiologia,* 120: 249–257.

Petipa, T.S. (1978). Matter accumulation and energy expenditure in planktonic ecosystems at different trophic levels. *Mar. Biol.,* 49: 285–293.

Petipa, T.S., E.V. Pavlova and G.N. Mironov (1970). The food web structure, utilization and transport of energy by trophic levels in the planktonic communities. In *Marine Food Chains,* (Ed.) J.H. Steele. Oliver and Boyd, Edinburgh, pp. 142–167.

Pimm, S.L. (1980). Properties of food webs. *Ecology,* 61: 219–225.

Pimm, S.L. (1982). *Food Webs.* Chapman and Hall, London.

Pimm, S.L. and J.C. Rice (1987). The dynamics of multispecies, multilife-stage models of aquatic food webs. *Theor. Pop. Biol.,* 32: 303–325.

Pimm, S.L. and J.H. Lawton (1977). Number of trophic levels in ecological communities. *Nature,* 268: 329–331.

Pimm, S.L. and J.H. Lawton (1978). On feeding on more than one trophic level. *Nature,* 275: 542–544.

Porter, K.G. (1984). Natural bacteria as food resources for zooplankton. In: *Current Perspectives in Microbial Ecology,* (Eds.) M.J. Klug and C.A. Reddy, pp. 340–345.

Porter, K.G. and J.D. Orcutt (1980). Nutritional adequacy, manageability and toxicity as factors that determine the food quality of green and blue-green algae for *Daphnia.* In *The Evolution and Ecology of Zooplankton Communities,* (Ed.) W.C. Kerfoot. New England, Hanover, N.H.

Pourriot, R. (1977). Food and feeding habits of Rotifera. *Arch. Hydrobiol. Beih.,* 8: 243–260.

Price, H.J. and G.A. Paffenhöfer and J.R. Strickler (1983). Modes of cell capture in calanoid copepods. *Limnol. Oceanogr.,* 28: 116–123.

Rao, T.R. and Ram Kumar (2002). Patterns of prey selectivity in the cyclopoid copepod, *Mesocyclops thermocyclopoides* Harada. *Aquatic Ecol.,* 36: 411–424.

Ricci, C. (1984). Culturing of some bdelloid rotifers. *Hydrobiologia,* 112: 45–51.

Roche, K.F. (1990). Some aspects of vulnerability to cyclopoid predation of zooplankton prey individuals. *Hydrobiologia,* 198: 152–162.

Roche, K.F. (1987). Post-encounter vulnerability of some rotifer prey types to predation by the copepod *Acanthocyclops robustus. Hydrobiologia,* 147: 229–233.

Rosenberg, G.G. (1980). Filmed observations of filter feeding in the marine planktonic copepod *Acartia clausii Limnol. Oecangr.,* 25: 738–742.

Rothhaupt, K.O. (1990a). Differences in particle size-dependent feeding efficiencies of closely related rotifer species. *Limnol. Oceanogr.,* 35: 16–23.

Rothhaupt, K.O. (1990b). Population growth rates of two closely related rotifer species: Effects of food quantity, particle size and nutritional quality. *Freshwater Biol.,* 23: 561–570.

Rubenstein, D.I. and M.A.R. Koehl (1977). Mechanisms of filter feeding: Some theoretical consideration AM. *Nature,* 111: 981–994.

Sanders, R.W., C.E. Williamson, P.L. Stutzman, R.E. Moeller, C.E. Goulden and R. Aoki-Goldsmith (1996). Reproductive success of "herbivorous" zooplankton fed algal and nonalgal food resources. *Limnol. Oceanogr.,* 41: 1295–1305.

Santer, B. (1993). Potential importance of algae in the diet of adult Cyclops vicinus. *Freshwater Biology,* 30: 269–278.

Santer, B. (1996). Nutritional suitability of the dinoflagellate *Ceratium furcoides* for four copepod species. *J. Plankton Res.,* 18: 323–333.

Santer, B. and F. van den Bosch (1994). Herbivorous nutrition of *Cyclops vicinus:* The effect of a pure algal diet on feeding, development, reproduction and life cycle. *J Plankton Res.*, 13: 789–799.

Saunders, P.T. (1978). Population dynamics and the length of food chains. *Nature*, 272: 189–190.

Schulze, P.C. and C.L. Folt (1989). Effects of conspecifics and phytoplankton on predation rates of the omnivorous copepods *Epischura lacustris* and *Epischura nordenskioldi. Limnol. Oceanogr.*, 34: 444–450.

Schulze, P.C. and C.L. Folt (1990). Food resources, survivorship and reproduction of the omnivorous calanoid copepod *Epischura lacustris. Ecology*, 71: 2224–2240.

Seaman, M.T., M. Gophen, B.Z. Cavari and B. Azoulay (1986). *Brachionus calyciflorus* Pallas as agent for the removal of *E. coli* in sewage ponds. *Hydrobiologia*, 135: 55–60.

Sherr, E. and B. Sherr (1988). Role of microbes in pelagic food webs: A revised concept. *Linmol. Oceanogr.*, 33: 1225–1227.

Sprules, W.G. and J.E. Bowerman (1988). Omnivory and food chain length in zooplankton food webs. *Ecology*, 69: 418–426.

Starkweather, P.L. (1980). Aspects of the feeding behaviour and trophic ecology of suspension feeding rotifers. *Hydrobiologia*, 73: 63–72.

Starkweather, P.L. and P.E. Kellar (1983). Utilization of cyanobacteria by *Brachionus calyciflorus*: *Anabaena flosaquae* (NRC-44-1) as a sole or complementary food source. *Hydrobiologia*, 104: 373–377.

Starkweather, P.L., J.J. Gilbert and T.M. Frost (1979). Bacterial feeding by the rotifer *Brachionus calyciflorus:* Clearance and ingestion rates, behaviour and population dynamics. *Oecologia*, 44: 26–30.

Stemberger, R.S. (1985). Prey selection by the copepod *Diacyclops thomasi. Oecologia*, 65: 492–497.

Sterner, R.W. (1986). Herbivores' direct and indirect effects on algal populations. *Science*, 231: 605–607.

Taghon, G.L. (1981). Beyond selection: optimal ingestion rate as a function of food value. *Am. Nat.*, 118: 202–214.

Taylor, W.D. and R.W. Sanders (1991). Protozoa. In: *Ecology and Classificaiton of North American Freshwater Invertebrate,* (Eds.) J.H. Thorp and A.P. Covich. Academic Press, pp. 37–93.

Tiselius, P. and P.R. Jonsson (1990). Foraging behaviour of six calanoid copepod: Observations and hydrodynamic analysis. *Mar. Ecol. Prog. Ser.*, 56: 49–56.

Toth, L.G. and N.P. Zankai (1985). Feeding of Cyclops *vicinus* (Uljanin) (Copepoda: Cyclopoida) in Lake Balaton on the basis of gut content analyses. *Hydrobiologia*, 122: 251–260.

Toth, L.G., N.P. Zankai and O.M. Messner (1987). Alga consumption of four dominant Planktoic Crustaceans in Lake Balaton (Hungary). *Hydrobiology*, 145: 323–332.

Vadstein, O., G. Oie and Y. Olsen (1993). Particle size dependent feeding by the rotifer *Brachionus plicatilis Hydrobiologia*, 255/256: 261–267.

Vanderploeg, H.A. and G.A. Paffenhöfer (1985). Modes of algal capture by the freshwater copepod *Diaptomus sicilis* and their relation to food-size selection. *Limnol. Oceanogr.*, 30: 871–885.

Vanni, M.J. and W. Lampert (1992). Food quality on life history traits and fitness in the generalist herbivore *Daphnia. Oecologia*, 92: 48–57.

Verity, P.G. and G.-A. Paffenhöfer (1996). On assessment of prey ingestion by copepod. *J. Plankton. Res.*, 18: 1767–1779.

Vijverberg, K. (1989). Culture techniques for studies the growth, development and reproduction of copepods and cladocerans under laboratory an *in situ* conditions: A review. *Freshwater Biol.,* 21: 317–373.

Walz, N. (1995). Rotifer populations in plankton communities: Energetics and life history strategies. *Experimentia,* 51: 437–453.

Wallace, R.L. and T.W. Snell (1991). Rotifera. In: *Ecology and Classification of North American Freshwater Invertebrates,* (Eds.) J.H. Thorp and A.P. Covich. Academic Press, pp. 187–248.

Waterman, T.H. (ed). (1960). *The physiology of Crustacea, Vo1. 1: Metabolism and Growth.* Academic Press, pp. 670.

Wiackowski, K., A.M. Venteä, M. Moilanen, V. Saarikari, K. Vuorio and J. Sarvala (2001). What factors control planktonic ciliates during summer in a highly eutrophic lake? *Hydrobiologia,* 443: 43–57.

Wickham, S. A. (1995). Trophic rebtions cyclopoid copepods and ciliated protists; Compley interactions link the microbial and classic food webs. *Limnol. Oceanogr.,* 40(6): 1173–1181.

Wickham, S.A. (1998). The direct and indirect impact of Daphnia and Cyclops on a freshwater microbial food web. *J. Plankton Res.,* 20: 739–755.

Wickham, S.A., J.J. Gilbert and U.G. Berninger (1993). Effects of rotifers and ciliates on the growth and survival of *Daphnia. J. Plankton Res.,* 15: 317–334.

Wiesse, T. and H. Muller (1990). Significance of heterotrophic nanoflagellates and ciliates in large lakes: evidence from lake Constance. In: *Large Lakes,* (Eds.) M.M. Tilzer and C. Serruya. Springer Verlag, Berlin, pp. 540–553.

Williamson, C.E. (1991). Copepoda. In: *Ecology and classification of North American Freshwater Invertebrates.* (Ed.) J. Throp. Academic Press San Diego, pp. 787–821.

Williamson, C.E. (1980). The predatory behavior of *Mesocyclops edx:* Predator preferences, Prey defences and starvation induced changes. *Limnol. Oceanogr.,* 25. 903–909.

Williamson, C.E. (1983). Invertebrate predation on planktonic rotifers. *Hydrobiologia,* 104: 385–396.

Williamson, C.E. (1989). Predator-prey interactions between omnivorous diaptomid copepods and rotifers: The role of prey morphology and behaviour. *Limnol Oceanogr.,* 32: 167–177.

Williamson, C.E. and H.A. Vanderploeg (1988). Predatory suspension-feeding in diaptomus: Prey defenses and the avoidance of cannibalism. *Bull. Mar. Sci.,* 43: 561–572.

Williamson, C.E. and J.J. Gilbert (1980). Variation among zooplankton predator the potential of *A. planchna, wsocyclops* and cyclops to attack, capture and eat various rotifer prey. In: *Evolution and Ecology of Zooplankton Communities,* (Ed.) W.C. Kerfoot, p. 205–217.

Williamson, C.E. and N.M. Butler (1986). Predation on rotifers by the suspension-feeding calanoid copepod *Diaptomus pallidus. Limnol. Oceanogr.,* 31: 393–402.

Wong, C.K. (1984). A study of the relationships between mouthparts and food habits in several species of freshwater calanoid copepods. *Can. J. Zool.,* 62: 1588–1595.

Work, K.A. and K.E. Havens (2003). Zooplankton grazing on bacteria and cyanobacteria in a eutrophic lake. *J. Plankton Res.* 25: 1301–1307.

Wylie, J.L. and D.J. Currie (1991). The relative importance of bacteria and algae as food sources for crustacean zooplankton. *Limnol. Oceanogr.,* 36: 708–728.

Xu, Z. and C.W. Burns (1991). Development, growth and survivorship of juvenile calanoid copepods on diets of cyanobacteria and algae. *Int. Revue. Ges. Hydrobiol.*, 76: 73–87.

Yodzis, P. (1981). The stability of real ecosystem. *Nature*, 289: 674–676.

Zaret, T.M. (1980). *Predation and Freshwater Communities*. Yale Univ. Press, New Have, pp. 187.

Chapter 28

Relation Between Water Circulation and Zooplankton Distribution in the Bay Environment of Kakinada, East Coast of India

N.V. Prasad

Division of Marine Biology, Department of Zoology, Andhra University, Visakhapatnam–530 003

ABSTRACT

Distribution of zooplankton in relation to hydrographical conditions and water circulation have been described from the bay environment of Kakinada. A total of 23 diverse groups of zooplankton were observed. The overall mean values of biomass and numerical abundance at three stations were 1.79, 1.80, 2.91 $ml.m^{-3}$ and 38482, 39269, 38447 $no.m^{-3}$ respectively. Distribution of major groups of zooplankton populations were governed by various behavioural and physiological adaptations of the plankton populations to ever changing hydrographical conditions. The prevailing northerly and southerly currents along the coast also influence the distribution of the planktonic organisms. In spite of these wide fluctuating conditions we can distinguish characteristic populations within the bay region which show regular annual cycles and colonization.

Keywords: *Zooplankton, Distribution, Water movements, Bay environment.*

Introduction

The Kakinada Bay (82 14′–82 22′ E and 16 50′–17 00′ N) is situated on the east coast of India, about 160 km southwest of Visakhapatnam. It is the main receptacle, for the river runoff of the Godavari estuarine system. The total area of the bay is 132 km^2 (Ramasarma and Ganapathi, 1968). Two major distributaries of the river Godavari namely, Coringa and Gaderu open into the Kakinada Bay. Other small channels *viz.,* Chollangi, Matlapalem and Pillavarava also open into the bay on the southwestern and southeastern corners respectively. On the eastern side of the bay, there is a long narrow sand bar continuous with the eastern tip of the Hope Island separating the bay from the sea. Due to the presence of the sand bar, the Kakinada Bay forms a semi-enclosed body of water where the water movements are unique. Two types of circulation patterns are seen in the bay. The water movements are coupled with the salinity distribution in the bay which in turn limits the plankton distribution. Studies made on the distribution of zooplankton in relation to water movements, in Sundays river estuary in Algoa Bay (Wooldrige and Erasmus, 1980) and Mandovi-Zuari estuarine system (Madhupratap and Rao, 1979) revealed that the zooplankton distribution is dependent on the water-mass movements. The objective of the present investigation was to describe the plankton distribution and variations in the numerical abundance due to the prevailing hydrographical conditions and in particular the circulation in the Kakinada Bay.

Materials and Methods

Surface water and zooplankton samples were collected monthly from three stations in the Kakinada Bay during July 1996–June1999. Station 1 is located at the northern most region of the Kakinada Bay and it has greater influence of coastal currents of Bay of Bengal. Station 2 is located at the central part of the bay and it presents a more or less stable environment due to the least disturbance due to the wave action and water movements. Whereas, Station 3 is located towards the southern part of the Bay and is very shallow. During most part of the year this station experiences brackish water conditions, due to the large scale influx of river water discharges from Gaderu, Coringa, Matlapalem, Pillavarava and Chollangi creeks. The zooplankton collections were made with a 120 μm mesh net fitted with a calibrated flow meter. Biomass was estimated by the displacement method (Wickstead, 1965) and the aliquot method was adopted for the enumeration of the zooplankton populations density. Hydrographical parameters such as temperature, salinity, dissolved oxygen and pH were measured by adopting standard methods (APHA, 1995).

Results and Discussion

Kakinada Bay waters exhibited wide range of physical and chemical factors in both time and space with a marked variation from station to station. The surface water temperature ranged from a minimum of 26.0 C (Station 1, December 1998) to a maximum of 34.4 C (Station 3, May 1997), the mean being 29.47 C (Table 28.1). Differences between atmospheric and surface water temperature were negligible and it ranged between 0.60 to 1.22 C. Salinity varied from estuarine conditions (15.12 per cent; Station 3) to a normal sea water (34.65 per cent; Station 1). In general, the salinity fluctuations are very high at Station 3, while at Station 1 marine-water conditions are prevailed during most part of the year, where neritic influence was more. Dissolved oxygen varied between 2.92 ml.l^{-1} (Station 3) and 7.39 ml.l^{-1} (Station 1). The mean pH values at Stations 1, 2 and 3 were 7.09, 7.12 and 7.14 respectively. Inter-station differences were negligible. Altogether 23 groups of zooplankton were observed. Copepods formed a major component (61.45 per cent) of total zooplankton, followed by decapods (8.42 per cent), bivalves (7.36 per cent), gastropods (5.02 per cent), chaetognaths (2.19 per cent) and lucifers (1.28 per cent). Other groups were occurred relatively in small numbers. Zooplankton density varied from

11,125 (Station 1) to 1,49,766 no.m^{-3} (Station 3) (Table 28.1). Biomass of zooplankton ranged from a minimum of 0.40 ml.m^{-3} (Station 1), to a maximum of 8.40 ml.m^{-3} (Station 3) (Table 28.1).

Table 28.1: Hydrographical and Biological Characteristics (range and mean) in Kakinada Bay (Values in Parentheses Denote Mean)

Parameter	*Station 1*	*Station 2*	*Station 3*
Temperature (°C)	26.0–33.0	26.2–33.0	26.6–33.4
	(29.9 ± 1.73)	(29.0 ± 1.83)	(29.52 ± 1.48)
Salinity (%)	21.09–33.4	18.95–34.65	15.12–34.60
	(27.88 ± 4.42)	(26.0 ± 4.42)	(24.85 ± 5.21)
Dissolved oxygen (ml.l^{-1})	3.24–7.39	3.30–6.27	2.92–7.22
	(4.72 ± 1.07)	(4.73 ± 0.94)	(4.65 ± 1.04)
pH	6.2–7.9	6.7–7.9	6.7–7.8
	(7.09 ± 1.22)	(7.12 ± 1.44)	(7.14 ± 1.38)
Numerical abundance	11,125–1,14,970	11,283–1,06,323	72,704–1,49,766
(no.m^{-3})	(38,482 ± 26,849)	(39,296 ± 26,323)	(37,447 ± 17,489)
Biomass (ml.m^{-3})	0.40–4.36	0.46–4.92	0.98–8.40
	(1.79 ± 1.42)	(1.80 ± 1.69)	(2.91 ± 1.95)
Holoplankton (%)	54.49–88.43	53.49–81.46	53.42–81.62
	(70.60)	(72.46)	(69.05)
Meroplankton (%)	11.57–45.57	8.12–46.51	17.12–46.58
	(29.39)	(27.42)	(30.90)

Water Movements and Zooplankton Distribution

Rochford (1951) defined circulation in estuaries as the process responsible for maintaining a seaward movement of freshwater component and for estuarine invasion by neretic waters. This process is able to limit the distribution of zooplankton. The Kakinada Bay by virtue of its position on the east coast of India as a semi-enclosed body of water, has a complex pattern of water movements. Among the factors that influence the movements of water in the bay mention may be made of the two coastal currents which run along the coast, the southerly and northerly currents operating during the periods August-December and January-July respectively. Due to these currents prevailing in the coast, the pattern of circulation in Kakinada Bay is clockwise from December to June and counter clockwise from July to November (Figures 28.1A&B).

The zooplanktonic organisms have various behavioural adaptations like the utilization of tidal currents, vertical migration, high reproductive rate and changes in the larval behaviour by which they have been successfully thriving well in the dynamic systems like estuaries and bays (Wooldrige and Erasmus, 1980). Another aspect relating to the zooplankton movement is that the species inhabiting a water mass will tend to have an overall distribution beyond the area where it reproduces and maintains itself successfully and it will be carried to greater or lesser degree outside its own area by drift (Raymont, 1983). From this study it is clear that the zooplankton populations dwelling in Kakinada Bay have special adaptations of their own to survive within the bay withstanding the water movements in the bay.

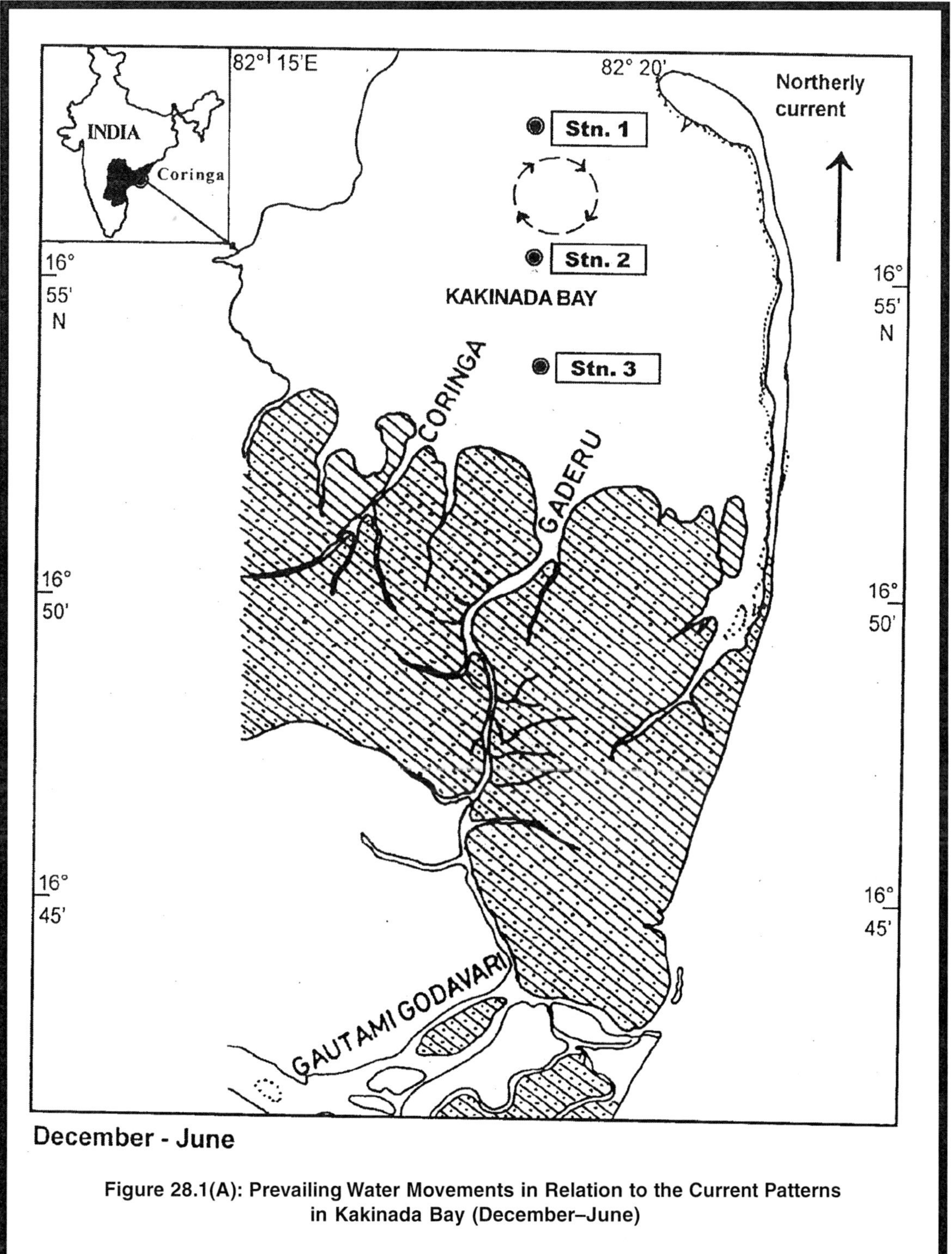

Figure 28.1(A): Prevailing Water Movements in Relation to the Current Patterns in Kakinada Bay (December–June)

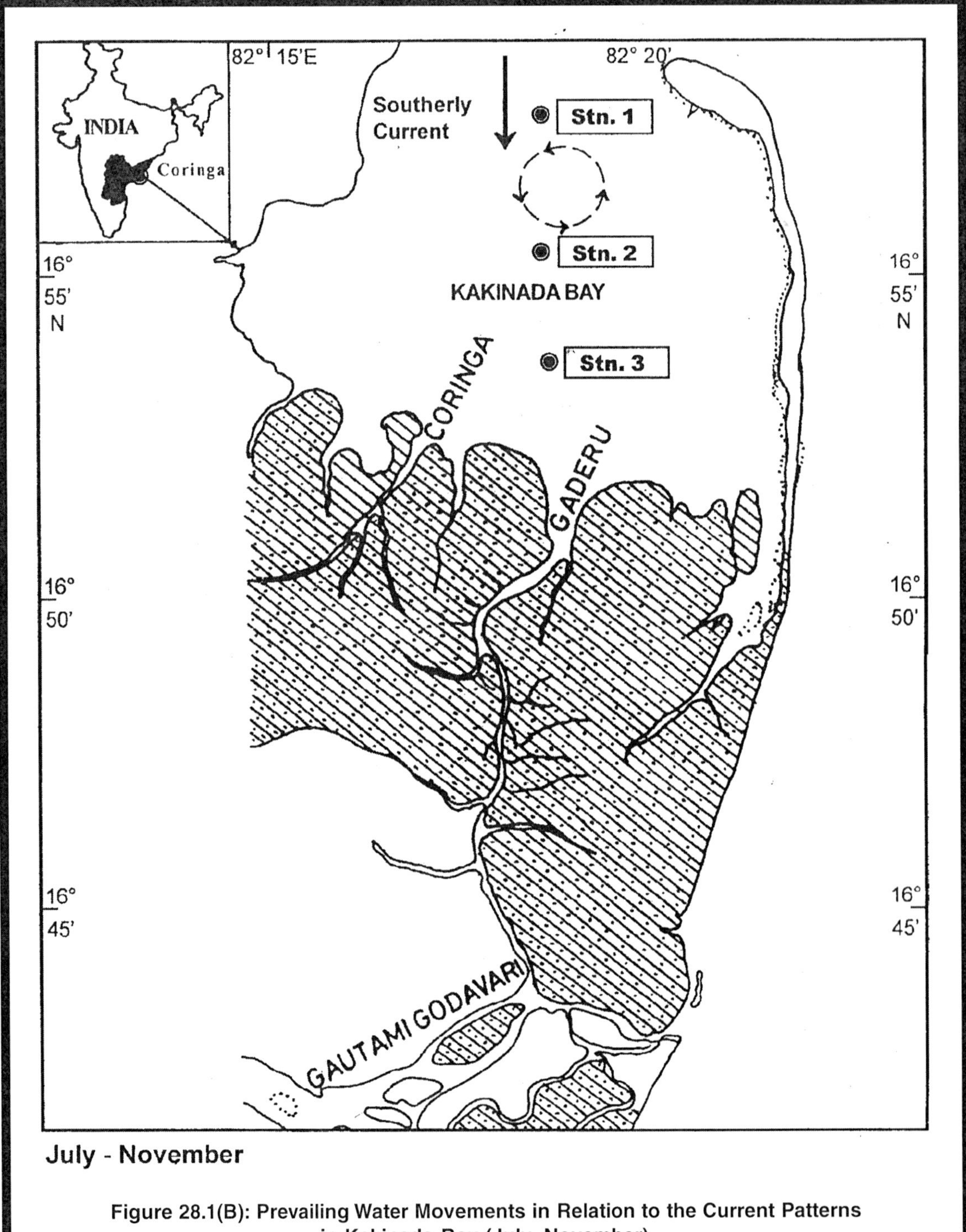

Figure 28.1(B): Prevailing Water Movements in Relation to the Current Patterns in Kakinada Bay (July–November)

The plankton populations were dominated by the copepods which represented to an average of 75–80 per cent of total zooplankton. Among the copepods *Paracalanus* sp., *Eucalanus* sp., *Acrocalanus* sp., *Oithona* sp., *Oncaea* sp., *Acartia* sp., *Corycaeus* sp., and *Macrosetella* sp., were most predominant. The other holoplanktonic groups were represented by chaetognaths, lucifers, pteropods, cladocerans, ostracods, doliolids, appendicularians, ctenophores and pleurobranchians which were seen occasionally and in few numbers in the plankton samples. The meroplanktonic component of the plankton population was dominated by the decapod larvae, bivalve veligers, gastropod veligers, polychaete larvae of which the bivalve and the gastropod veligers dominated the populations in good numbers. The present study is in conformity with the works carried out in the bay estuarine systems of temperate regions (Madhupratap and Rao, 1979; Wooldrige and Erasmus, 1980). The zooplankton of temperate bay and estuaries is typically higher in abundance and lower in diversity than zooplankton of adjacent neritic waters (Kimmer, 1993). Bay fauna are typically dominated (Ueda, 1991) by one or two species of ubiquitous species like *Acartia* sp., *Paracalanus* sp., *Oithona* sp., *Centropages* sp. etc.

Various factors aid the indigeneous zooplankton to remain within the estuary. A high reproductive rate may replace that component lost by way of the net seaward transport of mixed fresh and salt waters (Ketchum, 1954; Barlow, 1955; Prasad, 2003). Depending on the life history, distribution of zooplankton in accordance with the salinity along with the prevailing hydrographical conditions of the area the entire Kakinada Bay area can arbitrarily be divided into two zones:

Zone I

The marine dominant area representing the northern bay which is the zone where reproduction, growth and maturation of plankton populations takes place and representing mostly the holoplankton groups in the plankton collection (Station 1). Immigration of larvae such as crustaceans, bivalves and gastropod veligers from the southern region of Kakinada Bay was often encountered. In this zone the dominant holoplanktonic groups were copepods, chaetognaths, hydromedusae, siphonophores, appendicularians, ostracods, cladocerans and pteropods. Among the meroplankton the dominant groups were decapod larvae, bivalve veligers, gastropod veligers and polychaete larvae.

Zone II

This area was dominated by brackish water and is located in the southern part of the Kakinada Bay (Stations 1 and 2), which seems to be the main area for the seasonal recruitment and colonization. This zone supports mixed zooplankton populations. Among the meroplankton gastopod veligers, bivalve veligers, decapod larvae and polychaete larvae were abundant. In this zone there seem to exist typical seasonal pattern of distribution in the plankton abundance, and the abundance showed two major peaks during pre-monsoon and a minor peak during the post-monsoon period.

Distribution of major groups of zooplankton populations were governed by various behavioural and physiological adaptations of the plankton populations to ever changing hydrographical conditions. Wooldridge and Erasmus (1980) from the Sundays estuary (South Africa) reported that behavioural adaptations include utilization of tidal currents, vertical migration of zooplankton in well stratified estuaries. The meroplankton decapod larvae are able to utilize tidal currents for their horizontal distribution (Young and Carpenter, 1977; Cronin, 1982). The prevailing circulation and water mixing process in the semi-enclosed Kakinada Bay seems to influence the distribution of plankton populations. The distribution of various plankton groups was also influenced by rainfall to some extent during the monsoon period. Relatively high freshwater runoff during monsoon provides embayment with a large amount of nutrients which allowed the development of phytoplankton community (Marshall, 1982), this in turn may result in the production of herbivorous zooplankton, mostly meroplankton in nature. The characteristic phenomenon that enables the organisms to survive in particular zone or biotope is highly appreciable and it may be due to the mechanism that generates environmental variability for animals in estuaries. The first component corresponds to the presence of longitudinal and vertical gradients in physico-chemical parameters, and the second component is the dynamic environmental variability results from the water movements which determine the rate and direction of changes in abiotic conditions an individual experiences (Laprise and Dodson, 1993).

References

APHA (1995). *Standard Methods for the Examination of Water and Wastewater*, 19th Edn. American Public Health Association Publication, Washington D.C. pp. 874.

Barlow, J.P. (1955). Physical and biological process determining the distribution of zooplankton in a tidal estuary. *Ecol. Bull.,* 109: 211–225.

Cronin, T.W. (1982). Estuarine retention of larvae of the crab *Rhithropanopeus harrisii. Estuar. Coast. Shelf Sci.,* 15: 207–220.

Ketchum, R.H. (1954). Relation between circulation and plankton populations in estuaries. *Ecology,* 35: 191–200.

Kimmer, W. J. (1993). Distribution patterns of zooplankton in Tomales Bay, California. *Estuaries,* 16: 264–272.

Laprise, R. and J.J. Dodson (1993). Environmental variability a factor controlling spatial patterns in distribution and species diversity of zooplankton in St Lawrence estuary. *Mar. Ecol. Prog. Ser.,* 107: 67–81.

Madhupratap, M. and T.S.S. Rao (1979). Tidal and diurnal influence on estuarine zooplankton. *Indian J. Mar. Sci.,* 8: 9–11.

Marshall, H.G. (1982). The composition of phytoplankton within the Chesapeale Bay plume and adjacent waters of the Virginia coast USA. *Estuar. Coast. Shelf Sci.,* 15: 29–42.

Prasad, N.V. (2003). Composition,and abundance of meroplankton in Coringa mangrove ecosystem with reference to aquaculture. *J. Aqat. Biol.,* 18: 29–34.

Rachford, D.J. (1951). Graging by a river zooplankton community: Importance of microzooplankton. *Aust. J. Mar. Freshwat. Res.,* 2: 1–17.

Raymont, J.E.G. (1983). *Plankton and Productivity in the Oceans–Zooplankton.* Pergamon press, Paris, pp. 824.

Ramasarma, D.V. and P.N. Ganapathi (1968). Hydrography of the Kakinada Bay. *Bull. Nat. Inst. Sci.,* India, 38: 48–79.

Ueda, H. (1991). Horizontal distribution of planktonic copepods inlet waters. In: *Proceedings of the Fourth International Conference on Copepoda,* (Eds.) S.I. Uye, S. Nishida and J.S. Ho. Plankton Society of Japan, Hiroshima, pp. 143–160.

Wickstead. J.H. (1965). *An Introduction to the Study of Tropical Plankton.* Hutchinson Tropical Monogrpah, Hutchinson and Co. Pub, London, pp. 160.

Wooldrige, T.R. and T. Erasmus (1980). Utilization of tidal currents by estuarine zooplankton. *Estuar. Coast. Mar. Sci.,* 11: 107–114.

Young, P.C. and S.M. Carpenter (1977). Recruitment of post larval penaeid prawns to nursery areas in Moretene Bay, Queens Land. *Aust. J. Mar. Freshwat. Res.,* 28: 745–751.

Chapter 29

Limnological Studies of Mosam River of Maharashtra with Relation to Phytoplankton

N.H. Aher & S.N. Nandan***

**Department of Botany, Arts, Commerce and Science College, Sakri – 424 304, District Dhule (MS)*
***P.G. Department of Botany, S.S. V.P.S's L.K. Dr. P.R. Ghogrey Science College, Dhule – 424 005 (MS)*

ABSTRACT

Limnological study of algae was carried out on Mosam river flowing through Baglan Tahsil of Maharashtra from July 1998 to June 2000. Algal and water samples were collected at first week in monthly intervals from 3 stations of Mosam river during the period of two years of investigation. The most pollution tolerant genera and species of 4 groups of algae *viz.*, Chlorophyceae, Cyanophyceae, Bacillariophyceae and Euglenineae were recorded of 3 stations. The assessment of water quality of river was made for 3 stations by algal communities. In present study 16 pollution tolerant species were observed at 3 stations of Mosam river. The data of physico-chemical analysis is also supported for the study of pollution of Mosam river.

Keywords: *Limnology, Algae, Water pollution.*

Introduction

Algae are involved in water pollution in number of significant ways. Pollution may bring about on environment of algal nutrients in water. There are as few studies on algal exploitation as to biological indicators of water quality and pollution (Gonzalves and Joshi, 1946; Singh, 1960; Palmer, 1969;

Venkateswarlu, 1969; Kant Shashi, 1985; Weilgolaski, 1975; Gunale, 1991; Hosmani and Barati, 1980; Nandan and Patel, 1985; More and Nandan, 2000).

In Maharashtra state, very few workers had paid attention on ecology of algae with relation to pollution. So it was thought worth to study the algal flora of polluted waters of Mosam river of Baglan Tahsil. In present investigation, an attempt has been made to study the algal communities from polluted habitats of Mosam river.

Materials and Methods

Mosam river is one of the main source of Girna river. It is larger river of the Baglan Tahsil. The river is originated in the Sahydris north of Hanumangad hills (1062 meters above the MSL) or Salher fort with its rivulets or nala from deciduous forest of western ghat of Sahyadri mountains ranges. The Mosam river flows west-east from Haranbaree →Mulher →Antapur → Taharabad →Jayakheda and Nampur. It joins the Girna river about 3 km below Malegaon City (Chandanpuri gate).

Water samples were collected in plastic can with 5 litres capacity and algal samples were collected in acid washed glass bottles and presented in 4 per cent formalin at monthly intervals from 3 stations of the river from July 1998 to June 2000. Water samples were analyzed for studying of physico-chemical parameters according to the Methods given by APHA, 1985.

A quantitative and qualitative study of 4 groups of algae *viz.*, Chlorophyceae, Cyanophyceae, Bacillariophyceae and Euglenineae was made. The pollution tolerant genera and species were recorded for 3 stations of Mosam river. The assessment of water Quality of Mosam river was made by algal communities for 3 stations of river. The data of physico-chemical analysis is also supported for the study of pollution of Mosam river.

Results and Discussion

The pollution tolerant genera were recorded at 3 stations of river (Table 29.1). The pollution tolerant species were also observed in present study according to Palmer (1969) as shown in Table 29.2. The pollution tolerant algae could be used as indicators of organic pollution as reported by earlier workers (Hosmani and Bharati, 1980). Nandan and Patel, 1985; Gunale, 1991; More and Nandan, 2000). Twenty most frequent genera were taken account for pollution study.

Table 29.1: Algal Genera which are Pollution Tolerant Genera of Algae from 3 Stations of Mosam River in Order of Decreasing Emphasis (Palmer, 1969)

Sl.No.	*Genus*	*Group**	*Total Points*	*Stations*		
				SD–I	*SD–II*	*SD–III*
1.	*Euglena*	E	172	+	+	+
2.	*Oscillatoria*	B	161	+	+	+
3.	*Scenedemus*	G	112	–	+	+
4.	*Chlorella*	G	103	+	+	+
5.	*Nitzschia*	D	98	+	+	+
6.	*Navicula*	D	92	+	+	+
7.	*Stigeoclonium*	G	69	+	+	+
8.	*Synedra*	D	58	–	–	+

Contd...

Table 19.3–Contd...

Sl.No.	Genus	Group*	Total Points	Stations		
				SD–I	SD–II	SD–III
9.	*Phacus*	E	57	+	+	+
10.	*Phormidium*	B	52	–	+	–
11.	*Melosira*	D	50	+	+	+
12.	*Gymphonema*	D	48	+	+	+
13.	*Cyclotella*	D	47	+	+	+
14.	*Microcystis*	B	39	+	+	+
15.	*Spirogyra*	G	37	+	+	+
16.	*Pediastrum*	G	35	–	+	–
17.	*Trachelomonas*	E	34	+	+	+
18.	*Fragilaria*	D	33	+	+	+
19.	*Ulothrix*	G	33	+	+	+
20.	*Surivella*	D	33	–	–	+
21.	*Lyngbya*	B	28	+	–	+
22.	*Spirulina*	B	25	–	–	+
23.	*Cymeblla*	D	24	+	+	+
24.	*Coelastrum*	G	24	–	+	–
25.	*Cladophora*	G	24	+	+	+
26.	*Hantzschia*	D	21	+	+	–
27.	*Achnanthes*	D	19	+	+	+
28.	*Pinnularia*	D	18	–	+	+
29.	*Cocconeis*	D	17	–	+	+
30.	*Cosmarium*	G	17	+	+	+
31.	*Gonium*	G	17	–	+	–
32.	*Crucigenia*	G	14	+	+	+

Groups *: E: Euglenoids; B: Blue-green algae; G: Green algae; D: Diatoms

+: Present, –: Absent

S–1: Haranbaree; S–2: Mulher; S–3: Antapur.

Pollution index factor was assigned to each genus by determining the relative number of total points scored by each genus of algae (Palmer, 1969). The results are presented in Table 29.1. The total score of each station of river was greater than 20 indicating the confirmed high organic pollution (Figure 29.1). The station 3 showed the highest organic pollution as compared to other stations. This might be due addition of sewage water in the river. The trend of increase in organic pollution was observed from Station 1 onwards confirming the earlier observations (Weilgolaski, 1975; Hosmani 'and Bharati, 1980; Nandan and Patel, 1985; Gunale, 1991; More and Nandan, 2000). This was also supported by data of physico-chemical analysis of water samples.

Table 29.2: Algal Species which are Pollution Tolerant Species of Algae from 3 Stations of Mosam River in Order of Decreasing Emphasis (Palmer, 1969)

Sl.No.	Algal Species	Group*	Total Points	Stations		
				SD–I	SD–II	SD–III
1.	*Stigeoclonium tenue*	G	34	+	+	+
2.	*Synedra ulna*	D	33	–	–	+
3.	*Chlorella vulgaris*	G	29	+	+	+
4.	*Cyclotella meneghiana*	D	27	–	+	–
5.	*Oscillatoria princeps*	B	24	–	+	+
6.	*Hantzschia amphioxys*	D	23	–	+	–
7.	*Oscillatoria chalybea*	B	22	–	–	+
8.	*Euglena acus*	E	20	–	+	–
9.	*Surivella ovata*	D	20	–	–	+
10.	*Melosira granulata*	D	17	–	+	–
11.	*Phacus pleuronectus*	D	16	–	+	–
12.	*Cocconeis placentula*	D	14	–	+	+
13.	*Coelastrum microporum*	G	14	–	+	–
14.	*Cladophora glomerata*	G	11	+	+	+
15.	*Gonium pectorale*	G	10	–	+	–
16.	*Tetrahedorn muticum*	G	10	+	–	–

Groups*: E: Euglenoids (Euglenineae); G: Green algae (Chlorophyceae); B: Blue-green algae (Cyanophyceae); D: Diatoms (Bacillariophyceae).

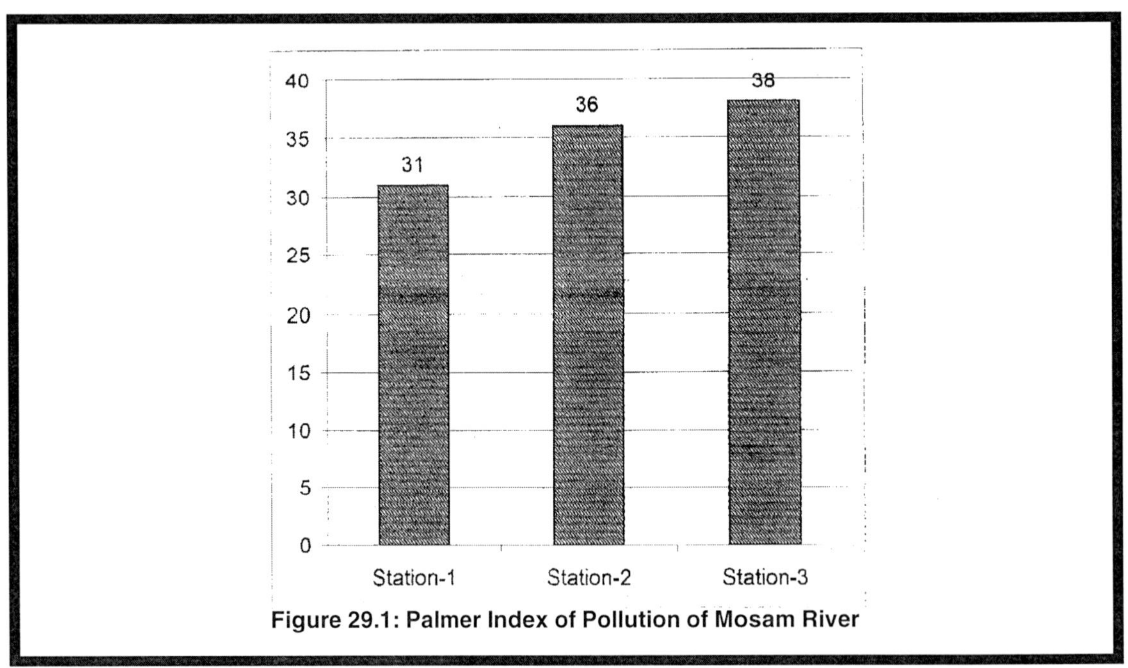

Figure 29.1: Palmer Index of Pollution of Mosam River

Table 29.3: Pollution Index of Algal Genera (Palmer, 1969) at 3 Stations of Mosam River

Sl. No.	*Group/Genera*	*Palmers Pollution Index Number*		
		S–1	*S–2*	*S–3*
I	**Chlorophyceae**			
	1. *Scenadesmus*	–	4	4
	2. *Chlorella*	3	3	3
	3. *Strigeoclonium*	2	2	2
	4. *Coelastrum*	–	1	–
	5. *Spirogyra*	1	1	1
	6. *Pediastrum*	–	1	1
II	**Cyanophyceae**			
	7. *Oscillotoria*	4	4	4
	8. *Phormidium*	1	1	1
	9. *Microcystis*	1	1	1
	10. *Lyngbya*	1	–	1
III	**Bacillariophyceae**			
	11. *Navicula*	3	3	3
	12. *Nitzchia*	3	3	3
	13. *Synedra*	–	–	2
	14. *Melosira*	1	1	1
	15. *Cyclotella*	1	1	1
	16. *Gymponema*	1	1	1
	17. *Fragilaria*	1	1	1
IV	**Euglenineae**			
	18. *Euglena*	5	5	5
	19. *Phacus*	2	2	2
	20. *Trachelomonas*	1	1	1
	Total Score	**31**	**36**	**38**

Acknowledgement

We are grateful to principal, Dr. S.N. Nandan, S.S.V.P.S's L.K. Dr. P.R. Ghogrey Science College, Dhule for providing the facility for investigation.

References

A.P.H.A. (1985). *Standard Methods for the Examination of Water and Wastewater,* 16th Edition. American Publ. Hlth. Asso., New York.

Gonzalves, E.A. and D.B. Joshi (1946). Freshwater algae near Bombay–I. The seasonal succession of algae in a tank of Bandra. *J. Bomb. Nat. Hist. Soc.,* 46: 144–176.

Gunale, V.R. (1991). Algal communities as indicators of Pollution. *J. Environ. Biol.,* p. 223–232.

Hosmani, S.P. and S.G. Bharati (1980). Algae as indicators of organic pollution. *Phykos.,* 19(1): 23–26.

Kant, Shashi (1985). Algae as indicators of organic pollution. *Proc. All India Applied Phycological Congress,* Kanpur, p. 77–86.

More, Y.S. and S.N. Nandan (2000). Hydrobiological study of algae of Panzara river (Maharashtra). *Ecol. Env. & Cons.,* 6(1): 99–103.

Nandan, S.N. and R.J. Patel (1986). Assessment of water quality of Vishwamitri river by algal analysis. *India. J. Env. Hlth.,* 29(2): 160–161.

Palmer, C.M. (1969). A composite rating of algae tolerating organic pollution. *J. Phycol.,* 5: 78–82.

Singh, V.P. (1960). Phytoplankton ecology of the Inland Water of U.P. *Proc. Symp. Algol.,* ICAR, New Delhi, p. 243–271.

Venkateswarlu, V. (1969). An ecological study of the algae of the. river Moosi, Hyderabad (India) with special reference to water pollution–III. The algal periodicity, *Hydrobiologia,* 33: 533–560.

Weilgolaski, F.E. (1975). Biological indicators of pollution, *Urban Ecol.,* 1: 63–69.

Chapter 30

Preliminary Observations on Breeding Behaviour of Freshwater Rotifer *Brachionus calyciflorus*

Sampada Tadphale and Madhuri Pejaver
Department of Zoology, B.N. Bandodkar College of Science, Thane

ABSTRACT

Preliminary observations on breeding behavior of *Brachionus calyciflorus* were recorded. *B. calyciflorus* reached maturity at 24 hrs. after hatching. Three eggs were produced per individual on an average. Embryonic development was observed to be completed within 7–9 hrs. Average size of neonate was recorded to be 96 µm 104 µm.

Keywords: *Brachionus calyciflorus, Breeding behaviour.*

Introduction

Rotifers are very common in Indian waters. They enjoy the position of quantitatively dominant group among the zooplankton community and comprise an integral part of aquatic food chain. They are the natural food of fishes, prawns and can be used as food for larvae of these animals in aquaculture practices. Hence the study of behavioral patterns will be helpful for fishery industries.

The present study deals with the preliminary observations on breeding pattern of fresh water rotifer *Brachionus calyciflorus.*

Materials and Methods

Zooplankton were originally collected from local fresh by the net (mesh 25, diameter of pore 60 µ). From the collected plankton, *B. calyciflorus* were identified with the help of key given by Battish (1992) and isolated. Isolated specimens were propagated under laboratory conditions. They were fed on microalgae (Jaseen *et al.,* 1993). Culture of these rotifers was maintained at room temperature (28 C–30 C).

To observe breeding behaviour, 0–2 hrs neonates were cultured individually and observed throughout their life span. 5 sets of experiments were conducted in sterile, 10 ml glass beakers. In every set of experiment one neonate of the size 96 µm 104 µm was introduced in a 10 ml glass beaker with 5 ml of distilled water. The breeding behaviour of this neonate was observed up to F3 generation by isolating the progeny of F1, F2 and F3 generation from the parent in separate beaker.

Each individual was checker after every 4 hrs. to observe number of eggs produced per organism, size of egg, time span of attachment of egg with the parent and size of neonate and were recorded. Average of five sets is represented in the Table 30.1.

Table 30.1

Generation	*Total No. of Eggs Produced*	*Egg No.*	*Egg Size*	*Time Span of Attachment of Egg with Organism*	*No. of Hatched Organisms*	*Size of Hatched Organism*
Parent	3	1st	73.6 m × 96 m	7–9 hrs	1	96 m × 104 m
		2nd	81.6 m × 99.2 m	4–5 hrs	–	–
		3rd	73.6 m × 96 m	4–5 hrs	–	–
F1	3	1st	76.8 m × 96 m	7–9 hrs	1	96 m × 104 m
		2nd	75.2 m × 96 m	4–5 hrs	–	–
		3rd	72 m × 97.6 m	4–5 hrs	–	–
F2	3	1st	73.6 m × 96 m	7–9 hrs	1	96 m × 97.6 m
		2nd	76.8 m × 94.4 m	4–5 hrs	–	–
		3rd	73.6 m × 91.2 m	4–5 hrs	–	–
F3	3	1st	81.6 m × 96 m	7–9 hrs	1	96 m × 104 m
		2nd	76.8 m × 94.4 m	4–5 hrs	–	–
		3rd	75.2 m × 96 m	4–5 hrs	–	–

Results and Discussion

During the present study, life span of *B. calyciflorus* was observed to be 4 days. While in other species, *B. plicatilis life* span was observed to be 5–7 days (Reddy, 1998). During the present study, it was observed that *B. calyciflorus* reached maturity at 24 hrs. after hatching. Marian Paul (2004) reported similar maturity period for *B. plicatilis. B. calyciflorus* produced three eggs in its life span, during the study. Among the eggs produced, some remained attached for less time span *i.e.* 4–5 hrs and some remained attached for longer time span *i.e.* 7–9 hrs. The eggs that were attached for less time span underwent the dormant condition. According to observations 66.66 per cent eggs remained in dormant condition.

During the present study, it was observed that the size of egg of *B. calyciflorus* varied from 72 µm 97.6 µm to 81.6 µm 96 µm. In this species embryonic development was completed within 7–9 hrs. while Marian Paul (2004) observed *B. plicatilis* completed its embryonic development within 11 hrs. The average size of neonate of *B. calyciflorus* was observed to be same as parent individual *i.e.* 96 µm 104 µm.

Acknowledgement

The authors wish to acknowledge Head, staff and research scholars of zoology department, B.N. Bandodkar College of Science for their constant encouragement.

References

Battish, S.K. (1992). *Freshwater Zooplankton of India.* Oxford and IBH Publishing Co. Pvt. Ltd., p. 69–79.

Janseen, C.R., M.D. Ferrando, Rodrigo and G. Persoon (1993). Ecotoxicological studies with the fresh water rotifer, *Brachionus calyciflorus. Hydrobiologia,* 255/256: 21–32.

Marian Paul (2004). Mass culture of rotifers, Abstract book of National Seminar on Perspectives of Aquaculture, pp. 49.

Reddy, A.K. (1998). *Culture of Rotifers, Training Manual on Culture of Live Food Organisms for Aquahatcharies.* CIFE Publication, p. 39–43.

Chapter 31

Biodiversity of Chlorophyceae in Haranbari Dam of Baglan (Maharashtra)

N.H. Aher and S.N. Nandan***

**Department of Botany, Arts, Commerce and Science College, Sakri – 424 304, District Dhule (MS)*
***P.G. Department of Botany, S.S.V.P.S's L.K. Dr. P.R. Ghogrey Science College, Dhule – 424 005 (MS)*

ABSTRACT

A critical study has been made on the diversity of Chlorophyceae in Haranbari dam of Baglan during the period from July 1998 to June 2000. Algal samples were collected at first week of monthly intervals from 3 stations of Haranbari dam during the period of two years of investigation. Algal samples preserved in 4 per cent formalin for taxonomic investigation. Line drawings of Chlorophyeae were made by camera lucida. Taxa of Chlorophyceae were identified with the help of standard monographs and recent literature. In present study this class consists of mainly species of *Gonium, Chlorococcum, Poediastrum, Tetradedron, Westella, Crucigenia, Ulothrix, Oedogonium, Zygnema, Spirogyra* and *Cosmarium.* The population of *Desmids* and *Chlorococcales* was dominant in winter and summer seasons. From Ulothrichales like *Ulothrix* and from Oedogoniales like *Oedononium,* all these filaments forms were observed from all 3 stations. In the present study 25 taxa of belonging to 11 genera of Chlorophyceae were studied taxonomically in algal community.

Keywords: *Biodiversity, Freshwater, Green algae, Dam.*

Introduction

Algae are natural inhabitants of water. It is important group of thallophytes. Algae constitute a part of food chain of aquatic ecosystem. They affect the population of Zooplankton and other aquatic organisms. The number of publications on limnology and ecology of freshwater rivers, dams etc. have been appeared in Maharashtra (Deore, 1978; Barhate and Tarar, 1981; Gunale and Balu, 1981; Jagdale *et al.,* 1987; Bodas, 1991; Ragothaman and Jaiswal, 1995; More, 1997; Ansari Ziya, 1997; Patil and Nandan, 1998; Nandan and More, 2001, 2002). Though many rivers of dam of Maharashtra were observed hydro-biologically very few rivers and dams carried out work with relation to algae. Biodiversity of 4 groups of algae and limnological studies of algae were made by selecting Haranbari dam of Baglan of Nasik district of Maharashtra. In present investigation Biodiversity of Chlorophyceae of Haranbari dam is explained.

Material and Methods

Haranbari dam is located on the Mosam river. The dam is situated towards the west direction of Taharabad and near the village Haranbari. Geographically Haranbari dam is loacated at latitude 20 51′2″ North and longitude 74 11′23″ East at 714.52 meters above the MSL. The catchment area of dam is about 5478 hectors. The water storage capacity of it is 34.78 million cubic meter. The Haranbari dam is the earthen dam. The wall length of the dam is 748.60 meters and height of dam wall is 44.44 meters. The dam is mainly constructed for irrigation and drinking water purpose.

Algal samples were collected in acid washed glass bottles and preserved in 4 per cent formalin at monthly intervals from 3 stations of Haranbari dam from July 1998 to June 2000.

For qualitative study of Chlorophyceae algae and identification of Chlorophyceae line drawings of different forms of taxa were made by Camera lucida. The members of Chlorophyceae algae were identified by relevant monographs and recent available literature. (Kant and Gupta, 1998; Ramanathan, 1964; Gonzalves, 1981; Randhawa, 1959; Bernard, 1908; Iyengar and Desikachary, 1981; Philipose, 1967; Nandan and Patel, 1985a).

Taxonomy of Chlorophyceae

Biodiversity of Chlorophyceae genera and species were recorded at 3 stations of Haranbari dam. Genes Gonium 1 taxon, Chlorococcum 1 taxon, Pediastrum 2 taxa, Tetradedron 3 taxa, Westella 1 taxon, Crucigenia 2 taxa, Ulothrix 3 taxa, Oedogonium 2 taxa, Zygnema 1 Taxon, Spirogyra 5 taxa, Cosmarium 4 taxa.

1. ***Gonium Pectorale* Muller (Plate 31.1, Figure 1)**

   Huber. Pestalozzi. G.1961, 16(5):624, F.877

   Colonies square with rounded corners, cells 15.6μ broad and 20.8μ long.

   Habitat: HD–II, October, 1999.

2. ***Chlorococcum infusionum* (Schrank) Meneghini (Plate 31.1, Figure 2)**

   J. Meneghini, 1842, P.27, Pl.II, F.3

   Cells usually spherical and variable dimensions 101.8μ in diameters.

   Habitat: HD–II, November, 1991.

1,4
A 6.34µ
2,3
B 17.39µ
1
2
3
4
1. *Gonium pectorale* Muller.
2. *Chlorococcum infusionum* Meneghini
3. *Pediastrum simplex* Mayen
4. *Pediastrum tricornutum* Borge

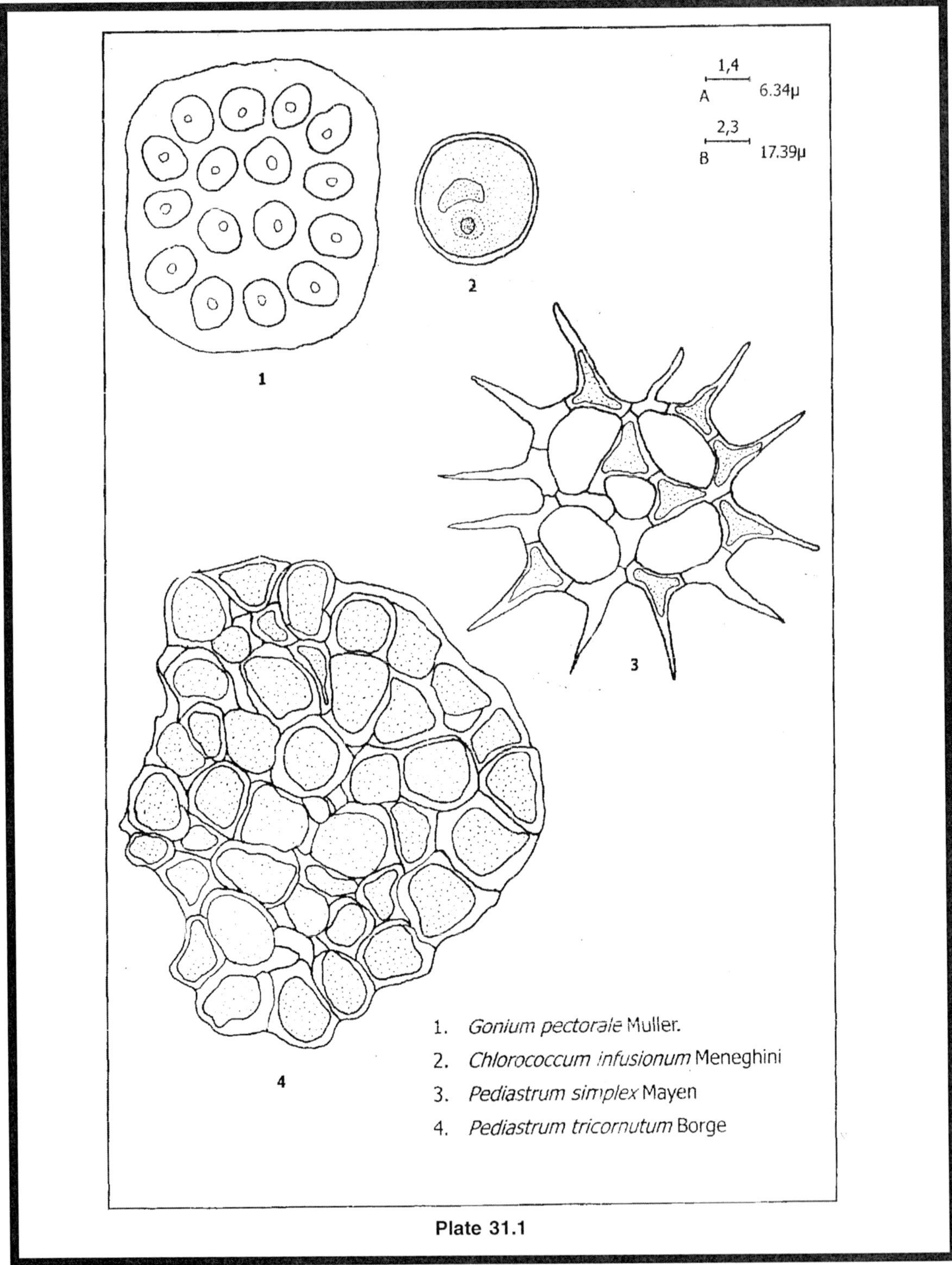

Plate 31.1

3. *Pediastrum simplex* Mayen (Plate 31.1, Figure 3)
   F.I.F. Mayen 1829, P.772, Pl.43, F.1.5
   Colonies circular, cells 8–13μ broad, 19–26μ long.
   Habitat: HD–I, December, 1999.

4. *Pediastrum tricornutum* Borge (Plate 31.1, Figure 4)
   Kant and Gupta, 1998, P.80, Pl.23, F.9.
   Outer face of Marginal cells with two projections, cells 9μ board and 13μ long.
   Habitat: HD–II, December, 1999.

5. *Tetradedron minimum* (A. Braun.) Hansgirg (Plate 31.2, Figure 1)
   A. Hansgirg, 1988, Pl.131, 1889, P.18.
   Cells small and quadrangular with the sides concave and angles rounded. Cells 12μ in diameter.
   Habitat: HD–I, August, 1999.

6. *Tetradedron scrobiculatum* (Lagartheim) De. Toni (Plate 31.2, Figure 2)
   C.B.De Toni, 1989, P.601.
   Cell wall coarsely scrobiculate, Cells 10.77μ in diameter.
   Habitat: HD–II, September, 1999.

7. *Tetradedron octaedrium* V. Spinosum (Rernsch) W.et. G.S. West (Plate 31.2, Figure 3)
   W. and G.S. West, 1901a, P.183.
   Cells octagonal with eight lateral planes. Cells 13.3μ diameter and the spines 3.17μ long.
   Habitat: HD–I, October, 1999.

8. *Westella botrayoides* (W. West) Dewildeman. (Plate 31.2, Figure 4)
   Dewildeman, 1897, P.532.
   Colonies of Irregular shape and about 40–80 cells. Cells 3-8μ in diameter.
   Habitat: HD–III, September, 1999.

9. *Crucigenia fenestrata* (Schmidle) Schmidle (Plate 31.2, Figure 5)
   W. Schmidle, 1900, P.234.
   Colonies rectangular 9.5μ in diameter.
   Habitat: HD–III, September, 1999.

10. *Crucigenia tetrapedia* (Kirchner) W.et.G.S. West (Plate 31.2, Figure 6)
    W. and G.S. West, 1920a, P.62, Pl.1, F.11–12.
    Colonies 4 celled, outersides of cells always concave, cells 3.17μ in diameter, four celled colonies 7.6μ diameter.
    Habitat: HD–I, July, 1999; HD–II, August, 1999; HD–III, July 1998.

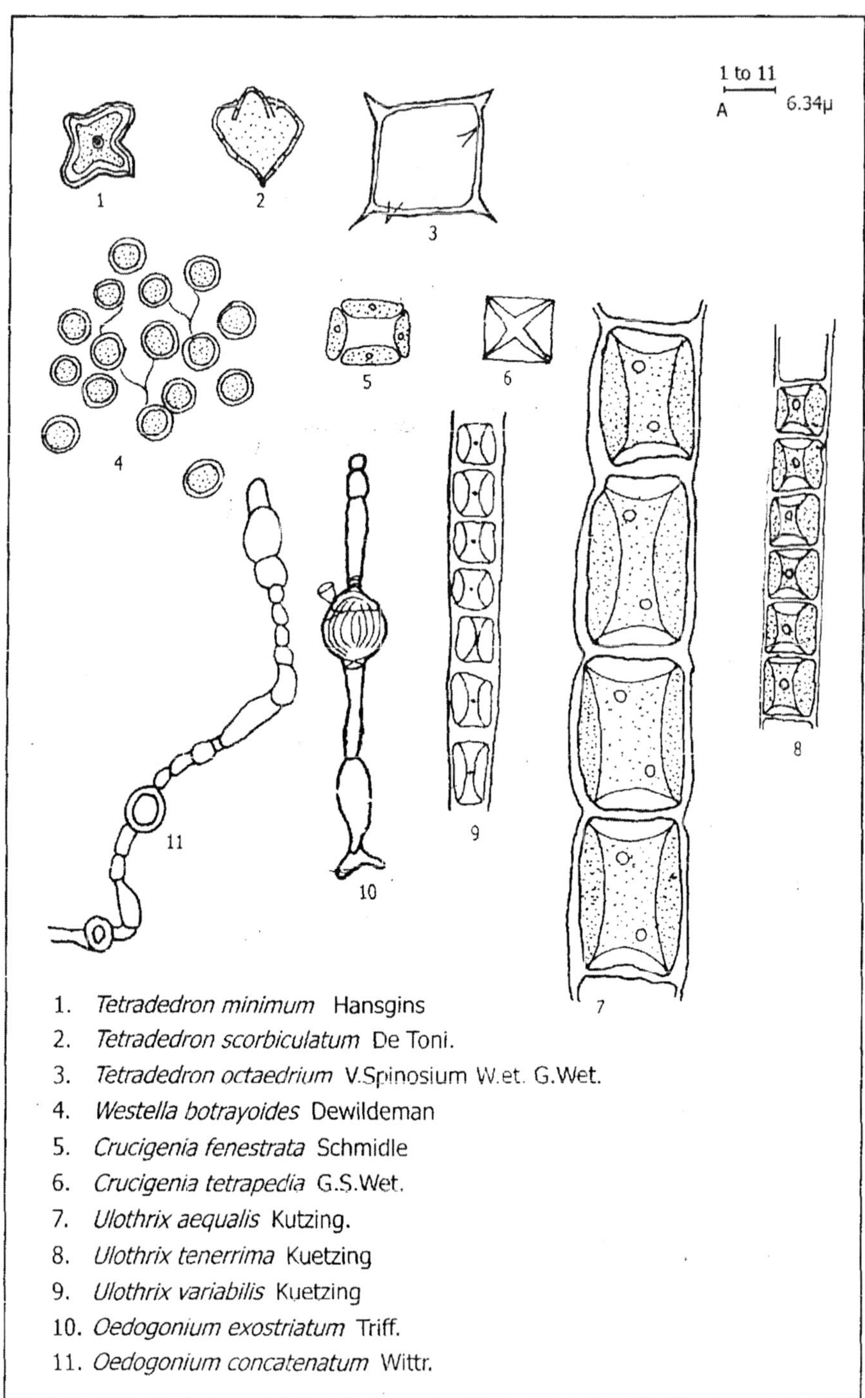

1. *Tetradedron minimum* Hansgins
2. *Tetradedron scorbiculatum* De Toni.
3. *Tetradedron octaedrium* V.Spinosium W.et. G.Wet.
4. *Westella botrayoides* Dewildeman
5. *Crucigenia fenestrata* Schmidle
6. *Crucigenia tetrapedia* G.S.Wet.
7. *Ulothrix aequalis* Kutzing.
8. *Ulothrix tenerrima* Kuetzing
9. *Ulothrix variabilis* Kuetzing
10. *Oedogonium exostriatum* Triff.
11. *Oedogonium concatenatum* Wittr.

Plate 31.2

11. *Ulothrix-aequalis* Kuetzing (Plate 31.2, Figure 7)
    Ramanathan, 1964. P.36, Plate 9, I–L.
    Cells generally cylindrical, 14.5μ broad and 1–2 time as long as broad.
    Habitat: HO–III, August, 1999.

12. *Ulothrix tenerrima* Kuetzing. (Plate 31.2, Figure 8)
    Ramanathan 1964, P.37, Plate 10 A–C.
    Cells 7.6μ thick, 2/3-1½ times as long as broad with one pyronoid.
    Habitat: HD–I, August, 1999.

13. *Ulothrix variabilis* Kuetzing (Plate 31.2, Figure 9)
    Ramanathan, 1964, P.39, Pl.10, D–F.
    Filaments sometimes with a pointed basal cells; Cells 5–7μ, ½-1½ times as long as broad.
    Habitat: HO–I, November, 1999; HO–II, August, 1999.

14. *Oedogonium exostriatum* Triff. (Plate 31.2, Figure 10)
    E.A. Gonzalves, 1981, P.378, F.9–271.
    Nannandrous, vegetative capitellate, 10μ in diameter, 39.30μ long, oogonia depressed-globose, 28μ in diameter 21μ long.
    Habitat: HO–III, November, 1999.

15. *Oedogonium concatenatum* Wittr. (Plate 31.2, Figure 11)
    E.A. Gonzalves, 1981, P.446, F.9.345A.
    Nannandrous, gynandrosporous, vegetative cell, Cylindrical, 25–40μ in diameter, 75–400 long, anthridia single or upto 4 seriate, 13–15μ in diameter, 12–25μ long.
    Habitat: HO–I, July, 1999.

16. *Zygnema collinsianum* Transeen (Plate 31.3, Figure 1)
    Randhawa, 1959, P.249, F.209 a–d.
    Vegetative cells 22 45μ chloroplast 2 in each cells, reproduction by conjugation.
    Habitat: HD–II, March, 1999.

17. *Spirogyra varians* (Hassall) Kuetzing (Plate 31.3, Figure 2)
    Randhawa, 1959, P.297, F.256.
    Vegetative cells 29 45μ end wall plane, Chloroplast 1, Conjugation scalariform, and lateral.
    Habitat: HD–II, July, 1999.

18. *Spirogyra sahnii* Randhawa (Plate 31.3, Figure 3)
    Randhawa, 1959, P.307, F.274 a–e.
    Vegetative cells 60.86 70.37μ usually they are broader than long. Conjugation lateral.
    Habitat: HD–III, July, 1999.

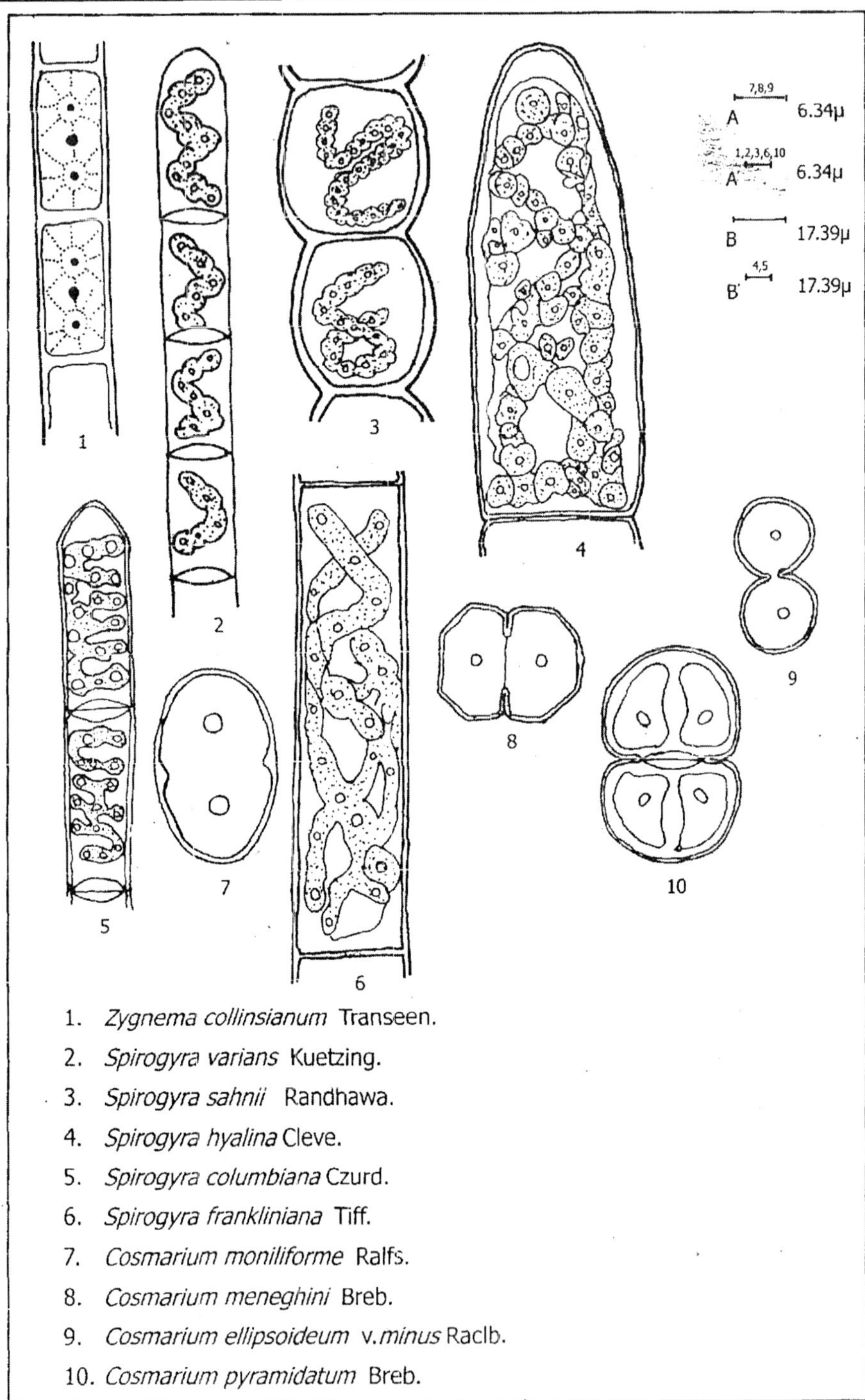

1. *Zygnema collinsianum* Transeen.
2. *Spirogyra varians* Kuetzing.
3. *Spirogyra sahnii* Randhawa.
4. *Spirogyra hyalina* Cleve.
5. *Spirogyra columbiana* Czurd.
6. *Spirogyra frankliniana* Tiff.
7. *Cosmarium moniliforme* Ralfs.
8. *Cosmarium meneghini* Breb.
9. *Cosmarium ellipsoideum* v.*minus* Raclb.
10. *Cosmarium pyramidatum* Breb.

Plate 31.3

19. *Spirogyra hyalina* Cleve (Plate 31.3, Figure 4)

    Randhawa, 1959, P.318, F.294 a–c.

    Vegetative cells, 59.12 126.94μ with plane endwalls. Tubes formed by both gametangia.

    Habitat: HD–I, January, 2000.

20. *Spirogyra columbiana* Czurd (Plate 31.3, Figure 5)

    Randhawa, 1959, P.318, F.295.

    Vegetative cells 53.9 92.16μ with plane end walls, chloroplast 1, conjugation scalariform, tubes formed both gametangia.

    Habitat: HD–I, January, 2000.

21. *Spirogyra frankliniana* Tiffany (Plate 31.3, Figure 6)

    Randhawa, 1959, P.413, Fig.500.

    Vegetative cells 32.96 84.95μ with plane end walls. Conjugation scaliform; tube formed by both gamentangia.

    Habitat: HD–II, July, 1999.

22. *Cosmarium moniliforme* (Turp) Ralf. (Plate 31.3, Figure 7)

    Ch. Bernarad, 1908, 2, P.90, F.58.

    Cells 30μ long, 18μ broad, Isthmus 2μ.

    Habitat: HD–III, February, 2000.

23. *Cosmarium meneghini* Breb. (Plate 31.3, Figure 8)

    Ch. Bernard, 1908, 2. P.90, F.64, 65.

    Cells 17μ long, 13μ broad, Isthmus 3μ.

    Habitat: HD–I and HD–II, September, 1999.

24. *Cosmarium ellipsoideum* Elfv. v. minus. Raclb. (Plate 31.3, Figure 9)

    Ch. Bernard, 1908, 2. P.106, F.65.

    Cells 17μ long, 9.51μ broad, Isthus, 2–2½μ.

    Habitat: HD–I, September, 1999.

25. *Cosmarium pyramidatum* Breb. (Plate 31.3, Figure 10)

    Ch. Bernard, 1908, 2, P.107, F.75–76.

    Cells 71μ long, 61μ broad, Isthmus 20–25μ.

    Habitat: HD–I and HD–II, September, 1999.

Acknowledgement

We are grateful to principal, Dr. S.N. Nandan, S.S.V.P.S's L.K. Dr. P.R. Ghogrey Science College, Dhule for providing the facility for investigation.